# An illustrated guide to British upland vegetation

# An illustrated guide to British upland vegetation

A M Averis and A B G Averis
*Haddington, East Lothian*

H J B Birks
*Botanical Institute, University of Bergen*

D Horsfield and D B A Thompson
*Advisory Services, Scottish Natural Heritage, Edinburgh*

M J M Yeo
*Joint Nature Conservation Committee, Peterborough*

Joint Nature Conservation Committee
Monkstone House
City Road
Peterborough
PE1 1JY
UK

ISBN 1 86107 553 7

# Contents

# Contents

# List of colour plates

**These plates appear after page 224**

# Preface

The purpose of this guide to upland vegetation is simple – to focus attention on the nature, diversity and importance of habitats which cover almost a third of Great Britain's land surface and create landscapes so characteristic of its wilder parts. But this book is much more than that; it is also a tribute to the sustained efforts of a small group of people who have studied the vegetation of the uplands.

Forty years ago two remarkable young ecologists introduced their classic volume on *Plant Communities of the Scottish Highlands* (McVean and Ratcliffe 1962) with the following statement: 'A vegetation monograph can be arranged in three ways – ecologically, or according to the main habitat types and altitudinal zonation; systematically, or according to an hierarchical classification of the units distinguished; and physiognomically, or according to the life-form of the dominant species'. It is a great tribute to the two authors, Donald McVean and Derek Ratcliffe, that their book is still a source of raw data on upland vegetation and an inspiring read. Yet in some ways the issues which these two scholars tackled are still taxing us today. Precisely how does one describe the vegetation seen in the field, and to what extent is it possible to classify this in a systematic way? And to what extent does what we see reflect broader environmental influences on plant growth and development?

With the publication of volumes 2 and 3 of *British Plant Communities* (in 1991 and 1992), edited by John Rodwell, and with substantial contributions from ten leading botanists, an important line was drawn under the classification and description of the bulk of Great Britain's upland vegetation. At long last we had an exhaustive treatment of upland vegetation, with a huge amount of detail on the taxonomy, habitats and ecology of each plant community.

But as is so often the way with science, time has brought new information to light. Over the past fifteen or so years, a huge amount of fieldwork by just a handful of ecologists has given new insights into the vegetation of the British uplands. While the basic classification of vegetation is much the same, our knowledge of the different communities, notably their distribution, ecology and conservation importance, has advanced considerably. This is due largely to the work of three of our co-authors – Ben and Alison Averis, and David Horsfield. They have greatly improved our knowledge of upland vegetation, and in their different ways have made significant contributions to what we know today about the importance of different parts of the uplands.

These days, however, the politics of conservation seem to conspire to eclipse the efforts of those working in the field. We are fortunate, therefore, in having Marcus Yeo as a further co-author. As Head of the Habitats Advice Team in the Joint Nature Conservation Committee he had responsibility for advising on the UK's implementation of the Habitats Directive (Council Directive 92/43/EEC) – the major European statutory instrument for protecting habitats and species.

Working with our colleagues, we have tried to produce a fresh overview of the distribution and ecology of British upland vegetation. This builds substantially on the classification and content of *British Plant Communities*, which has provided a solid foundation for our endeavours. We owe an enormous debt to John Rodwell and his co-workers for what is now widely referred to as the NVC – The National Vegetation

Classification. Our work also echoes many of the ideas in *Plant Communities of the Scottish Highlands*. It is therefore a special pleasure for us to dedicate this guide to our two mentors – Derek Ratcliffe and Donald McVean. We hope that we have done something to solve what they described as the 'essential problem' of representing adequately the total range of variation in vegetation.

D B A Thompson and H J B Birks
February 2004

# Acknowledgments

Many people have commented on various parts of this work and made helpful suggestions, improving the text considerably. In particular, we would like to thank Brian and Sandy Coppins, Andrew Coupar, Claire Geddes, John Gordon, Richard Jefferson, Barbara Jones, Peter Jones, Lyndsey Kinnes (now Lyndsey Duncan), Jim Latham, Phil Lusby, Magnus Magnusson, Sandy Payne, Derek Ratcliffe, Mick Rebane, Ian Strachan, David Stevens, Alex Turner and Marion Whitelaw. The following provided assistance with the collation of records for maps of NVC communities: Alan Brown, Andrew Coupar, Lynne Farrell, Hector Galbraith, Ken Graham, Stuart Hedley, Geoff Johnson, Jane MacKintosh, Eliane Reid, Jim Reid, Richard Lindsay, Alex Turner, Mike Webb and Stan Whitaker. Figures 2, 3, 4, 5, 8 and 9 are reproduced with permission from Harley Books. In addition to those mentioned in the Preface, we owe a special debt to Chris Sydes for managing much of the fieldwork which contributed to the earlier mapping of NVC communities throughout upland Great Britain. We are also grateful to Sally Johnson, Alex Geairns, Kirsty Meadows and Ian Kingston for help with the final stages of producing the book.

# Introduction

If one divides Great Britain into broad-scale zones based on vegetation and environment, including climate, a particularly important division is seen between the cooler, wetter north and west (upland Great Britain), and the warmer, drier south and east (lowland Great Britain). The difference between these areas is shown, for example, by the greater extent of peat and the greater quantity and diversity of humidity-demanding bryophytes and ferns in upland Great Britain, and the greater quantity of thermophilous or heat-demanding plant species in lowland Great Britain. An upland type of climate, and hence an upland environment generally, is restricted to the highest ground in south-west England, but descends to sea-level in the cooler north-west. For the purpose of this book, upland Great Britain is defined as those areas of the country which have an upland type of environment, regardless of their altitude (Figure 1). As well as being wetter and cooler than lowland environments, these upland areas are generally more windy, and their soils are generally less productive than in the lowlands (Pearsall 1968; Ratcliffe and Thompson 1988).

Various other definitions of upland have been made, such as land above a certain altitude (e.g. 300 m), land above the upper limit of enclosed farmland, or land classed as Less Favoured Areas because of its low agricultural productivity. In south-west England and Wales some types of vegetation are recognised as upland types because they occur there mainly on higher ground, but in the colder north of Scotland the same types of vegetation are common near sea-level. Enclosure and agricultural improvement may alter the soils and vegetation, but do not affect the climate. Rather, agricultural improvement simply produces modified vegetation within a generally upland environment.

About a third of Great Britain is upland, and here we find many of the wildest and most beautiful parts of our countryside and the largest areas of natural-looking vegetation (Ratcliffe and Thompson 1988). Upland Great Britain encompasses a tremendous variety of habitats and vegetation types, including heaths, bogs, grasslands, woods, scrub, cliffs, screes, snow-beds and high rocky summits. The plant species composition of much of the vegetation here, and also in the Irish uplands, is unique in Europe. The different regions of upland Great Britain are distinctive in their geology, terrain, climate, land use and vegetation. For example, there are the rounded grassy and boggy hills of central Wales; the steep, craggy mountains of north-west Wales, the Lake District and the western Scottish Highlands; the limestone pavements of the Craven Pennines and south Cumbria; the heathy and boggy stepped basalt landscapes on Mull and Skye; the rolling heather moors of the eastern Highlands; the knob-and-lochan terrain of north-west Sutherland and the Outer Hebrides; and the expansive, pool-studded bog landscapes of the Flow Country in Caithness and Sutherland.

## The upland environment

### Geology and landforms
The landforms which constitute upland Great Britain are very varied, because of the great range in the type and age of the bedrock and the varying effects of glacial activity.

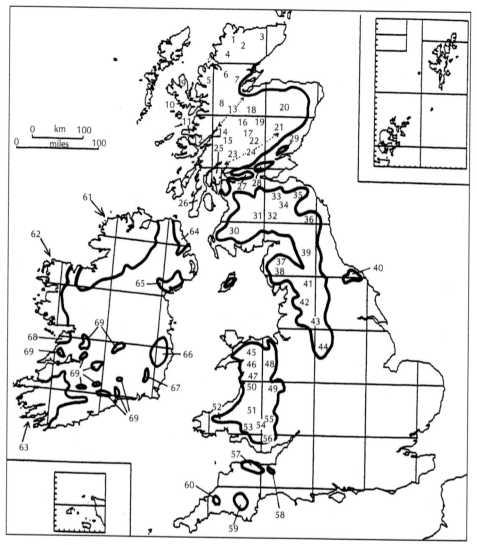

**Figure 1** Upland areas of Great Britain and Ireland, showing the 100 km grid lines of the Ordnance Survey national grid.

Key to code numbers and letters

1  Foinaven and Cranstackie
2  Ben Klibreck
3  Flow Country
4  Ben More Assynt
5  Beinn Eighe, Liathach and other Torridon hills
6  Beinn Dearg and Fannich Hills
7  Ben Wyvis
8  Glen Affric area
9  Trotternish
10 Cuillin Hills
11 Ardnamurchan
12 Morvern
13 Great Glen
14 Ben Nevis
15 Glen Coe and Rannoch Moor
16 Creag Meagaidh
17 Ben Alder
18 Monadhliath Mountains
19 Cairngorms
20 Ladder Hills
21 Lochnagar and Caenlochan
22 Ben Lawers
23 Ben Lui
24 Breadalbanes
25 Cowal
26 Kintyre
27 Campsie Fells
28 Ochil Hills
29 Sidlaw Hills
30 Merrick
31 Lowther Hills
32 Tweedsmuir-Moffat Hills
33 Pentland Hills

(continued opposite)

In south-west England the granite moorlands of Bodmin Moor and Dartmoor and the Devonian Old Red Sandstone of Exmoor were not covered by ice in the Pleistocene and their relief is gentle, with long smooth ridges crowned by jagged tors. The Brecon Beacons and the Black Mountain in south Wales are also made of Old Red Sandstone, but these hills were glaciated and the scenery is more spectacular, with the hillsides carved out into great precipices and corries. In mid-Wales the vast undulating bog-clothed ranges of the central Welsh plateaux extend north to the wild and dramatic ice-etched peaks of Snowdonia, where the soft Silurian and Ordovician slate and shale give way to hard volcanic andesite, diorite and tuff.

Further north still is the Lake District: a mass of jagged, sharp-pointed, high-ridged hills crammed into an area about the same size as Greater London. Most of the Lake District hills are made of the same volcanic rocks as Snowdonia: this was another centre of intense volcanic activity. And here again they are butted up against the Silurian and Ordovician slate and shale which forms the massive rounded plateaux of Skiddaw and the northern Buttermere fells. Carboniferous limestone forms a ring of lower, smooth-edged hills around the margins of the Lake District; on some of these hills there are large areas of bare limestone pavement.

The Pennines form the high spine of northern England. Broad, flat-topped, peat-covered Millstone Grit fells eroded into craggy 'edges' extend from Edale in Derbyshire to the Geltsdale Fells just short of the Scottish border. This hard acid rock is interrupted by the Carboniferous limestone of the Craven Pennines: Ingleborough, Pen-y-Ghent, Fountains Fell, Malham, and the hills around Upper Teesdale. Further east still are the Jurassic rocks of the North York Moors, with gentle rolling heather-clad moorlands.

Silurian and Ordovician sedimentary rocks of the types which occur in Wales and parts of the Lake District form the Southern Uplands of Scotland: gently undulating ridges with little spectacular scenery, although they are dissected in places by dramatic cliff-lined glens. In Galloway in the west of this area, granite intrusions form more rugged hills around Merrick, Cairnsmore of Fleet and Criffel.

To the north of the Silurian and Ordovician hills of the Southern Uplands are the hills of central Scotland, composed mainly of extrusive igneous rocks of Devonian and Carboniferous age. Some of these hills form steep-sided plateaux: the Campsie Fells, the

**Figure 1** Key (*continued*)

| | |
|---|---|
| 34 Moorfoot Hills | 52 Mynydd Preseli |
| 35 Lammermuir Hills | 53 Black Mountain |
| 36 The Cheviot | 54 Brecon Beacons |
| 37 Skiddaw | 55 Black Mountains |
| 38 Scafell | 56 South Wales coalfield |
| 39 Cross Fell and Upper Teesdale | 57 Exmoor |
| 40 North York Moors | 58 Quantock Hills |
| 41 Craven District | 59 Dartmoor |
| 42 Forest of Bowland | 60 Bodmin Moor |
| 43 Dark Peak | 61 NW Ireland uplands |
| 44 White Peak | 62 Mid-W Ireland uplands |
| 45 Snowdonia | 63 SW Ireland uplands |
| 46 Rhinog | 64 Antrim Mountains |
| 47 Cadair Idris | 65 Mourne Mountains |
| 48 Berwyn | 66 Wicklow Mountains |
| 49 South Shropshire Hills | 67 Blackstairs Mountains |
| 50 Pumlumon | 68 The Burren |
| 51 Central Wales plateau | 69 Other southern Irish uplands |

Ochils and the hills between Glasgow and the Firth of Clyde. The Pentland Hills have a sharper outline and are made partly of Devonian Old Red Sandstone.

The upland scenery of the Scottish Highlands is enormously diverse. The most obvious difference is between the eastern and western Highlands. In the west the hills tend to have sharp summits and pinnacled ridges, exemplified by the Glen Coe hills, An Teallach, Beinn Eighe, Liathach, and, at the most spectacular, shattered and splintered extreme, the Black Cuillin of Skye. These jagged hills of the west, with their high, narrow ridges and sharp summits, give way eastwards to the main West Highland water-shed. These huge ranges of hills extend from Foinaven and Ben More Assynt in the north to Ben Nevis in the mid-west Highlands. These are not so dramatic as the western hills and their wildest scenes are hidden from the roads, but even so they are carved into great corries and deep spectacular glens. Further east the hills are smoother in outline: the vast plateaux of the Cairngorms, Caenlochan and Lochnagar and the massive rounded bulky skyline of the Monadhliath and Drumochter hills. These eastern hills have been less dissected by glaciation than those in the west and therefore have a larger area of high plateau.

Most of the hills in the central and northern Highlands are made of Moine Schist, a hard crystalline rock which is locally base-rich. From Islay, Jura and southern Argyll, running in a diagonal band north-east through the Breadalbanes to the Banff coast are the Dalradian rocks. They are renowned for bands of highly calcareous mica-schist and limestone with their astonishing herb-rich vegetation and calcareous mires, especially well developed on Ben Lui, Ben Lawers and Caenlochan.

In the north-west of Scotland, the edge of the Moine rocks (the Moine thrust plane) lies against the older rocks of the north-west Highlands. The Lewisian gneiss, the oldest of all, is a deeply-folded, banded rock. It is mostly acid but is locally base-rich – strongly so on Ben More Assynt where at 998 m the Lewisian gneiss reaches its highest elevation in Great Britain and Ireland. The Lewisian gneiss occurs from Lewis and Sutherland south to Coll, Tiree, Iona and Islay. Glacial action has moulded hillocks and scoured out lines of weakness in the Lewisian rocks to give intervening small lochs, and is responsible for the distinctive 'knob-and-lochan' topography. This wild landscape of rock, heath, bog and water is typical of the Outer Hebrides and north-west Sutherland. Lying over the Lewisian gneiss, from Islay, Colonsay, Iona, Rum and south-eastern Skye north to Sutherland, is the Torridonian sandstone. This rock forms the tiered precipices of Beinn Bhàn, the Torridon hills, An Teallach, Cul Mór and Cul Beag, and, standing out on the ancient eroded Lewisian foreland, the dramatic isolated peaks of Suilven, Stac Polly, Quinag and Canisp. Some of these hills have a white cap of quartzite, formed during the Cambrian period and looking like snow from a distance. The Durness limestone, with its special vegetation and flora, also occurs along the line of the Moine thrust, outcropping in many places from south-eastern Skye north to near Durness in Sutherland.

Granite intrusions are scattered widely in the Highlands, from Arran, the Ross of Mull and the Red Cuillin of Skye eastwards through Morvern, north Argyll and Rannoch Moor to the Cairngorms and Lochnagar. This hard, coarse-grained, acidic rock produces hills which are smooth rather than jagged in outline, but with prominent outcrops and screes of a distinctive pinkish colour.

In the Inner Hebrides and Morvern there is much basalt of Tertiary age. This produces a stepped, 'trap' landscape which is the nearest British equivalent to the basalt islands of the Faroes and Iceland. Many of the lines of basalt cliffs have weathered into distinctive columns, well-seen in parts of Skye and Mull. In north Argyll there are extensive out-crops of Older Devonian basalt and other extrusive igneous rocks. They have weathered into a series of low hills with very complex topography.

Eastern Caithness and Orkney are composed of Devonian Old Red Sandstone and most of the ground is low and gently undulating, though with some spectacular sea-cliffs. Shetland is also low-lying but is rockier, has more complicated topography, and an extraordinarily varied geology.

## Climate

Upland Great Britain is generally cooler, wetter and windier than lowland Great Britain. However, this is very much a generalisation because the uplands encompass considerable variation in climate, as is evident from climate maps and from the regional accounts in Wheeler and Mayes (1997).

The most obvious variation is that in temperature, which decreases with increasing altitude and with increasing distance north. The temperature decreases by about 0.65 °C with every 100 m increase in altitude, though this figure is a rough guide only and the

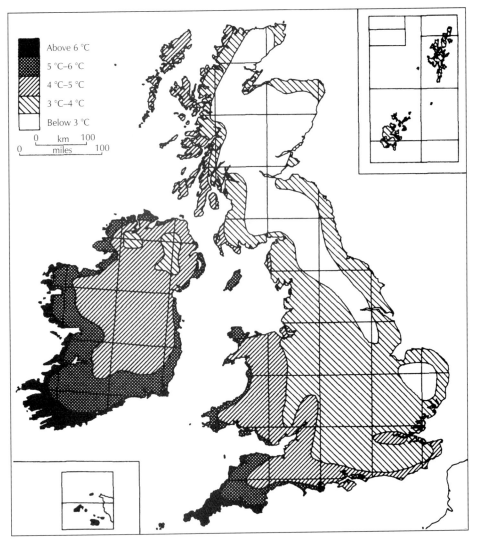

**Figure 2** January mean temperature. Values are corrected to sea level. Isopleths are spaced at intervals of 1 °C, except for the extreme south-west of Ireland and England, where small areas have a January mean in excess of 7 °C. From Hill *et al.* (1991); *Atlas of the Bryophytes of Britain and Ireland. Vol. 1.*

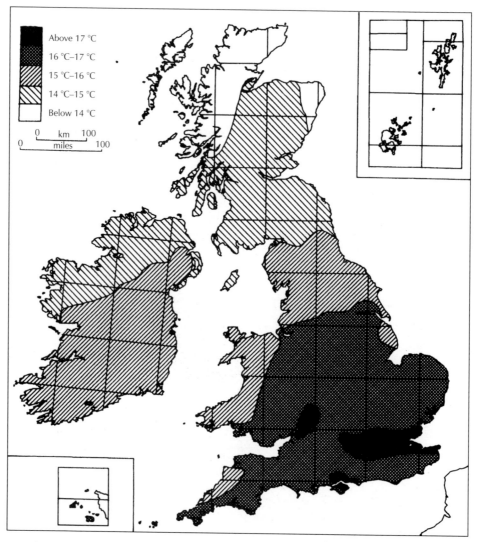

**Figure 3** July mean temperature. Values are corrected to sea level. Isopleths are spaced at intervals of 1 °C. Source as for Figure 2.

actual lapse rate varies in relation to weather conditions (Wheeler and Mayes 1997). Temperature inversions can occur during periods of still, cold weather: cold air flows down slopes and settles in hollows or valley bottoms which become colder than the adjacent slopes. The pattern of decrease in temperature northwards depends on the season: in winter the south-west is mildest and the north-east is coldest (Figure 2), whereas in summer the south-east is warmest and the north-west coolest (Figure 3).

There is generally more rain at higher altitudes, but this pattern is overshadowed by the broader-scale distribution of rainfall. The west receives heavier and more frequent rain than the east (Figure 4). The highest amount of rainfall is consistently recorded in the mid-western and south-western Highlands, the Lake District and Snowdonia. Frequency of rainfall shows a slightly different pattern, with the highest average annual number of 'wet days' occurring in the mid-western and north-western Highlands and the Hebrides (Figure 5). Thus there are parts of north-western Scotland (and western Ireland) where the amount of rain is less than in the Lake District and Snowdonia, but it is

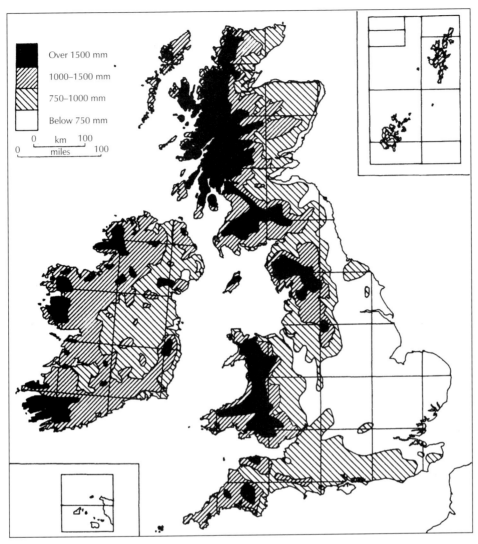

**Figure 4** Annual rainfall. The map shows total precipitation, of which a small amount falls as snow, especially in the higher Scottish mountains. Source as for Figure 2.

more evenly spread through the year so that there are more wet days and the climate is effectively wetter.

As with rainfall, snowfall is more frequent at higher altitudes but also shows a pronounced broad-scale pattern of variation, in this case from the south-west, where it is least, to the north-east, where it is greatest (Meteorological Office 1952). More significant for vegetation is the length of time during which snow covers the ground, and this depends not only on the amount of snowfall but also on the topography. In exposed situations, fallen snow may soon be blown away so that snow-lie is not prolonged, even though snowfall may be frequent. In contrast, north-facing to east-facing slopes, hollows and corries are less windy and experience less sunlight so they tend to hold snow for longer periods, especially in the Scottish Highlands where there can be a build-up of snow blown in from adjacent extensive and exposed summit plateaux.

The windiest areas are in the far west (Meteorological Office 1952), although, as with snow-lie, wind varies considerably depending on the topography. Within a small distance there can be marked contrasts between bleak and windswept exposed slopes and

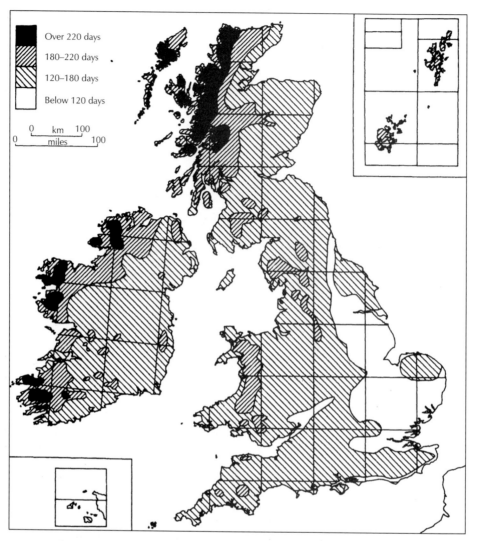

**Figure 5** A wet day is defined here as a day with at least 1 mm of precipitation. Source as for Figure 2.

summit plateaux, and sheltered hollows, valleys and ravines. Wind direction is very variable but is most frequently from between the south and the north-west. This has an effect on the vegetation; for example, on the islands of Skye and Mull, trees can grow much taller in sheltered habitats on eastern coasts than in more exposed western places.

Two or more climatic variables can be combined to form a climatic index, and two such indices are useful to summarise the broad gradients of climatic variation in Great Britain and Ireland.

The first index is one of climatic severity (Figure 6), calculated as the mean annual number of wet days (Ratcliffe 1968) divided by the annual accumulated temperature (Gregory 1954). Low accumulated temperature and frequent rainfall both contribute to high index values. This index was calculated to show the pattern of variation from the most strongly upland (cold and wet) climate to the most strongly lowland (warm and dry) climate; wet days and accumulated temperature were used because wetness and temperature are particularly important elements contributing to the contrast in vegetation between lowland and upland environments. Higher index values indicate a colder

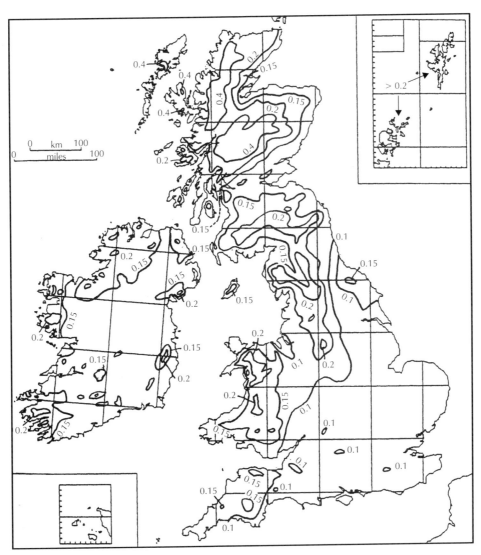

**Figure 6** Index of climatic severity. This index is calculated as the mean annual number of wet days (>1 mm) (Ratcliffe 1968) divided by the annual accumulated temperature (Gregory 1954). Higher values indicate a more severe (colder and/or wetter) climate. The line for the index value 0.15 is a good general guide to the boundary between upland and lowland Great Britain.

and/or wetter climate. Obviously, a particular index value does not always indicate exactly the same levels of both wetness and temperature; for example, places with an index value of 0.15 in south-west England are milder but wetter than places with a similar index value in north-east Scotland. Areas with an index value of at least 0.15 correspond fairly well with the distribution of upland types of vegetation. In these areas bryophytes, ferns and montane plant species are generally more common, and peat cover more extensive, than in lowland areas; this reflects the cooler and wetter climate. The index value of 0.15 corresponds better with the division between upland and lowland environments than do individual measures of altitude, temperature, wetness or wind. The climate is most severe (index value of >0.4) in parts of the north-western, central and eastern Scottish Highlands; these are the areas with the best-developed montane vegetation and the richest montane floras.

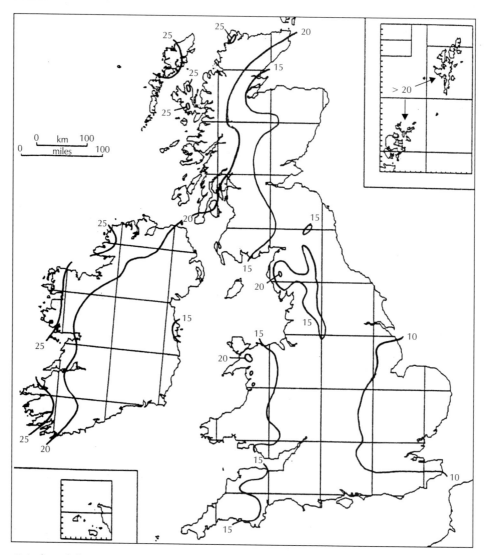

**Figure 7** Index of climatic oceanicity. This index is calculated as the mean annual number of wet days (>1 mm) (Ratcliffe 1968) divided by the range of monthly mean temperatures (°C) (Meteorological Office 1952). Higher values indicate a more oceanic (wetter and/or more equable) climate.

The second index is one of climatic oceanicity (Figure 7) calculated as the mean annual number of wet days (Ratcliffe 1968) divided by the range of annual monthly mean temperatures (Meteorological Office 1952). A small temperature range and frequent rainfall both contribute to high index values. This index was calculated to show the pattern of variation from the most strongly oceanic (wet/equable) climate to the most strongly continental (dry/less equable) climate. There are some great contrasts in vegetation in relation to oceanicity. Many plant species, especially certain bryophytes, lichens and ferns, are strongly associated with wet and equable conditions and grow mainly in extreme western upland areas of Great Britain and Ireland with an oceanicity index higher than 20. The map of an oceanicity index for Europe (using rain days (>0.01 mm) instead of wet days (>0.1 mm) because data for wet days are available only for Great Britain and Ireland) shows very clearly that north-western Scotland and western Ireland have by far the most oceanic climate in Europe. This is reflected in the fact that these areas are the European headquarters for many western 'oceanic' bryophytes, lichens and ferns.

## Soils

The soil is one of the most important determinants of which plants will grow where. Almost all vascular plants extract nutrients and water directly from the soil, and their distributions are limited by the distribution of soils and the balance of nutrients and water.

Soils, like the rocks from which most of them are derived, can be classed according to whether they are basic or acid in chemical reaction. Basic soils have more of the elements calcium, magnesium, sodium and potassium, whereas acid soils are deficient in these elements and have few plant nutrients. Base-rich soils are generally derived from rocks such as limestone, calcareous mica-schist, basalt and serpentine. There are also bands of basic rock within generally acid formations, such as the Borrowdale Volcanic rocks, Moine schist, Lewisian gneiss and Torridonian sandstone, and these also weather into base-rich soils. Acid soils come from acid rocks, such as granite, quartzite, gneiss, many schists and many sandstones, including Millstone Grit.

Acid soils predominate in the uplands. This is partly because most of the rocks themselves are acid; calcareous rocks are very localised (Figure 8). The cool wet upland climate also plays a part. The excess of precipitation over evaporation favours two things: leaching of mineral soils on slopes and the accumulation of peat on level ground. This is why so many upland soils are predominantly acid and poor in nutrients, even where the rocks are moderately base-rich. They have not always been so. Soils were undoubtedly more base-rich and fertile before the original forests were cleared. Tree felling, leaching of soils without a protective cover of trees, and nutrient extraction in timber, wool and meat have all combined to reduce the fertility and carrying capacity of upland soils over the centuries. Plants themselves can also acidify soils; *Calluna vulgaris*, *Betula pubescens* and *Pinus sylvestris* are known to have especially acidifying effects. This does, however, depend on the starting point; for example, *B. pubescens* can have an acidifying effect as it colonises herb-rich *Agrostis-Festuca* grassland, but will enrich the soil as it colonises very acidic *Calluna* heath.

Paradoxically, base-rich soils are often maintained by leaching, where they are irrigated by water running down from basic rocks higher up the slope. They are also maintained by physical and chemical weathering of the parent rock.

Soils are classified by their profiles: the layers of material or horizons which they comprise and which are revealed by digging and examining the exposed vertical faces of soil.

The most extensive type of soil in the uplands is peat (Figure 9). This is an organic soil made of the partly decomposed and compacted remains of plants such as *Sphagnum* mosses and *Eriophorum vaginatum*. Bogs and wet heaths are the typical vegetation here, but *Nardus stricta* grasslands and *Juncus squarrosus* grasslands also occur over peat where the original bog vegetation has been destroyed by too much burning, grazing and draining. Peat clothes the granite plateaux of south-west England and much of the gently rolling hill ground of central Wales. It also covers the gentle dip slopes of the Pennines and the North York Moors. It is quite common in the Lake District and Snowdonia, but is not very extensive there because most of the hills are too steep for peat to accumulate. Vast areas of land in the Highlands are blanketed with peat. Particularly notable expanses of peat are in Lewis at the northern end of the Outer Hebrides; Sutherland, Caithness and Rannoch Moor.

Podsols and podsolised soils are common too, especially on steep slopes. These acid leached soils have developed under pine and birch woodland, heaths and acid grassland. They have a thin top layer of raw humus over a leached and, in many cases, light-coloured upper horizon. The minerals and organic particles leached out of this horizon accumulate in a dark brown or yellowish layer at its base.

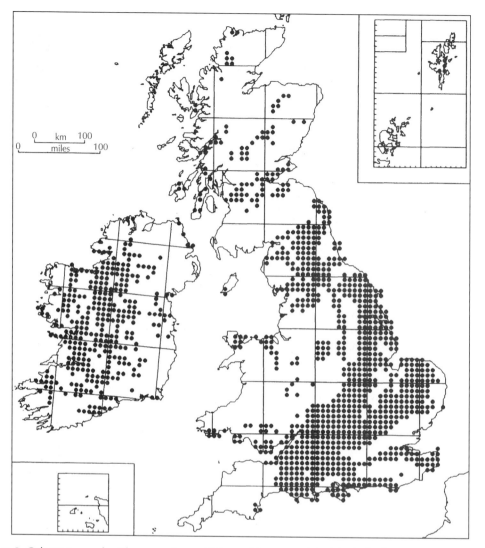

**Figure 8** Calcareous rocks. The map shows 10 km grid squares (•) with chalk, limestone or metamorphic calcareous rock underlying at least 5% of the land area of the square. Source as for Figure 2.

Brown earths, also known as brown forest soils, developed under the original deciduous woodlands and persist under herb-rich woodlands, richer forms of grassland and herb-dominated communities.

Rendzinas develop over limestone and are dark brown or reddish with high concentrations of certain plant nutrients. The usual vegetation here is species-rich grassland or woodland and, in the north-west Highlands, *Dryas octopetala* heath.

Gley soils are permanently waterlogged and are commonly peaty or mottled with shades of grey and brown. They are the usual soils under rush-mires and sedge-mires and can also occur under acid grasslands, including *Molinia caerulea* grasslands.

Many of the soils on steep slopes in the montane zone above the tree-line are classed as rankers or lithosols: shallow accumulations of raw humus lying directly over scree or bare rock. Montane dwarf-shrub heaths are common on this type of soil. These soils can be fragile and prone to erosion, especially when the slopes are frequently crossed by large numbers of deer and sheep.

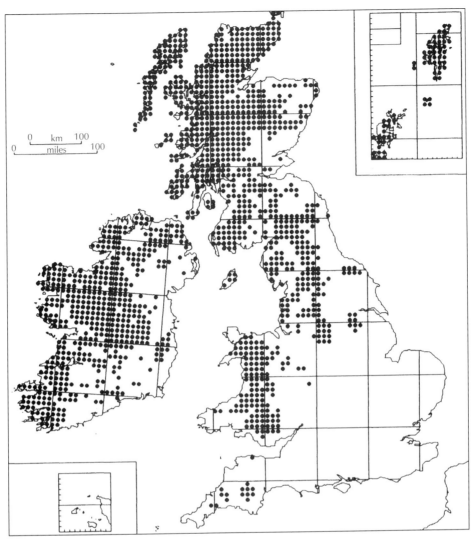

**Figure 9** Bog peat. The map shows 10 km grid squares (•) with at least 1% of their land surface covered by peat at least 50 cm deep. Source as for Figure 2.

On the higher summits and ridges the soils are repeatedly disturbed by frost-heaving and redistributed by solifluction, which is the slow movement of saturated soil and fragments of rock down a slope. Solifluction may be amorphous where a saturated superficial layer of soil flows over the frozen layers beneath, or structured where stones, gravel and soil are sorted into patterns by the churning action of frost. A great variety of patterns can develop, depending on the slope of the ground and the texture of the soil. The patterns include alternating stripes of large and small stones, polygons comprising a regular network of larger stones filled in with finer material, hummocks and ridges of soil, and terraces or steps. Many of the larger features were formed when the climate was much colder than it is today, but some are undoubtedly still active. *Racomitrium lanuginosum* moss-heath and bryophyte-dominated snow-bed communities commonly cover these solifluction soils. South of the Highlands, moss-heath may be replaced by montane *Festuca-Agrostis* grasslands.

Although many of the soils of high montane ground are very fragmentary – little more than greasy accumulations of decaying moss and humus over shattered stone – there are many places where *Racomitrium lanuginosum* heath conceals deep deposits of wind-blown or aeolian sand. There are good examples on Conival in Assynt, Beinn Alligin in Wester Ross, Aonach Mór to the east of Ben Nevis, and the Drumochter hills. In some exposed places along the north-western coast of Scotland the soils are enriched by wind-blown calcareous shell-sand, as at Invernaver in Sutherland. This can allow calcicolous vegetation to develop on otherwise acid and unfavourable soils. Some new studies in the uplands are examining interactions between earth science and biological processes (e.g. Thompson *et al.* 2001; Burt *et al.* 2002).

## History, land use and management

Prior to human activity beginning about 5000–6000 years ago, birch (*Betula* spp.), pine (*Pinus sylvestris*), oak (*Quercus* spp.) and hazel (*Corylus avellana*) woodland covered much of what is now bare, treeless upland (Tipping 2003). In north Wales, the Lake District and Scotland there was ground above the altitudinal limit of trees (the tree-line). Although it is difficult to tell where the woodlands were continuous and where there were just scattered trees, especially in the north and west of Scotland, there were probably very few places in the uplands from which no trees were visible (Tipping 2003). The Pennines, the North York Moors, Bodmin Moor, Exmoor and Dartmoor were probably completely covered by forest, except on cliffs, screes, areas of shallow soils, mires, springs, flushes and the most exposed summits (Birks 1988).

The landscape of upland Great Britain has been greatly modified by human influence over many centuries, with extensive deforestation beginning about 2000–4000 years ago. Centuries of deforestation have reduced woodland to a tiny fraction of its original extent. In the more open landscape, regenerating trees and shrubs are more vulnerable to browsing by herbivores, such as native deer and introduced cattle (*Bos taurus*), sheep (*Ovis aries*) and goats (*Capra hircus*). This has kept woodland regeneration in check and has led to the maintenance of large areas of open grassland and heath (e.g. Ratcliffe and Thompson 1988). Without the annual addition of nutrients from tree and shrub leaf litter the previously rich forest soils have become leached and degraded over time. There has been further removal of nutrients from the ecosystem in the form of wool and meat from sheep, and milk and meat from cattle. Deliberate and repeated burning of moorland vegetation has caused further modification, including drying out of peat, and, often in combination with grazing, the associated conversion of bog vegetation to heath or grassland. The composition of upland vegetation has also been influenced by changes in the type of grazing animal over the past few centuries. Historically, most of the uplands were grazed by mixtures of cattle, sheep (including wethers), ponies and goats, but now sheep (mainly ewes and lambs) are the most common animals in almost all upland areas. The selective grazing behaviour of ewes and lambs appears to have encouraged the spread of unpalatable species, such as *Nardus stricta* and *Juncus squarrosus*.

However, land use in upland Great Britain has not been as intensive as in the lowlands, and its influence has not completely overwhelmed that of variation in natural physical factors, such as rock type, topography, soils and climate. The result of the combination of natural variation and anthropogenic land use is a richly varied landscape. The main forms of land use in upland Great Britain in recent decades have been sheep grazing, deer forest, grouse moor, cattle grazing, water abstraction, recreation, and afforestation with non-native conifers. The pattern of land use varies between different parts of upland Great Britain. For example, centuries of intensive sheep grazing have led to the development of extensive grasslands through much of the hill ground in Wales, the Lake District and parts of the Southern Uplands and the Breadalbanes, while moor-burning and light grazing associated with grouse shooting have maintained extensive heather-dominated heaths in the Pennines, North York Moors, eastern Highlands and parts of the Southern Uplands.

Throughout most of upland Great Britain, grazing by native and introduced herbivores has until recent years prevented woodland regeneration in all but the most inaccessible places. For example, McVean and Ratcliffe (1962), discussing potential oak wood regeneration in the Scottish Highlands, wrote: 'In the absence of any semi-natural hill oak woods which are fenced against the grazing animal it is impossible to say how this potential regeneration would develop'. Only four decades later the situation is quite different. Since the 1980s many areas of deciduous woodland and open ground have been fenced to exclude grazing animals. Some enclosures are stock-fenced, some are deer-fenced, some are quite temporary or mobile, and others are more permanent. The result is that a substantial and increasing amount of native woodland regeneration is now taking place in many parts of upland Great Britain. In many other places native tree and shrub species are being planted on open ground to create new woodlands which are nearer to natural communities than are the all-too-familiar dense conifer plantations.

Patterns of human settlement in upland Great Britain vary from crofting in the western Highlands and Hebrides to concentrated villages among a general scatter of farms in the English and Welsh uplands. The human population of much of upland Great Britain was once greater than it is today, although in some areas, such as the Yorkshire Dales, it is believed to be about the same now as it was in medieval times. This change in population is especially marked in the Highlands, where the population was greatly reduced during the infamous Highland clearances and where today one can find many old ruined houses and settlements in remote places. Some of the more remote dwellings were never permanently occupied. Known as 'shielings' in Scotland and northern England (*airidh* in the Gaelic) and *hafotai* in Wales, they were places where people lived in summer when the cattle were taken up the hills to graze the high pastures.

## Pollution

Various human activities over recent centuries have released chemical pollutants into the air, and this has affected upland vegetation. Early industries, such as lead smelting, would have generated sulphur dioxide and other pollutants, but atmospheric pollution increased greatly at the time of the Industrial Revolution.

During and since the Industrial Revolution the two main pollutants affecting the uplands have been sulphur dioxide and nitrogen oxides. Both of these have acidifying effects on the environment; their effects can be hard to separate from each other and from those of other pollutants. Sulphur dioxide emissions from industrial activities were high within and close to industrial areas in the nineteenth and twentieth centuries, but they have decreased markedly over the past 30 years (National Expert Group on Transboundary Air Pollution 2001). Sulphur dioxide is probably one of the main pollutants responsible for the dramatic impoverishment of bog vegetation in the southern Pennines, where there have been great losses of *Sphagnum* species (Lee *et al.* 1988). Slightly further afield the effects of pollution are less dramatic but still apparent. For example, in the extensive heaths and bogs of the Forest of Bowland the moss *Hylocomium splendens* (a very common species in heaths, bogs and grasslands throughout most of the British uplands) is very rare or absent except in small areas of limestone grassland where the base-rich soils compensate for the acidity of the rain. This species is considered to be sensitive to atmospheric pollution (Farmer *et al.* 1992).

Nitrogen deposition has increased substantially in the twentieth century. For example, in the southern Pennines it increased approximately fourfold between about 1868 and 1988 (Lee *et al.* 1988). Nitrogen oxide emissions have doubled since 1990, but total nitrogen deposition has remained more or less constant over the same period, and nitrites (salts of nitrous acid $HNO_2$) are now more prevalent than nitrates (salts of nitric acid $HNO_3$) (National Expert Group on Transboundary Air Pollution 2001). The acidifying effect of nitrogen deposition is now greater than that of sulphur.

In recent years atmospheric pollution in the form of nitrites, nitrates, ammonium and other pollutants in cloud and mist droplets might have been damaging the vegetation even on some of our more remote hills, perhaps especially some of the higher cloud-shrouded mountains (Lee *et al.* 1988). On some hills, the loss of *Racomitrium lanuginosum* from summit heaths has been attributed to this (Thompson and Baddeley 1991; Thompson and Brown 1992; Baddeley *et al.* 1994).

The bryophytes and lichens in snow-bed vegetation are susceptible to the effects of polluted snow. Snow can be so polluted that it actually looks dark grey rather than white, because it is loaded with particles of carbon and ash as well as with a range of chemicals (Woolgrove 1994). Much of the nitrogen and sulphur deposited in snow-beds in the eastern Scottish Highlands has been found to come from eastern Europe and the Baltic countries as well as from industrial parts of Scotland and the rest of Great Britain and Ireland (Woolgrove 1994). The west-central Highlands suffer most, because they receive very high precipitation with moderate amounts of pollutants (Fowler and Irwin 1989). Snow-patches can also scavenge pollutants from mist and cloud as they pass over the hills (Woolgrove 1994). Once the snow in a snow-bed is polluted, the effect of continual melting and re-freezing within the mass of snow is to move the dissolved ions into the lower part of the snow where they become concentrated. As the snow melts, it releases periodic flushes of polluted water onto the vegetation beneath it, and into springs and streams on the slopes below. This leads to a great increase in the amount of nitrogen in the cells of at least some bryophytes (Woolgrove and Woodin 1995), which can cause such physiological damage that the plants die.

Another effect of atmospheric pollution in the nineteenth and twentieth centuries is the acidification of bark and rock surfaces, and a corresponding change in the species composition of bryophyte and lichen assemblages. One of the best-known results has been the loss of many species of lichens, especially large foliose and fruticose species of *Lobaria*, *Sticta*, *Pseudocyphellaria*, *Pannaria*, *Degelia* and *Usnea*. Many of these species were once widespread in Great Britain but they are now scarce outside the least-polluted areas in the far west and north. Several bryophytes, including species of the genera *Ulota*, *Orthotrichum* and *Frullania*, appear to be vulnerable to the effects of acidification by atmospheric pollution (Hill *et al.* 1991, 1992, 1994). Ratcliffe (1968) considered that the larger particles of certain airborne pollutants could interfere adversely with the metabolism of some of the smaller and more delicate liverworts, such as the small Lejeuneaceae.

Densely planted non-native conifers can have an acidifying effect on soils, and water run-off can cause acidification of ground and watercourses downslope from plantations (Gee and Stoner 1988). From various studies (e.g. Tyler and Ormerod 1994) it appears that this has depleted populations of riparian invertebrates and birds within and along upland watercourses.

Acid deposition and afforestation have led to widespread acidification of the water of lakes and streams in north and central Wales, the Pennines, the Lake District, Galloway and parts of the Scottish Highlands. This has caused the loss of fish and other aquatic organisms, the loss of submerged vascular plants, and the spread of *Juncus bulbosus* and aquatic *Sphagnum* species such as *S. denticulatum*.

Dust blown from quarries is thought to have an effect on vegetation and to be deleterious for certain plants, including *Sphagnum*. Effects probably vary depending on the parent rock type. For example, deposition of limestone dust in the vicinity of limestone quarries can lead to the development, on oak and alder, of lichen assemblages typical of trees and shrubs with more basic bark, such as ash, elm and elder.

Some upland streams and rivers, especially in parts of Wales, northern England and southern Scotland, have become polluted by run-off from adjacent lead and zinc mines or spoil heaps.

# Upland vegetation

### Studies of upland vegetation

Many studies have been made of the vegetation in different parts of the British uplands over the last 100 years or so. Up until the 1950s there were a number of localised studies, for example of the Cairngorms, the Breadalbanes and Upper Teesdale. There are too many such studies to name here; several of them have been reviewed in more general works such as Tansley (1939) and Poore and McVean (1957). Some other studies covered larger areas but focused only on particular types of vegetation; for example, the study of native pine woodland by Steven and Carlisle (1959).

The first study to deal with the vegetation of a large area in a systematic way was that carried out by Donald McVean and Derek Ratcliffe in the Scottish Highlands. During the late 1950s they recorded quadrat samples of different vegetation types from all over the Highlands, finally presenting their data and descriptions of the vegetation types in the Nature Conservancy monograph *Plant Communities of the Scottish Highlands* (McVean and Ratcliffe 1962). Birks (1973) did similarly detailed work on the vegetation of the island of Skye as part of his studies on the past and present vegetation of Skye.

Geographical coverage was later extended to the Southern Uplands of Scotland by Eric Birse and Jim Robertson, using slightly different field methods and analytical procedures. They described plant communities from both upland and lowland Scotland as part of the Soil Survey of Scotland (Birse 1980, 1984; Birse and Robertson 1976; Robertson 1984).

The National Vegetation Classification (NVC) began in 1975 as a comprehensive attempt to sample, classify and describe British vegetation types. A total of about 35,000 quadrat samples was assembled from fieldwork and previous studies, such as those of McVean and Ratcliffe (1962) and Birks (1973). Analysis of these quadrats produced a scheme in which vegetation was classified at three levels: broad groups, communities and sub-communities. The broad groups are as follows: woodland and scrub (25 communities), mires (38 communities), heaths (22 communities), calcicolous grasslands (14 communities), mesotrophic grasslands (13 communities), calcifugous grasslands and montane communities (21 communities), swamps and tall-herb fens (28 communities), aquatic communities (24 communities), saltmarshes (28 communities), shingle, strandline and sand-dune communities (19 communities), maritime cliff communities (12 communities), and vegetation of open habitats (42 communities). The NVC deals with the whole of Great Britain (upland and lowland) and most upland vegetation types published in previous work are accommodated in this scheme. The work was finally presented as a series of five volumes: *British Plant Communities* (Rodwell 1991a, 1991b, 1992, 1995, 2000), with descriptions of the flora and habitat of each vegetation type, and floristic tables. The NVC is now widely accepted as the standard system for the classification of vegetation in Great Britain, for use in surveys, site evaluation, and so on.

By 1980 it was clear that the NVC would not be available for some years, and that in the meantime some kind of classification was required for vegetation surveys being carried out in the British uplands. To this end Birks and Ratcliffe (1980) drew up a *Classification of upland vegetation types in Britain* based mainly on the findings of McVean and Ratcliffe (1962) and Birks (1973).

During and since the 1980s many surveys of vegetation were carried out in upland Great Britain. The Nature Conservancy Council Upland Vegetation Survey team carried out surveys of vegetation of upland Sites of Special Scientific Interest (SSSI) and a selection of undesignated areas in Scotland and England during the 1980s and early 1990s. The vegetation was classified using the system of Birks and Ratcliffe (1980), incorporating the NVC as parts of this became available. The information was presented as a series of overlays onto 1:25,000 monochrome aerial photographs. Some of this was published by Ratcliffe and Thompson (1998), Thompson and Baddeley (1991), Thompson and Brown (1992), Brown *et al.* (1993a,b), and Thompson *et al.* (1995).

The Nature Conservancy Council Welsh Field Unit conducted similar surveys of the Welsh uplands during the 1980s. For each site a report was prepared, including descriptions of the site and its different vegetation types, and vegetation maps were drawn onto Ordnance Survey sheets, mostly at a scale of 1:25,000.

During the mid- to late 1980s the Nature Conservancy Council Scottish Field Unit carried out surveys of most of the semi-natural deciduous woodland in Argyll, Lochaber, Speyside, Nairn, Moray, Banff, Angus and Galloway. For each site the woodland vegetation was mapped and described using the NVC scheme. The data for all woods were summarised into a series of reports, one for each district (e.g. MacKintosh 1988, 1990; Tidswell 1988, 1990; and various other reports).

During the 1990s the Scottish Natural Heritage Peatland Survey Team surveyed large areas of bog, mainly in Sutherland and Caithness, using both the NVC and the classification system of British ombrotrophic bogs by Lindsay (1995).

Large areas of upland Great Britain have been surveyed since the early 1990s using the NVC. These surveys were carried out by various people on contracts commissioned by Scottish Natural Heritage, the Countryside Council for Wales, English Nature and other organisations. They have generally been written up as detailed reports, typically including descriptions of vegetation types, site descriptions and evaluations, quadrat data, site species lists, and mapped information drawn onto Ordnance Survey maps.

No classification of vegetation can be expected to be totally comprehensive. Although the NVC is the most thorough and detailed classification of British vegetation to date, new information continues to accumulate. This provides further insights into the flora and distribution of previously recognised vegetation types, and also indicates the presence of additional, newly recognised vegetation types. This information was summarised by Rodwell *et al.* (2000).

In recent years much work has been done to look at the effects of different types of land management on upland vegetation. This has resulted in the publication of the detailed two-volume guide *Upland habitats: surveying land management impacts* by MacDonald *et al.* (1998). The information presented in this guide has helped vegetation surveyors to assess the ways in which upland sites are affected by management. In particular, the guide has been used in surveys designed specifically to look at the effects of land management on vegetation, rather than the type of vegetation. The government conservation agencies are now undertaking comprehensive work across the uplands (and other habitats) to determine the condition of protected sites.

## General characteristics of upland vegetation

British upland vegetation is immensely variable. Semi-natural vegetation predominates, in contrast to the lowlands which are mostly arable farmland and agriculturally-improved pasture, with parks and gardens in the towns and cities. Bryophytes, lichens and ferns play a larger part in upland vegetation than they do in the lowlands. Acidophilous species predominate in the uplands, because of the relative scarcity of base-rich rocks and soils.

At the very simplest level the patterns of vegetation over the ground are determined largely by altitude and slope. Woodland on lower slopes gives way with increasing altitude to dwarf-shrub heaths, which, in turn, are replaced by grasslands and eventually, on the highest ground, by moss-heaths and snow-beds. Bogs cover level ground where peat has been able to accumulate, and springs, flushes and irrigated mires occur throughout the altitudinal sequence wherever water comes to the surface of the ground. However, in most places this simple pattern has become greatly modified by a history of deforestation, grazing and burning, which has produced more complex patterns of vegetation.

Heathland and unimproved grassland are the most extensive vegetation types in the uplands. Grassland often occurs above the upper altitudinal limit of heaths, as part of a natural altitudinal sequence, but it is also very common at lower altitudes where dwarf shrubs have been eliminated by grazing and burning. In areas with a long history of intensive grazing by sheep, for example in parts of Wales and the Lake District, grassland covers very large areas and heathland is comparatively scarce, being restricted mainly to places which are difficult of access for sheep. Many Scottish hills are grassy too, especially where the rocks are more base-rich and the vegetation can sustain high numbers of livestock, but in general the hills of the Highlands are heathery. Bracken is common around the lower fringes of the uplands, especially in the Lake District, Wales, the Southern Uplands and the south-west Highlands. Bog vegetation is widespread and common on deep, usually more or less level, waterlogged peat at all altitudes, and can be very extensive. Soligenous and topogenous mires are also common at all altitudes, although individual stands are mostly small. There is still much native woodland and scrub at lower altitudes, but this makes up only a small fraction of the original extent. Most of this woodland is on steeper and less accessible ground. Large tracts of ground in the uplands have been planted with exotic conifers, most commonly Sitka spruce and lodgepole pine. Tall-herb vegetation and montane willow scrub occur chiefly in small stands on ungrazed cliff ledges. Summit heaths and snow-beds dominated by bryophytes occur on the highest ground and are most extensive in the Scottish Highlands.

Bryophytes and ferns generally favour humid, equable conditions and are therefore more common on sheltered, shaded, cool and humid slopes, facing between north-west and east, than on warmer or drier slopes, facing between south-east and west. This pattern can be seen particularly well in heaths, grasslands and woodlands.

### The recognition of vegetation types

The science and art of classifying vegetation is based on the premise that plants tend to grow in recognisable and repeated patterns, assemblages or communities, and that these assemblages can be sampled and described. In reality, upland vegetation varies as an immensely complicated multi-dimensional continuum (far too complex to be understood in all its detail) within which recognised vegetation types, such as NVC communities, are artificially defined noda or points of reference. Variation can be recognised at different scales too. For example, in some places a stand of vegetation recognised as belonging to a single particular type can cover many hectares, but in other places there are complex mosaics of small patches of different vegetation types between which there are many intergradations.

Just as a map is not the territory that it represents, a vegetation classification is not the vegetation itself. Vegetation does not fall neatly into NVC sub-communities, but varies in many dimensions and through time. Classifications such as the NVC work at the human scale, but the boundaries that seem obvious to us may mean less to other animals, or indeed to the plants which actually make up the vegetation. For the plants themselves, life is a matter of the opportunity to establish initially, the suitability of environmental conditions, and competition with one's immediate neighbours for light, water and nutrients. A danger of using any system of classification is that it can encourage the recognition only of predetermined patterns. Other important variation is either not noticed or not considered because it is not dealt with in the classification. It is often more informative to look at vegetation first in a more general or unbiased way, and then to see how it fits within the framework of the classification used.

The published tables in the NVC are synthesised from quadrat samples recorded all over the country and over many years. Given the variation among the quadrat samples which make up the published floristic table for any particular NVC type, one cannot expect any individual stand of vegetation to conform exactly to the published table. It is often, therefore, inappropriate and misleading to refer to a stand of vegetation as atypical just because its species composition differs in some way from the summary NVC table.

The vegetation of upland Great Britain contains so much that is of interest and of national or international significance that the identification of different vegetation types is obviously a very important part of the study of upland vegetation in general. The national status of different vegetation types can help in the evaluation of the conservation importance of a site, and in the definition of the boundaries of sites which are to be afforded special protection.

The identification of upland vegetation types is an important element in the study of mammals, birds and insects, all of which ultimately depend on plants. For example, certain types of vegetation are selected as nest sites by breeding birds because they provide, directly or indirectly, a food supply, or because they provide safety from predators. Eagles and other large predatory birds feed on animals which in turn feed on or in vegetation. Many butterfly species feed on a very narrow range of plant species. The survey of breeding dotterel in the late 1980s (Galbraith *et al.* 1993) owed much of its success to knowledge of the combinations of vegetation types used by these birds.

The recognition of different vegetation types is important when assessing the state of the uplands in relation to land management and other anthropogenic influences. For example, the separation of different types of bogs and heaths can help to clarify past and present draining and burning; heathland and grassland types can show up grazing practices; some types of bog and montane moss-heath can suggest the effects of atmospheric pollution or the effects of nutrients in the urine and droppings of grazing animals; and the distribution of snow-bed types might be useful in monitoring the effects of global warming. Certain types of vegetation are more resistant to trampling than others and this is an essential part of deciding how to manage footpaths in the hills. A knowledge of vegetation types informs decisions about future management. For example, a site may contain certain heathland types which are scarce and highly vulnerable to burning. Where it is planned to plant new woodland of native tree species, identification of the precursor vegetation type is important when deciding which tree and shrub species to plant and which areas to leave unplanted.

### British upland vegetation in a European context
It is valuable to consider distribution patterns on a wider geographical scale, especially as this highlights the international importance of Great Britain and Ireland for many plant species and vegetation types.

British and Irish plant species have been grouped into European phytogeographical classes by various authors: Salisbury (1932), Matthews (1937, 1955), Birks (1973, 1976), and Preston and Hill (1997) for vascular plants, and Greig-Smith (1950), Ratcliffe (1968), Schuster (1983), Düll (1983, 1984, 1985) and Hill and Preston (1998) for bryophytes. The classifications by Preston and Hill (1997), and Hill and Preston (1998) are the most comprehensive and are based on the most up-to-date distributional data. Their phytogeographical classes are based on a combination of two patterns of distribution. One pattern is that of latitude and altitude, with the most northern or montane species at one extreme and the most southern and thermophilous species at the other. The other is that of restriction to the oceanic west, with the most strongly western, oceanic species at one extreme and the very widespread circumpolar species at the other.

The phytogeographical elements in the flora of upland vegetation vary considerably across Great Britain. Northern European species are best represented in northern and montane communities, southern European species are better represented in south-western communities, and western European species are commonest in western communities. This is illustrated in Table 1, which shows the percentage representation of each phytogeographical class in selected upland vegetation types of contrasting British distribution. Recent studies have drawn out contrasts between the vegetation of different parts of Europe (e.g. Nagy *et al.* 2003 a,b).

It is more difficult to set up a European phytogeographical classification for British vegetation types. Unlike species, they are not distinct taxonomic units. However, much information is available on the European distributions of vegetation types, and from this it is possible to obtain some indication of the European status of many types of British upland vegetation. Much of our upland vegetation is distinctly western and oceanic in character and distribution. Indeed, many vegetation types are more common in the uplands of Great Britain and Ireland than anywhere else in Europe. These are discussed under the individual accounts of vegetation, but particularly notable are the blanket bogs, wet heaths, *Calluna* heaths, *Racomitrium* moss-heaths and bryophyte-rich rocky woodlands. Some oceanic upland vegetation types are dominated by oceanic species, such as *Agrostis curtisii*, *Erica cinerea*, *Ulex europaeus* and *U. gallii*. Others are dominated by more widespread species, such as *Calluna vulgaris*, *Molinia caerulea*, *Pteridium aquilinum* and *Racomitrium lanuginosum*, which in Europe attain dominance only in the more western oceanic areas (Ratcliffe and Thompson 1988; Rodwell 1997). Further information on the European context of our vegetation can be found in the Phytosociological conspectus of British plant communities in Volume 5 of *British Plant Communities* (Rodwell 2000), which places all the NVC communities within a pan-European framework of vegetation types, as developed by the European Vegetation Survey project.

Given that Great Britain's climate is extremely wet and equable compared with that of mainland Europe, it is not surprising that many of our western, oceanic species are bryophytes and ferns – plants which are generally better adapted to humid and equable environments. Some of our upland vegetation types are outstanding for these plants; for example, bryophyte-rich and fern-rich rocky western woodlands and bryophyte-rich heaths on north-facing hillsides. The floras of these vegetation types show fascinating links with tropical parts of the world. Many oceanic bryophytes have markedly disjunct world distributions (e.g. Scotland, Himalaya and British Columbia) or are scattered among tropical mountain areas and reach their northern world limits in Great Britain. A few heaths in north-western Scotland have the northernmost records in the world of bryophytes growing epiphyllously on the living leaf tissue of vascular plants; this life form occurs mainly in warm, humid tropical and sub-tropical forests.

## Nature conservation in the British uplands

The most extensive tracts of semi-natural and near-natural habitat in Great Britain are in the uplands. Many of these habitats, together with their flora and fauna, are internationally important. Because of this, their value for nature conservation has long been realised, and various efforts have been made to safeguard their important and distinctive elements both within protected sites and across the wider upland environment.

# Introduction

**Table 1** Breakdown of the flora of selected British upland vegetation types according to European phytogeographical classes.

Vegetation types are NVC communities: H4 *Ulex gallii-Agrostis curtisii* heath, H10 *Calluna vulgaris-Erica cinerea* heath, H13 *Calluna vulgaris-Cladonia arbuscula* heath, H14 *Calluna vulgaris-Racomitrium lanuginosum* heath, U4 *Festuca ovina-Agrostis capillaris-Galium saxatile* grassland, U7 *Nardus stricta-Carex bigelowii* grass-heath, and U12 *Salix herbacea-Racomitrium heterostichum* snow-bed. Phytogeographical classes are those of Preston and Hill (1997) for vascular plants and Hill and Preston (1998) for bryophytes. Lichens and a very few vascular and bryophyte species are not classified. Values in cells are percentages of the total flora in the published NVC tables (Rodwell 1991b, 1992).

| NVC type: | H4 | H10 | U4 | H14 | H13 | U7 | U12 |
|---|---|---|---|---|---|---|---|
| **Broad vegetation type:** | heath | heath | grass-land | heath | heath | grass-land | snow-bed |
| **British distribution:** | SW | W | wide-spread | NW | NE | N | N |
| **Mean altitude (m):** | 235 | 295 | 319 | 429 | 683 | 756 | 1046 |
| 11 Oceanic Arctic-montane | 0.0 | 0.0 | 0.0 | 0.0 | 0.0 | 0.0 | 0.0 |
| 12 Suboceanic Arctic-montane | 0.0 | 0.0 | 0.0 | 0.0 | 0.0 | 0.0 | 0.0 |
| 13 European Arctic-montane | 0.0 | 0.0 | 0.0 | 2.9 | 3.2 | 4.0 | 16.3 |
| 14 Eurosiberian Arctic-montane | 0.0 | 0.0 | 0.0 | 0.0 | 1.6 | 1.3 | 2.3 |
| 15 Eurasian Arctic-montane | 0.0 | 0.0 | 0.0 | 0.0 | 0.0 | 0.0 | 0.0 |
| 16 Circumpolar Arctic-montane | 0.0 | 0.0 | 0.0 | 4.4 | 4.8 | 4.0 | 11.6 |
| 21 Oceanic Boreo-arctic montane | 0.0 | 0.0 | 0.0 | 0.0 | 0.0 | 0.0 | 0.0 |
| 22 Suboceanic Boreo-arctic montane | 0.0 | 0.0 | 0.0 | 0.0 | 0.0 | 0.0 | 0.0 |
| 23 European Boreo-arctic montane | 0.0 | 0.0 | 0.0 | 0.0 | 0.0 | 1.3 | 0.0 |
| 24 Eurosiberian Boreo-arctic montane | 0.0 | 0.0 | 0.0 | 0.0 | 0.0 | 0.0 | 0.0 |
| 26 Circumpolar Boreo-arctic montane | 1.6 | 7.1 | 2.4 | 7.4 | 11.3 | 14.7 | 16.3 |
| 32 Suboceanic Wide-boreal | 0.0 | 0.0 | 0.0 | 0.0 | 0.0 | 0.0 | 0.0 |
| 34 Eurosiberian Wide-boreal | 0.0 | 1.4 | 0.0 | 0.0 | 0.0 | 0.0 | 0.0 |
| 35 Eurasian Wide-boreal | 0.0 | 0.0 | 1.2 | 0.0 | 0.0 | 0.0 | 0.0 |
| 36 Circumpolar Wide-boreal | 7.8 | 8.6 | 7.2 | 5.9 | 6.5 | 9.3 | 9.3 |
| 41 Oceanic Boreal-montane | 0.0 | 0.0 | 0.0 | 0.0 | 0.0 | 2.7 | 0.0 |
| 42 Suboceanic Boreal-montane | 0.0 | 0.0 | 0.0 | 1.5 | 0.0 | 1.3 | 0.0 |
| 43 European Boreal-montane | 0.0 | 0.0 | 1.2 | 0.0 | 0.0 | 1.3 | 0.0 |
| 44 Eurosiberian Boreal-montane | 1.6 | 1.4 | 1.2 | 1.5 | 1.6 | 1.3 | 0.0 |
| 45 Eurasian Boreal-montane | 0.0 | 0.0 | 0.0 | 0.0 | 0.0 | 0.0 | 0.0 |
| 46 Circumpolar Boreal-montane | 1.6 | 1.4 | 0.0 | 2.9 | 3.2 | 2.7 | 2.3 |
| 51 Oceanic Boreo-temperate | 0.0 | 1.4 | 0.0 | 1.5 | 0.0 | 1.3 | 0.0 |
| 52 Suboceanic Boreo-temperate | 1.6 | 2.9 | 1.2 | 4.4 | 3.2 | 4.0 | 4.7 |
| 53 European Boreo-temperate | 9.4 | 11.4 | 10.8 | 10.3 | 8.1 | 10.7 | 9.3 |
| 54 Eurosiberian Boreo-temperate | 6.3 | 4.3 | 10.8 | 5.9 | 3.2 | 5.3 | 0.0 |
| 55 Eurasian Boreo-temperate | 1.6 | 4.3 | 6.0 | 4.4 | 3.2 | 2.7 | 2.3 |
| 56 Circumpolar Boreo-temperate | 6.3 | 7.1 | 4.8 | 4.4 | 3.2 | 4.0 | 2.3 |
| 63 European Wide-temperate | 0.0 | 0.0 | 0.0 | 0.0 | 0.0 | 0.0 | 0.0 |
| 64 Eurosiberian Wide-temperate | 1.6 | 1.4 | 2.4 | 0.0 | 0.0 | 0.0 | 0.0 |
| 65 Eurasian Wide-temperate | 0.0 | 0.0 | 0.0 | 0.0 | 0.0 | 0.0 | 0.0 |

**Table 1** (*continued*)

| NVC type: | H4 | H10 | U4 | H14 | H13 | U7 | U12 |
|---|---|---|---|---|---|---|---|
| Broad vegetation type: | heath | heath | grass-land | heath | heath | grass-land | snow-bed |
| British distribution: | SW | W | wide-spread | NW | NE | N | N |
| Mean altitude (m): | 235 | 295 | 319 | 429 | 683 | 756 | 1046 |
| 66 Circumpolar Wide-temperate | 1.6 | 2.9 | 4.8 | 1.5 | 0.0 | 1.3 | 0.0 |
| 70 Hyperoceanic Temperate | 0.0 | 1.4 | 0.0 | 1.5 | 0.0 | 0.0 | 0.0 |
| 71 Oceanic Temperate | 6.3 | 2.9 | 2.4 | 2.9 | 1.6 | 0.0 | 0.0 |
| 72 Suboceanic Temperate | 9.4 | 10.0 | 3.6 | 5.9 | 6.5 | 4.0 | 2.3 |
| 73 European Temperate | 14.1 | 11.4 | 22.9 | 2.9 | 1.6 | 4.0 | 0.0 |
| 74 Eurosiberian Temperate | 0.0 | 1.4 | 6.0 | 1.5 | 0.0 | 0.0 | 0.0 |
| 75 Eurasian Temperate | 0.0 | 0.0 | 1.2 | 0.0 | 0.0 | 0.0 | 0.0 |
| 76 Circumpolar Temperate | 1.6 | 1.4 | 1.2 | 0.0 | 0.0 | 0.0 | 0.0 |
| 80 Hyperoceanic Southern-temperate | 0.0 | 1.4 | 0.0 | 1.5 | 0.0 | 1.3 | 0.0 |
| 81 Oceanic Southern-temperate | 4.7 | 0.0 | 0.0 | 0.0 | 0.0 | 0.0 | 0.0 |
| 82 Suboceanic Southern-temperate | 0.0 | 0.0 | 0.0 | 0.0 | 0.0 | 0.0 | 0.0 |
| 83 European Southern-temperate | 1.6 | 0.0 | 3.6 | 0.0 | 0.0 | 0.0 | 0.0 |
| 84 Eurosiberian Southern-temperate | 1.6 | 1.4 | 3.6 | 0.0 | 0.0 | 0.0 | 0.0 |
| 85 Eurasian Southern-temperate | 1.6 | 1.4 | 1.2 | 0.0 | 0.0 | 0.0 | 0.0 |
| 86 Circumpolar Southern-temperate | 0.0 | 0.0 | 0.0 | 0.0 | 0.0 | 0.0 | 0.0 |
| 91 Mediterranean-Atlantic | 0.0 | 0.0 | 0.0 | 0.0 | 0.0 | 0.0 | 0.0 |
| 92 Submediterranean-Subatlantic | 0.0 | 0.0 | 0.0 | 0.0 | 0.0 | 0.0 | 0.0 |
| 93 Mediterranean-montane | 0.0 | 0.0 | 0.0 | 0.0 | 0.0 | 0.0 | 0.0 |
| Unclassified: vascular plants and bryophytes | 0.0 | 0.0 | 0.0 | 2.9 | 1.6 | 1.3 | 0.0 |
| Unclassified: lichens | 17.2 | 11.4 | 0.0 | 22.1 | 35.5 | 16.0 | 20.9 |
| Introduced | 1.6 | 0.0 | 0.0 | 0.0 | 0.0 | 0.0 | 0.0 |
| Total no. of species in NVC table | 64 | 70 | 83 | 68 | 62 | 75 | 43 |

The identification and protection of important wildlife sites has a long history in the UK. The first systematic attempt to prepare a national list of sites came to fruition in 1915, when the Society for the Promotion of Nature Reserves (a forerunner of the Royal Society for Nature Conservation) published a provisional list of 284 potential nature reserves. This had little immediate effect, but paved the way for the Government publication *Conservation of nature in England and Wales* in 1947, which envisaged a series of National Nature Reserves (NNRs) in Great Britain. The National Parks and Access to the Countryside Act 1949 established the Nature Conservancy as the statutory body responsible for establishing and managing NNRs and for identifying additional high-quality wildlife areas as Sites of Special Scientific Interest (SSSIs). The legal protection given to SSSIs has since been enhanced by the Wildlife and Countryside Act 1981, and, in England and Wales, by the Countryside and Rights of Way Act 2000. The statutory nature conservation agencies (English Nature, Scottish Natural Heritage and the Countryside Council for Wales) are responsible for notifying SSSIs, and work in partnership with owners and occupiers to protect and manage the sites. Common standards for site selection and monitoring are maintained by the Joint Nature Conservation Committee.

The national series of biological SSSIs is intended to represent the diversity of important wildlife and habitats in Great Britain. In general terms, the number and extent of sites should be sufficient to conserve the range of variation shown by semi-natural and natural habitats and their associated plants and animals. The SSSI series contains representative examples of upland ecosystems extending from south-west England to the north of Scotland. Because the uplands contain the most extensive tracts of remnant semi-natural habitat in Great Britain, they comprise a substantial proportion of the total area of SSSIs. Special mention should be made of the *Nature Conservation Review* (Ratcliffe 1977) which described the wildlife and habitats of Great Britain, and singled out 735 of the best examples of all habitats (graded according to national and international importance). This remains a remarkable repository of information, and the chapters in Volume 1 dealing with peatlands and upland grasslands and heaths provide rich descriptions of the range of ecological variation and of nature conservation matters.

The international importance of the British uplands is recognised in European legislation. In particular, many upland habitat types and some species are protected under Council Directive 92/43/EEC on the conservation of natural habitats and of wild fauna and flora (commonly known as the Habitats Directive). The relationship between upland NVC communities and habitats listed under Annex I of the Habitats Directive is shown in Table 2. The Directive includes a range of measures to conserve biodiversity, the most important of which is the requirement for Member States to protect important examples of habitats and species listed on the annexes of the Directive within Special Areas of Conservation (SACs). The UK network of candidate SACs currently comprises nearly 600 sites, and includes many important upland areas, including Eryri/Snowdonia and Berwyn in Wales, Dartmoor, the North Pennine Moors and the Lake District High Fells in England, and Ben Lawers, Ben Nevis, the Cairngorms and Inchnadamph in Scotland. Site details and an account of the selection rationale are provided in McLeod *et al.* (2002). Much of the work of the conservation agencies, in particular, focuses on conservation, management and advisory activities to reduce adverse effects of human activities on habitats and wildlife (e.g. Thompson *et al.* 1995, Thompson 2002).

Many upland areas in Great Britain are also classified as Special Protection Areas (SPAs) under Council Directive 79/409/EEC on the conservation of wild birds (commonly known as the Birds Directive) (Stroud *et al.* 2001). Together, SACs and SPAs comprise the Natura 2000 network – a series of wildlife conservation sites that extends across the European Union. Several examples of upland mires or water bodies have been designated as Ramsar sites under the Convention on Wetlands of International Importance especially as Waterfowl Habitat (the 'Ramsar Convention').

Over the past two decades there has been an increasing realisation that wildlife conservation cannot be achieved solely by the protection of special sites. Efforts have therefore been made to develop conservation measures across the wider environment, for example by providing farmers with incentives for positive habitat management through agri-environment schemes. Substantial areas of upland are now subject to management agreements through schemes such as Environmentally Sensitive Areas, the Countryside Stewardship Scheme (in England), the Countryside Premium Scheme (in Scotland), and Tir Gofal (in Wales). Some schemes focus specifically on the conservation management of protected sites, such as Natural Care in Scotland (which includes many moorland sites).

The UK Biodiversity Action Plan (BAP) (UK Government 1994) was published in response to the 1992 Convention on Biological Diversity. It provides a framework for the

**Table 2** Correspondence between upland NVC communities and habitats listed under Annex I of the Habitats Directive. The table only includes Annex I habitats that occur in the uplands and NVC communities that have accounts in this publication. Freshwater habitats have been excluded.

| Annex I habitat name | NVC name | Comments |
|---|---|---|
| Northern Atlantic wet heaths with *Erica tetralix* | M15 *Trichophorum cespitosum-Erica tetralix* wet heath *p.p.*<br>M16 *Erica tetralix-Sphagnum compactum* wet heath *p.p.* | In the lowlands, H5 *Erica vagans-Schoenus nigricans* heath and some examples of M14 *Schoenus nigricans-Narthecium ossifragum* mire also correspond to this Annex I type. |
| European dry heaths | H4 *Ulex gallii-Agrostis curtisii* heath<br>H8 *Calluna vulgaris-Ulex gallii* heath *p.p.*<br>H9 *Calluna vulgaris-Deschampsia flexuosa* heath<br>H10 *Calluna vulgaris-Erica cinerea* heath *p.p.*<br>H12 *Calluna vulgaris-Vaccinium myrtillus* heath *p.p.*<br>H16 *Calluna vulgaris-Arctostaphylos uva-ursi* heath *p.p.*<br>H18 *Vaccinium myrtillus-Deschampsia flexuosa* heath *p.p.*<br>H21 *Calluna vulgaris-Vaccinium myrtillus-Sphagnum capillifolium* heath *p.p.* | This Annex I type also includes various lowland dry heath communities.<br>High-altitude forms of *Calluna-Erica* heath, *Calluna-Vaccinium* heath, *Calluna-Arctostaphylos uva-ursi* heath, *Vaccinium-Deschampsia* heath and *Calluna-Vaccinium-Sphagnum* heath (above the natural tree-line) conform to the Annex I type Alpine and Boreal heaths. |
| Alpine and Boreal heaths | H10 *Calluna vulgaris-Erica cinerea* heath *p.p.*<br>H12 *Calluna vulgaris-Vaccinium myrtillus* heath *p.p.*<br>H13 *Calluna vulgaris-Cladonia arbuscula* heath<br>H14 *Calluna vulgaris-Racomitrium lanuginosum* heath<br>H15 *Calluna vulgaris-Juniperus communis* ssp. *nana* heath<br>H16 *Calluna vulgaris-Arctostaphylos uva-ursi* heath *p.p.*<br>H17 *Calluna vulgaris-Arctostaphylos alpinus* heath<br>H18 *Vaccinium myrtillus-Deschampsia flexuosa* heath *p.p.*<br>H19 *Vaccinium myrtillus-Cladonia arbuscula* heath<br>H20 *Vaccinium myrtillus-Racomitrium lanuginosum* heath<br>H21 *Calluna vulgaris-Vaccinium myrtillus-Sphagnum capillifolium* heath *p.p.*<br>H22 *Vaccinium myrtillus-Rubus chamaemorus* heath | Only high-altitudes examples of *Calluna-Erica* heath, *Calluna-Vaccinium* heath, *Calluna-Arctostaphylos uva-ursi* heath, *Vaccinium-Deschampsia* heath and *Calluna-Vaccinium-Sphagnum* heath are referable to this Annex I habitat. |
| Sub-Arctic *Salix* spp. scrub | W20 *Salix lapponum-Luzula sylvatica* scrub | This Annex I type also includes *Salix myrsinites* scrub, which is not described in the NVC, and stands of montane willows in a variety of other vegetation types. |

| Annex I habitat name | NVC name | Comments |
|---|---|---|
| *Juniperus communis* formations on heaths or calcareous grasslands | W19 *Juniperus communis* ssp. *communis-Oxalis acetosella* woodland *p.p.* | Stands of *Juniperus-Oxalis* woodland correspond to this Annex I type only if they are dominated by juniper. Stands of *Juniperus communis* ssp. *communis* on various other NVC grassland and heath types are also included.

Patches of *Juniperus-Oxalis* woodland within stands of W18 *Pinus-Hylocomium* woodland are referable to the Annex I type Caledonian forest.

Dwarf juniper heath (H15 *Calluna-Juniperus* heath) is referable to the Annex I type Alpine and Boreal heaths. Some lowland examples of this Annex I habitat are referable to W21 *Crataegus monogyna-Hedera helix* scrub. |
| *Calaminarian* grasslands of the *Violetalia calaminariae* | OV37 *Festuca ovina-Minuartia verna* community | As well as the *Festuca-Minuartia* community, this Annex I habitat includes related vegetation types that are not described in the NVC, and are characterised by metallophyte species or races of vascular and/or lower plants. |
| Siliceous alpine and boreal grasslands | U7 *Nardus stricta-Carex bigelowii* grass-heath
U8 *Carex bigelowii-Polytrichum alpinum* sedge-heath
U9 *Juncus trifidus-Racomitrium lanuginosum* rush-heath
U10 *Carex bigelowii-Racomitrium lanuginosum* moss-heath
U11 *Polytrichum sexangulare-Kiaeria starkei* snow-bed
U12 *Salix herbacea-Racomitrium heterostichum* snow-bed
U14 *Alchemilla alpina-Sibbaldia procumbens* dwarf-herb community | |
| Alpine and subalpine calcareous grasslands | CG12 *Festuca ovina-Alchemilla alpina-Silene acaulis* dwarf-herb community
CG13 *Dryas octopetala-Carex flacca* heath *p.p.*
CG14 *Dryas octopetala-Silene acaulis* ledge community | |

| Annex I habitat name | NVC name | Comments |
|---|---|---|
| Semi-natural dry grasslands and scrubland facies on calcareous substrates (Festuco-Brometalia) | CG9 Sesleria caerulea-Galium sterneri grassland p.p. CG10 Festuca ovina-Agrostis capillaris-Thymus polytrichus grassland p.p. | This Annex I type also includes various calcareous grassland NVC communities that occur in the lowlands. Some stands of Festuca-Agrostis-Thymus grassland on limestone with a significant representation of Mesobromion species are referable to this type, but most examples of this NVC community are referable to the Annex I type Species-rich Nardus grassland, on siliceous substrates in mountain areas (and submountain areas in continental Europe). |
| Species-rich Nardus grassland, on siliceous substrates in mountain areas (and submountain areas in continental Europe) | CG10 Festuca ovina-Agrostis capillaris-Thymus polytrichus grassland p.p. CG11 Festuca ovina-Agrostis capillaris-Alchemilla alpina grass-heath p.p. U4 Festuca ovina-Agrostis capillaris-Galium saxatile grassland p.p. U5 Nardus stricta-Galium saxatile grassland p.p. | Stands of Festuca-Agrostis-Thymus grassland and Festuca-Agrostis-Alchemilla grassland are only considered to be referable to this Annex I type where they occur on siliceous substrates; stands on limestone are excluded. Some species-rich examples of Festuca-Agrostis-Galium grassland and Nardus-Galium grassland are referable to this type, but most stands of these communities do not correspond to any Annex I habitat. |
| Molinia meadows on calcareous, peaty or clayey-silt-laden soils (Molinion caeruleae) | M26 Molinia caerulea-Crepis paludosa mire | Most forms of M24 Molinia caerulea-Cirsium dissectum fen-meadow also correspond to this Annex I type. |
| Hydrophilous tall herb fringe communities of plains and of the montane to alpine levels | U17 Luzula sylvatica-Geum rivale tall-herb community | |
| Mountain hay meadows | MG3 Anthoxanthum odoratum-Geranium sylvaticum grassland | |
| Active raised bogs | M1 Sphagnum denticulatum bog pool community p.p. M2 Sphagnum cuspidatum/fallax bog pool community p.p. M18 Erica tetralix-Sphagnum papillosum raised and blanket mire p.p. M19 Calluna vulgaris-Eriophorum vaginatum blanket mire p.p. | The NVC communities listed form the core of active raised bog in the UK but the list is not exhaustive. 'Active' is defined as supporting a significant area of vegetation that is normally peat-forming; this may include Sphagnum mosses, Eriophorum species or, in certain circumstances, Molinia. |

| Annex I habitat name | NVC name | Comments |
| --- | --- | --- |
| Degraded raised bogs still capable of natural regeneration | M3 *Eriophorum angustifolium* bog pool community *p.p.*<br>M15 *Trichophorum cespitosum-Erica tetralix* wet heath *p.p.*<br>M16 *Erica tetralix-Sphagnum compactum* wet heath *p.p.*<br>M18 *Erica tetralix-Sphagnum papillosum* raised and blanket mire *p.p.*<br>M20 *Eriophorum vaginatum* blanket and raised mire *p.p.*<br>M25 *Molinia caerulea-Potentilla erecta* mire *p.p.* | In degraded raised bogs the structure and function of the peat body has been disrupted and the bog surface is no longer 'active'. The NVC communities listed form the core of degraded raised bog in the UK but the list is not exhaustive.<br><br>*Trichophorum-Erica* and *Erica-Sphagnum compactum* wet heaths can occur on raised bogs, but are more common on shallow peats (< *ca* 0.5 m) where they are referable to the Annex I type Northern Atlantic wet heaths with *Erica tetralix*. |
| Blanket bog | M1 *Sphagnum denticulatum* bog pool community *p.p.*<br>M2 *Sphagnum cuspidatum/fallax* bog pool community *p.p.*<br>M3 *Eriophorum angustifolium* bog pool community *p.p.*<br>M15 *Trichophorum cespitosum-Erica tetralix* wet heath *p.p.*<br>M17 *Trichophorum cespitosum-Eriophorum vaginatum* blanket mire *p.p.*<br>M18 *Erica tetralix-Sphagnum papillosum* raised and blanket mire *p.p.*<br>M19 *Calluna vulgaris-Eriophorum vaginatum* blanket mire *p.p.*<br>M20 *Eriophorum vaginatum* blanket and raised mire *p.p.*<br>M25 *Molinia caerulea-Potentilla erecta* mire *p.p.* | The NVC communities listed form the core of blanket bog in the UK but the list is not exhaustive.<br><br>Stands of *Trichophorum-Erica* wet heath on shallow peats (< ca 0.5 m) are generally referable to the Annex I type Northern Atlantic wet heaths with *Erica tetralix*. |
| Transition mires and quaking bogs | M2 *Sphagnum cuspidatum/fallax* bog pool community *p.p.*<br>M4 *Carex rostrata-Sphagnum fallax* mire<br>M5 *Carex rostrata-Sphagnum squarrosum* mire<br>M8 *Carex rostrata-Sphagnum warnstorfii* mire<br>M9 *Carex rostrata-Calliergonella cuspidata/Calliergon giganteum* mire *p.p.*<br>M29 *Hypericum elodes-Potamogeton polygonifolius* soakway *p.p.*<br>S27 *Carex rostrata-Potentilla palustris* tall-herb fen | The NVC communities listed form the core of transition mire in the UK but the list is not exhaustive.<br><br>*Carex-Calliergonella* mire in more base-rich conditions or in association with other rich fen communities may be referable to the Annex I type Alkaline fen. Stands of *Carex-Calliergonella* mire containing *Cladium mariscus* usually conform to the Annex I habitat Calcareous fens with *Cladium mariscus* and species of the *Caricion davallianae*. |

| Annex I habitat name | NVC name | Comments |
|---|---|---|
| Depressions on peat substrates of the *Rhynchosporion* | M1 *Sphagnum denticulatum* bog pool community *p.p.*<br>M2 *Sphagnum cuspidatum/fallax* bog pool community *p.p.*<br>M15 *Trichophorum cespitosum-Erica tetralix* wet heath *p.p.*<br>M16 *Erica tetralix-Sphagnum compactum* wet heath *p.p.*<br>M17 *Trichophorum cespitosum-Eriophorum vaginatum* blanket mire *p.p.*<br>M18 *Erica tetralix-Sphagnum papillosum* raised and blanket mire *p.p.*<br>M21 *Narthecium ossifragum-Sphagnum papillosum* valley mire *p.p.*<br>M29 *Hypericum elodes-Potamogeton polygonifolius* soakway *p.p.* | This Annex I type occurs in hollows and depressions on blanket bogs, raised bogs, valley mires and heaths, and is usually characterised by an abundance of *Rhynchospora alba*. The relationship between this habitat and NVC categories is not straightforward.<br>In the lowlands, some forms of M14 *Schoenus nigricans-Narthecium ossifragum* mire correspond to this Annex I type. |
| Calcareous fens with *Cladium mariscus* and species of the *Caricion davallianae* | M9 *Carex rostrata-Calliergonella cuspidata/Calliergon giganteum* mire *p.p.* | This habitat type comprises species-rich examples of *Cladium mariscus* fen, particularly those stands enriched with elements of the *Caricion davallianae* (i.e. small-sedge fen composed of open low-growing vegetation).<br>In the lowlands, a number of additional NVC communities conform to this type. |
| Petrifying springs with tufa formation (*Cratoneurion*) | M37 *Palustriella commutata-Festuca rubra* spring<br>M38 *Palustriella commutata-Carex nigra* spring | |
| Alkaline fens | M9 *Carex rostrata-Calliergonella cuspidata/Calliergon giganteum* mire *p.p.*<br>M10 *Carex dioica-Pinguicula vulgaris* mire *p.p.* | The NVC communities listed form the core of alkaline fen in the UK but the list is not exhaustive. In the lowlands, stands of M13 *Schoenus nigricans-Juncus subnodulosus* mire may also correspond to this Annex I type.<br>Stands of *Carex-Calliergonella* mire containing much *Cladium mariscus* are referable to the Annex I type Calcareous fens with *Cladium mariscus* and species of the *Caricion davallianae*, and less base-rich examples of this NVC community may conform to the Annex I type Transition mires and quaking bogs.<br>High-altitude examples of *Carex-Pinguicula* mire that contain arctic-alpine species are referable to the Annex I type Alpine pioneer formations of *Caricion bicoloris-atrofuscae*. |

| Annex I habitat name | NVC name | Comments |
|---|---|---|
| Alpine pioneer formations of *Caricion bicoloris-atrofuscae* | M10 *Carex dioica-Pinguicula vulgaris* mire *p.p.* M11 *Carex viridula* ssp. *oedocarpa-Saxifraga aizoides* mire M12 *Carex saxatilis* mire M34 *Carex viridula* ssp. *oedocarpa-Koenigia islandica* flush | Stands of *Carex-Pinguicula* mire are only referable to this Annex I type if they occur at high altitude and contain arctic-alpine species. |
| Siliceous scree of the montane to snow levels (*Androsacetalia alpinae* and *Galeopsietalia ladani*) | U18 *Cryptogramma crispa-Athyrium distentifolium* snow-bed *p.p.* U21 *Cryptogramma crispa-Deschampsia flexuosa* community *p.p.* | This Annex I type comprises screes of siliceous rocks, generally at high altitude. As well as forms characterised by *Cryptogramma crispa*, it includes screes dominated by bryophytes or lichens, which are not covered by the NVC. |
| Calcareous and calcschist screes of the montane to alpine levels (*Thlaspietea rotundifolii*) | OV38 *Gymnocarpium robertianum-Arrhenatherum elatius* community *p.p.* | This Annex I type includes screes of calcareous or other base-rich rocks, generally at high altitude. It also includes a variety of plant communities not covered by the NVC. |
| Calcareous rocky slopes with chasmophytic vegetation | OV39 *Asplenium trichomanes-Asplenium ruta-muraria* community *p.p.* OV40 *Asplenium viride-Cystopteris fragilis* community *p.p.* | This Annex I type includes crevice vegetation of calcareous or other base-rich rocks. It includes a number of vegetation types with characteristic ferns, bryophytes and lichens, which are only partly covered by the NVC. |
| Siliceous rocky slopes with chasmophytic vegetation | U18 *Cryptogramma crispa-Athyrium distentifolium* snow-bed *p.p.* U21 *Cryptogramma crispa-Deschampsia flexuosa* community *p.p.* | This Annex I type includes crevice vegetation of siliceous rocks. It includes a number of vegetation types with characteristic ferns, bryophytes and lichens, which are poorly covered by the NVC. |
| Limestone pavements | W9 *Fraxinus excelsior-Sorbus aucuparia-Mercurialis perennis* woodland *p.p.* M10 *Carex dioica-Pinguicula vulgaris* mire *p.p.* M26 *Molinia caerulea-Crepis paludosa* mire *p.p.* M27 *Filipendula ulmaria-Angelica sylvestris* tall-herb fen *p.p.* MG5 *Cynosurus cristatus-Centaurea nigra* grassland *p.p.* CG9 *Sesleria caerulea-Galium sterneri* grassland *p.p.* CG10 *Festuca ovina-Agrostis capillaris-Thymus polytrichus* grassland *p.p.* CG13 *Dryas octopetala-Carex flacca* heath *p.p.* OV38 *Gymnocarpium robertianum-Arrhenatherum elatius* community *p.p.* | A range of calcicolous rock, heath, grassland, scrub and woodland NVC types occur on limestone pavement. The communities listed do not represent an exhaustive list. |

| Annex I habitat name | NVC name | Comments |
|---|---|---|
| Limestone pavements (*continued*) | OV39 *Asplenium trichomanes-Asplenium ruta-muraria* community *p.p.*<br>OV40 *Asplenium viride-Cystopteris fragilis* community *p.p.* | |
| *Tilio-Acerion* forests of slopes, screes and ravines | W9 *Fraxinus excelsior-Sorbus aucuparia-Mercurialis perennis* woodland *p.p.* | The relationship between this Annex I type and NVC categories is not straightforward. Essentially it incorporates stands of *Fraxinus-Sorbus-Mercurialis* woodland (and in the lowlands some forms of W8 *Fraxinus excelsior-Acer campestre-Mercurialis perennis* woodland) on rocky ground, including ravines, screes and other rocky slopes. |
| Old sessile oak woods with *Ilex* and *Blechnum* in Britain and Ireland | W11 *Quercus petraea-Betula pubescens-Oxalis acetosella* woodland *p.p.*<br>W17 *Quercus petraea-Betula pubescens-Dicranum majus* woodland *p.p.* | Oak need not necessarily be dominant in the stand for the woodland to conform to this Annex I type, but birch woods beyond the range of oak are excluded.<br>The habitat also includes some forms of W10 *Quercus robur-Pteridium aquilinum-Rubus fruticosus* woodland and W16 *Quercus species-Betula species-Deschampsia flexuosa* woodland. |
| Caledonian forest | W4 *Betula pubescens-Molinia caerulea* woodland *p.p.*<br>W17 *Quercus petraea-Betula pubescens-Dicranum majus* woodland *p.p.*<br>W18 *Pinus sylvestris-Hylocomium splendens* woodland *p.p.*<br>W19 *Juniperus communis* ssp. *communis-Oxalis acetosella* woodland *p.p.* | The majority of this Annex I habitat corresponds to *Pinus-Hylocomium* woodland. Stands of *Juniperus-Oxalis* woodland are only included where they occur within stands of *Pinus-Hylocomium* woodland; in other situations they should be considered under the Annex I type *Juniperus communis* formations on heaths or calcareous grasslands. Caledonian forest also includes some birch-dominated stands of *Quercus-Betula-Dicranum* woodland and *Betula-Molinia* woodland<br>*Pinus-Hylocomium* woodland may also include pine bog woodland which conforms to the Annex I type Bog woodland. |

| Annex I habitat name | NVC name | Comments |
|---|---|---|
| Bog woodland | W3 *Salix pentandra-Carex rostrata* woodland *p.p.* <br> W4c *Betula pubescens-Molinia caerulea* woodland, *Sphagnum* sub-community *p.p.* <br> W18 *Pinus sylvestris-Hylocomium splendens* woodland *p.p.* | The relationship between this Annex I type and the NVC is not straightforward. Pine types may be intermediate between *Erica-Sphagnum papillosum* mire or *Calluna-Eriophorum* mire and *Pinus-Hylocomium* woodland. Birch/willow/alder types may be close to *Betula-Molinia* woodland, *Salix-Carex* woodland or other wet woodland types. <br> Secondary birch woods on damaged raised bogs do not conform to this Annex I type. |
| Alluvial forests with *Alnus glutinosa* and *Fraxinus excelsior* (*Alno-Padion, Alnion incanae, Salicion alvae*) | W7 *Alnus glutinosa-Fraxinus excelsior-Lysimachia nemorum* woodland *p.p.* | Only woodlands on river floodplains conform to this Annex I type. <br> In the lowlands other forms of woodland dominated by alder or willows may also conform to this type. |

conservation and sustainable use of the UK's biodiversity, most notably through the preparation of action plans for 45 habitats and 391 species that are considered to be particularly at risk in the UK (available on the website http://www.ukbap.org.uk/). The action plans cover extensive upland habitats, such as blanket bog and upland heathland, as well as scarcer types, such as upland hay-meadows (see Table 3); several upland species, including Black Grouse *Tetrao tetrix*, Yellow Marsh Saxifrage *Saxifraga hirculus*, and the moss *Bryoerythrophyllum caledonicum*, are also represented. Each action plan contains biological objectives, and lists the actions required to meet these objectives, ranging from site safeguard to policy measures, research and publicity. A major success of the BAP approach has been the development of a collaborative approach between Government, statutory agencies and non-governmental organisations. Progress in implementing the action plans is reviewed in UK Biodiversity Group (2001).

## The aims of this book

The main aim of this book is to provide concise descriptions of all currently-recognised British upland vegetation types in a single volume. It is for everyone who wants to study, survey, manage, conserve or just enjoy the British uplands. It is written primarily for field surveyors, environmental consultants, students, teachers, researchers, field staff in conservation organisations, rangers and nature reserve wardens. It is a guide to help to identify, understand and appreciate the diversity of vegetation in the uplands.

Although earlier authors (see above) gave invaluable data on the species composition and structure of upland vegetation, some of their descriptions of vegetation are not very easy to understand, especially where they use phytosociological terminology. Many people have felt a need for concise and clear descriptions of vegetation which provide accurate botanical detail and also indicate what the vegetation looks like, where it occurs and how it can be distinguished from related types. This book attempts to provide such descriptions. It complements three earlier guides to specific volumes of British plant communities, namely grasslands and montane communities (Cooper 1997), woodlands (Hall *et al.* 2001) and mires and heaths (Elkington *et al.* 2001).

Vegetation surveys are continually adding to the existing body of information on upland vegetation types. New publications are being produced on many aspects of upland vegetation; for example, Scottish Natural Heritage have published a detailed guide to the effects of land management on upland habitats (MacDonald *et al.* 1998). However, this information is scattered through a great many reports and publications, so we have synthesised in this single volume our current understanding of the ecology, distribution, conservation and management of different vegetation types.

## Structure of the book

### Scope

This book adopts the NVC as a framework, because the NVC is the most detailed, comprehensive and widely-used classification of British vegetation at present. Where appropriate, the names of NVC communities and sub-communities have been updated to reflect recent changes in plant nomenclature (see p. 55). The published NVC volumes contain a huge amount of information that we have not been able to include here, so this book is best used in conjunction with *British Plant Communities* and is in no way intended as a replacement for it. Upland NVC types are spread between the five volumes of *British Plant Communities*, so here we have brought them all together in a single volume. We have included as much as possible of the recent information not included in

**Table 3** Correspondence between upland NVC communities and UK Biodiversity Action Plan (BAP) priority habitats.

The table only includes BAP priority habitats which occur in the uplands and NVC communities which have accounts in this publication. NVC communities which do not correspond to any priority habitat in the uplands (e.g. montane heaths and flushes) have been excluded. The term 'lowland' in several of the priority habitat names generally refers to land below the upper limit of agricultural enclosure. Priority habitats which are particularly characteristic of upland regions are indicated in bold.

| NVC community | BAP priority habitat | | | | | | | | | | | | | | | | | | |
|---|---|---|---|---|---|---|---|---|---|---|---|---|---|---|---|---|---|---|---|
| | **Upland oakwood** | **Upland mixed ashwoods** | Wet woodland | Lowland wood-pasture and parkland | **Native pine woodlands** | Lowland meadows | **Upland hay meadows** | Purple moor grass and rush pastures | Lowland calcareous grassland | **Upland calcareous grassland** | Lowland dry acid grassland | Lowland heathland | **Upland heathland** | Fens | Lowland raised bog | **Blanket bog** | Mesotrophic lakes | Eutrophic standing waters | **Limestone pavements** |
| W3 *Salix pentandra-Carex rostrata* woodland | | | x | | | | | | | | | | | | | | | | |
| W4 *Betula pubescens-Molinia caerulea* woodland | | | x | | x | | | | | | | | | | | | | | |
| W7 *Alnus glutinosa-Fraxinus excelsior-Lysimachia nemorum* woodland | | | x | | | | | | | | | | | | | | | | |
| W9 *Fraxinus excelsior-Sorbus aucuparia-Mercurialis perennis* woodland | | x | | | | | | | | | | | | | | | | | |
| W11 *Quercus petraea-Betula pubescens-Oxalis acetosella* woodland | x | | | x | x | | | | | | | | | | | | | | |
| W17 *Quercus petraea-Betula pubescens-Dicranum majus* woodland | x | | | | x | | | | | | | | | | | | | | |
| W18 *Pinus sylvestris-Hylocomium splendens* woodland | | | | | x | | | | | | | | | | | | | | |
| W19 *Juniperus communis* ssp. *communis-Oxalis acetosella* woodland | | | | | x | | | | | | | | | | | | | | x |

**BAP priority habitat**

| NVC community | Upland oakwood | Upland mixed ashwoods | Wet woodland | Lowland wood-pasture and parkland | Native pine woodlands | Lowland meadows | Upland hay meadows | Purple moor grass and rush pastures | Lowland calcareous grassland | Upland calcareous grassland | Lowland dry acid grassland | Lowland heathland | Upland heathland | Fens | Lowland raised bog | Blanket bog | Mesotrophic lakes | Eutrophic standing waters | Limestone pavements |
|---|---|---|---|---|---|---|---|---|---|---|---|---|---|---|---|---|---|---|---|
| M1 *Sphagnum denticulatum* bog pool community | | | | | | | | | | | | | | x | x | x | | | |
| M2 *Sphagnum cuspidatum/fallax* bog pool community | | | | | | | | | | | | | | x | x | x | | | |
| M3 *Eriophorum angustifolium* bog pool community | | | | | | | | | | | | | | x | x | x | | | |
| M4 *Carex rostrata-Sphagnum fallax* mire | | | | | | | | | | | | | | x | | | | | |
| M5 *Carex rostrata-Sphagnum squarrosum* mire | | | | | | | | | | | | | | x | | | | | |
| M6 *Carex echinata-Sphagnum fallax/denticulatum* mire | | | | | | | | | | | | | | x | | | | | |
| M9 *Carex rostrata-Calliergonella cuspidata/Calliergon giganteum* mire | | | | | | | | | | | | | | x | | | | | |
| M10 *Carex dioica-Pinguicula vulgaris* mire | | | | | | | | | | | | | | x | | | | | x |
| M15 *Trichophorum cespitosum-Erica tetralix* wet heath | | | | | | | | | | | | x | x | | x | x | | | |
| M16 *Erica tetralix-Sphagnum compactum* wet heath | | | | | | | | | | | | x | x | | x | | | | |
| M17 *Trichophorum cespitosum-Eriophorum vaginatum* blanket mire | | | | | | | | | | | | | | | | x | | | |
| M18 *Erica tetralix-Sphagnum papillosum* raised and blanket mire | | | | | | | | | | | | | | | x | x | | | |

**BAP priority habitat**

| NVC community | Upland oakwood | Upland mixed ashwoods | Wet woodland | Lowland wood-pasture and parkland | Native pine woodlands | Lowland meadows | Upland hay meadows | Purple moor grass and rush pastures | Lowland calcareous grassland | Upland calcareous grassland | Lowland dry acid grassland | Lowland heathland | Upland heathland | Fens | Lowland raised bog | Blanket bog | Mesotrophic lakes | Eutrophic standing waters | Limestone pavements |
|---|---|---|---|---|---|---|---|---|---|---|---|---|---|---|---|---|---|---|---|
| M19 *Calluna vulgaris-Eriophorum vaginatum* blanket mire | | | | | | | | | | | | | | | x | x | | | |
| M20 *Eriophorum vaginatum* blanket and raised mire | | | | | | | | | | | | | | | x | x | | | |
| M21 *Narthecium ossifragum-Sphagnum papillosum* valley mire | | | | | | | | | | | | | | x | | | | | |
| M23 *Juncus effusus/acutiflorus-Galium palustre* rush-pasture | | | | | | | | x | | | | | | x | | | | | |
| M25 *Molinia caerulea-Potentilla erecta* mire | | | | | | | | x | | | | | | x | x | x | | | |
| M26 *Molinia caerulea-Crepis paludosa* mire | | | | | | | | x | | | | | | x | | | | | x |
| M27 *Filipendula ulmaria-Angelica sylvestris* tall-herb fen | | | | | | | | | | | | | | x | | | | | x |
| M28 *Iris pseudacorus-Filipendula ulmaria* mire | | | | | | | | | | | | | | x | | | | | |
| M29 *Hypericum elodes-Potamogeton polygonifolius* soakway | | | | | | | | | | | | | | x | | | | | |
| M37 *Palustriella commutata-Festuca rubra* spring | | | | | | | | | | | | | | x | | | | | |
| H4 *Ulex gallii-Agrostis curtisii* heath | | | | | | | | | | | | x | x | | | | | | |
| H8 *Calluna vulgaris-Ulex gallii* heath | | | | | | | | | | | | x | x | | | | | | |
| H9 *Calluna vulgaris-Deschampsia flexuosa* heath | | | | | | | | | | | | x | x | | | | | | |

**BAP priority habitat**

| NVC community | Upland oakwood | Upland mixed ashwoods | Wet woodland | Lowland wood-pasture and parkland | Native pine woodlands | Lowland meadows | Upland hay meadows | Purple moor grass and rush pastures | Lowland calcareous grassland | Upland calcareous grassland | Lowland dry acid grassland | Lowland heathland | Upland heathland | Fens | Lowland raised bog | Blanket bog | Mesotrophic lakes | Eutrophic standing waters | Limestone pavements |
|---|---|---|---|---|---|---|---|---|---|---|---|---|---|---|---|---|---|---|---|
| H10 *Calluna vulgaris-Erica cinerea* heath | | | | | | | | | | | | x | x | | | | | | |
| H12 *Calluna vulgaris-Vaccinium myrtillus* heath | | | | | | | | | | | | x | x | | | | | | |
| H16 *Calluna vulgaris-Arctostaphylos uva-ursi* heath | | | | | | | | | | | | | x | | | | | | |
| H18 *Vaccinium myrtillus-Deschampsia flexuosa* heath | | | | | | | | | | | | | x | | | | | | |
| H21 *Calluna vulgaris-Vaccinium myrtillus-Sphagnum capillifolium* heath | | | | | | | | | | | | | x | | | | | | |
| MG3 *Anthoxanthum odoratum-Geranium sylvaticum* grassland | | | | | | | x | | | | | | | | | | | | |
| MG5 *Cynosurus cristatus-Centaurea nigra* grassland | | | | | | x | | | | | | | | | | | | | |
| MG8 *Cynosurus cristatus-Caltha palustris* grassland | | | | | | x | | | | | | | | | | | | | |
| CG9 *Sesleria caerulea-Galium sterneri* grassland | | | | | | | | | x | x | | | | | | | | | x |
| CG10 *Festuca ovina-Agrostis capillaris-Thymus polytrichus* grassland | | | | | | | | | | x | | | | | | | | | x |
| CG11 *Festuca ovina-Agrostis capillaris-Alchemilla alpina* grass-heath | | | | | | | | | | x | | | | | | | | | |
| CG12 *Festuca ovina-Alchemilla alpina-Silene acaulis* dwarf-herb community | | | | | | | | | | x | | | | | | | | | x |

**BAP priority habitat**

| NVC community | Upland oakwood | Upland mixed ashwoods | Wet woodland | Lowland wood-pasture and parkland | Native pine woodlands | Lowland meadows | Upland hay meadows | Purple moor grass and rush pastures | Lowland calcareous grassland | Upland calcareous grassland | Lowland dry acid grassland | Lowland heathland | Upland heathland | Fens | Lowland raised bog | Blanket bog | Mesotrophic lakes | Eutrophic standing waters | Limestone pavements |
|---|---|---|---|---|---|---|---|---|---|---|---|---|---|---|---|---|---|---|---|
| CG13 Dryas octopetala-Carex flacca heath | | | | | | | | | | x | | | | | | | | | x |
| CG14 Dryas octopetala-Silene acaulis ledge community | | | | | | | | | | x | | | | | | | | | |
| U1 Festuca ovina-Agrostis capillaris-Rumex acetosella grassland | | | | | | | | | | | x | | | | | | | | |
| U2 Deschampsia flexuosa grassland | | | | | | | | | | | x | | | | | | | | |
| U3 Agrostis curtisii grassland | | | | | | | | | | | x | | | | | | | | |
| U4 Festuca ovina-Agrostis capillaris-Galium saxatile grassland | | | | | | | | | | | x | | | | | | | | |
| S9 Carex rostrata swamp | | | | | | | | | | | | | | x | | | x | | |
| S10 Equisetum fluviatile swamp | | | | | | | | | | | | | | x | | | x | x | |
| S11 Carex vesicaria swamp | | | | | | | | | | | | | | x | | | x | | |
| S19 Eleocharis palustris swamp | | | | | | | | | | | | | | x | | | x | x | |
| S27 Carex rostrata-Potentilla palustris tall-herb fen | | | | | | | | | | | | | | x | | | | | |
| OV38 Gymnocarpium robertianum-Arrhenatherum elatius community | | | | | | | | | | | | | | | | | | | x |
| OV39 Asplenium trichomanes-Asplenium ruta-muraria community | | | | | | | | | | | | | | | | | | | x |
| OV40 Asplenium viride-Cystopteris fragilis community | | | | | | | | | | | | | | | | | | | x |

the published NVC volumes; for example, information from Rodwell *et al.* (2000) and from vegetation surveys undertaken since the NVC was published. We have also included a key which enables the user to assign a stand of vegetation first to one of the principal broad habitat formations and then to a particular vegetation type. Although we have not specifically referred to the use of quadrat data in this book, it is worth recording quadrat samples of vegetation for future reference. This applies particularly to vegetation that is difficult to classify in the field. We do not intend this book to be a guide to classifying vegetation into existing NVC types in a way which follows rigidly the details in the published NVC tables. Rather, it is intended to be a more general guide, which helps to provide an understanding of upland NVC types but also incorporates some of our own views on the classification of upland vegetation.

Some lichenologists are dissatisfied with the NVC because so few of the quadrat samples were recorded by persons properly able to identify lichens. Much work has been done on montane lichens by Fryday (1997) and where his studies can be related to NVC types his results have been incorporated here. Additional information on lichens has come from Brian and Sandy Coppins and from Oliver Gilbert's (2000) *New Naturalist* volume dealing specifically with lichens.

Accounts are presented here for 99 NVC communities: 34% of the total number of communities described in the five published NVC volumes. The 99 accounts cover all of the NVC types which occur mainly or exclusively in upland Great Britain, together with a few vegetation types which are principally lowland but which are sufficiently common in the uplands to be included here. Aquatic, sand dune, saltmarsh and sea-cliff communities are excluded. Types of lowland vegetation which occur rarely in the uplands are not included in the accounts, but are mentioned in the key where they are likely to be confused with upland vegetation types. Forms of upland vegetation that are not currently included in the NVC are described briefly in a section following the main vegetation accounts.

## Nature and design of the key
The key works in two stages. Just as a key to the identification of plant species takes the user first to genera and then to species, this key takes the user first to the broad formations of vegetation, such as heath, mire, grassland and swamp, and then within each of these it separates out the individual vegetation types (usually as NVC communities or sub-communities, but also including other vegetation types not described in the NVC). Lowland communities which are rare in the uplands but which can be confused with upland communities are included (with brief descriptions) in the key. The key is designed to be used to identify stands of vegetation in the field, and so it relies on characters such as dominant species and the structure, appearance and colours of the vegetation. The structure of the key differs in some ways from the broad habitat groupings used in the NVC (Heaths, Mires, Calcicolous Grasslands, etc.). So, for example, the *Dryas octopetala* communities key out as heaths, because *D. octopetala* is a dwarf shrub and these heaths can contain other woody species, although they are classed as calcicolous grasslands in the NVC.

## Nature and structure of each account
The accounts are intended to give a clear picture of each type of vegetation. There is one account for each NVC community. The descriptions avoid terms such as constancy, frequency, cover, abundance and preferential which are used in phytosociological texts in a way which can be confusing to those who know them as ordinary words with a different meaning in everyday English.

Each account has the same structure. At the beginning there is a list of synonyms, so that the vegetation types presented here can be related to the findings of earlier workers.

The published NVC volumes provide exhaustive lists of synonyms, but we have been more selective, confining our choice to the works that are most relevant to the uplands and with which our readers are likely to be most familiar: the monograph on Scottish Highland vegetation by McVean and Ratcliffe (1962), the account of the vegetation of Skye by Birks (1973), the studies of Scottish vegetation by Birse and Robertson (1976) and Birse (1980, 1984), the classification of woodlands by Peterken (1981), and the classification of British upland vegetation by Birks and Ratcliffe (1980).

The synonyms are followed by a description of the vegetation which aims to provide a concise and evocative picture of what it actually looks like and to be more than merely a list of species. Where appropriate, other characters are mentioned, such as the distinctive smell of base-rich mires. Subdivisions of NVC communities are described briefly; these are mainly NVC sub-communities but also include some other sub-types which are not described in the NVC.

Next there is a section on how to distinguish the vegetation type from similar communities which share many of the same species. This is information which is not always easy to glean from the NVC without an exhaustive comparison of the tables.

Then there is a section on the ecology of each type of vegetation, giving such details as the altitudinal range, the base-status of the soils, and the habitats in which it occurs. The landscape setting and habitat are illustrated in an accompanying drawing for each community.

The British distribution of the community is described, and illustrated by an updated distribution map in which the original NVC quadrat records are augmented by records from additional survey work throughout the country. Brief details of each community's European distribution are also provided.

The penultimate paragraph outlines the nature conservation interest of the vegetation type. Previous accounts of British upland vegetation have generally included some mention of the rare species which grow in different vegetation types. This has been done most recently in *British Plant Communities*. Since then, additional records have been made in some vegetation types, and in this section we have attempted to mention as many of these as possible. Although all semi-natural vegetation has a value for our indigenous fauna, some plant communities are particularly notable for the invertebrates, birds or other animals which make their homes or breed or feed in them. These notable faunas are mentioned in the text. Internationally important communities are identified here; many of these are included in the EC Habitats Directive (see Table 2).

Finally, management issues are dealt with, and in particular how the vegetation may best be managed for nature conservation purposes.

## Drawings

For each NVC community a black-and-white line drawing is provided. These drawings, done by Ben Averis, show the most typical habitats of the community, depicted at whatever scale is most appropriate for this. They do not attempt to show the complete range of habitats of each community because this would necessitate such distant or unrealistic views that the feel of the habitat and landscape would be lost. In some drawings, especially those of woodland and scrub, the vegetation itself is clearly depicted, at least in the foreground. However, many drawings are of large areas of ground comprising a range of different NVC types, usually with the particular type being dealt with indicated diagrammatically using stippling. This is done especially for vegetation types whose appearance and habitats would be difficult to portray in individual drawings or closer views; it seems better to show how they fit into more distant views of vegetation mosaics. Where two or more NVC communities share very similar habitats a single drawing is provided for each one, with minor modifications where necessary.

## Maps

The maps give the 100 km grid lines and show the known British distribution of each NVC community within the 10 × 10 km squares of the Ordnance Survey national grid. The records shown in each map are the original NVC samples (Rodwell 1991a, 1991b, 1992, 1995, 2000) augmented by more recent records. These later records are mainly from NVC survey and mapping work undertaken by the statutory conservation agencies, dating from 1981 to the present: the Scottish Natural Heritage (SNH) upland vegetation database, SNH peatland database, SNH lowland grassland inventory, SNH lowland heathland inventory, Nature Conservancy Council Upland Survey Project, North Pennine Project, Fenbase (Shaw and Wheeler 1990), West Yorkshire Plant Atlas (Lavin and Wilmore 1994), Countryside Council for Wales (CCW) upland vegetation database, Forestry Commission pinewood database (Forestry Commission 1998), various internal SNH reports, and some other records made by the authors. Available lowland records of these plant communities are also included for completeness.

These maps show a noticeably wider and more consolidated coverage across Great Britain than those presented in *British plant communities*, but they are still incomplete, and future surveys will certainly yield new records. In some cases variation in the density of records across the range of a community is the result of geographical variation in the intensity of survey work. Examples of intensive studies of localised areas include those of peatland communities in Lewis, Harris and Sutherland by the SNH Peatlands Project Team, and more general NVC mapping of Mull (Averis and Averis 1995a,b, 1999b).

We have not been able to trace all records of NVC communities to particular 10 × 10 km squares and so a few occurrences mentioned in the text are not mapped.

## Photographs

A selection of colour photographs is included in this book. Most of these are photographs of contrasting upland landscapes in different parts of the country. There are also a few photographs of selected, characteristic habitats and plant species, as well as of some of the effects of land-use on these.

## Critical taxa

The so-called 'critical taxa' *Euphrasia*, *Rubus*, *Hieracium* and *Taraxacum* are treated here as aggregates. There does not appear to be much value in distinguishing individual taxa within these genera for the purpose of classifying or understanding upland vegetation. Some bryophytes, for example *Hypnum cupressiforme*, are treated as *sensu lato* or *s.l.* (in the broad sense) where the taxonomy has been revised and it is not clear which of the new taxa are meant by earlier authors.

## Plant nomenclature

This book follows Stace (1997) for vascular plants, Blockeel and Long (1998) for bryophytes, and Purvis *et al.* (1992) for lichens. The nomenclature used in the NVC volumes is now superseded, especially for lichens; some lichen names in the NVC volumes have been obsolete for almost 40 years and cannot be found in the current literature (Fryday 1997). Where the current nomenclature has meant a change in the title of an NVC community, the former name has been listed as one of the synonyms.

## Literature

In the accounts, for brevity, we refer primarily to the NVC volumes (Rodwell 1991a,b, 1992, 1995, 2000) which cite the large primary literature on the flora and ecology of

British vegetation. Where appropriate, we also cite recent publications and reports that post-date the NVC volumes. These reports are available from the relevant nature conservation country agencies or the authors.

# Key to upland vegetation types

## Introduction

This key is intended to be a guide to upland plant communities using characteristics observable in single stands in the field. For this reason it does not rely on the number of times a species occurs over many stands in different locations (the frequency classes contained in the published NVC tables), but on easily-recognised features of the vegetation, such as its structure, dominant species and other common species. Other features, such as the habitat and the colour of the vegetation, are also mentioned where appropriate, to help distinguish different vegetation types.

The key takes the standard dichotomous form. The first part of the key groups the total range of upland vegetation into principal habitat formations, such as woodlands, dwarf-shrub heaths and grasslands. These habitat groupings do not correspond exactly to the framework used in the NVC. The published NVC volumes do not include a key to broad habitat categories; some grasslands are in the mires volume, some heaths are included with the calcareous grasslands, and so on. This key is intended to remedy this problem.

Subsequent sections of the key are designed to distinguish the communities, and in some cases sub-communities, that comprise the vegetation. As vegetation is so complicated and variable, brief descriptions of each community or sub-community are often included in order to reduce confusion. Nevertheless, because some stands of vegetation do not fit well into the NVC, and others are intermediate between one or more types, the key cannot be an infallible guide to the identity of any stand of vegetation. It is therefore important to read the relevant community accounts in this book, and where necessary the published NVC volumes, to be sure that vegetation has been correctly classified.

The main upland communities and sub-communities are shown in **bold type** in the key and have full accounts in this book. Those in standard type are communities which do not have a separate account in the book; generally these are lowland types which occur in parts of upland Great Britain where they might be confused with some forms of upland vegetation. Vegetation types which are not described in the published volumes of the NVC but which have been described by other authors are given the relevant reference.

## Key to principal upland vegetation formations

1   Vegetation dominated by woody species: trees, tall shrubs or dwarf shrubs, including *Dryas octopetala* but not *Thymus polytrichus*

                                                             **2**

   Vegetation dominated by grasses, rushes, sedges, forbs, ferns or bryophytes; if dwarf shrubs occur they are subordinate to other species

                                                             **5**

2    Vegetation dominated by trees or tall shrubs, or scrub of *Ulex europaeus*, *Cytisus scoparius*, tall *Juniperus communis* or montane willows, or underscrub of bramble

**3**

Heathy vegetation dominated by *Calluna vulgaris*, *Vaccinium* species, *Ulex gallii*, *Empetrum nigrum*, *Erica* species, *Arctostaphylos* species, *Myrica gale*, prostrate *Juniperus communis* or *Dryas octopetala*, with associated graminoids, forbs, bryophytes or lichens

**4**

3    Vegetation dominated by trees, including pine, birch, oak, ash, elm, alder, rowan, holly, hazel, *Salix cinerea* and *S. pentandra*

**Woodland (p. 59)**

Vegetation dominated by the shrubs *Crataegus monogyna*, *Prunus spinosa*, *Ulex europaeus*, *Cytisus scoparius*, *Juniperus communis* ssp. *communis*, *Salix aurita*, *S. arbuscula*, *S. lapponum*, *S. lanata*, *S. myrsinites*, *S. reticulata* or bramble

**Scrub (p. 63)**

4    Heathy vegetation on mineral soil or shallow peat, dominated by dwarf shrubs; Sphagna may be present but *S. papillosum* and *S. magellanicum* are absent or scarce, as is *Eriophorum vaginatum*

**Dwarf-shrub heaths (p. 64)**

Bog vegetation on deep peat, with much *Eriophorum vaginatum* and/or *Sphagnum papillosum*, *S. magellanicum* or *S. capillifolium*; usually comprising a low canopy of ericoid shrubs and in many places *Molinia caerulea* or *Trichophorum cespitosum*; erosion and hags may be present, typically showing peat depths of one or more metres

**Blanket, raised and valley mires (p. 70)**

5    Vegetation dominated by grasses, sedges or rushes

**6**

Vegetation dominated by ferns, forbs or bryophytes

**10**

6    Vegetation dominated by grasses (other than *Phragmites australis* or *Phalaris arundinacea*), *Juncus squarrosus*, *J. trifidus* or *Carex bigelowii* on damp or dry soils; dwarf shrubs may play a minor part in the sward

**7**

Vegetation dominated by sedges (other than *Carex bigelowii*), *Juncus effusus*, *J. acutiflorus*, *J. inflexus*, *Equisetum fluviatile*, *Eleocharis palustris*, *Schoenoplectus lacustris*, *Phragmites australis*, *Phalaris arundinacea*, *Eriophorum* species or *Trichophorum cespitosum*; usually on wet ground or in standing water; *Molinia caerulea* can be common but not dominant

**8**

7    Vegetation dominated by one or more of *Festuca* species, *Agrostis* species, *Deschampsia* species, *Nardus stricta*, *Arrhenatherum elatius*, *Dactylis glomerata*, *Anthoxanthum odoratum*, *Holcus lanatus*, *Molinia caerulea*, *Lolium perenne* or *Sesleria caerulea*

**Grasslands (p. 72)**

Vegetation dominated by *Juncus squarrosus, J. trifidus* or *Carex bigelowii*
**Rush- and sedge-heaths (p. 78)**

8   Vegetation composed of *Eriophorum* species, dwarf shrubs such as *Calluna vulgaris* and *Erica* species, and in many places *Trichophorum cespitosum, Molinia caerulea* and/or *Myrica gale* on deep peat, usually with bog mosses such as *Sphagnum papillosum* and *S. capillifolium*
**Blanket, raised and valley mires (p. 70)**

Vegetation dominated by sedges (other than *Carex bigelowii*), *Juncus effusus, J. acutiflorus, J. inflexus, Equisetum fluviatile, Eleocharis palustris, Schoenoplectus lacustris, Phragmites australis* or *Phalaris arundinacea* in standing water or on very wet and often flushed ground
         **9**

9   Vegetation dominated by *Carex rostrata, C. vesicaria, C. paniculata, Equisetum fluviatile, Eleocharis palustris, Schoenoplectus lacustris, Phragmites australis* or *Phalaris arundinacea*, commonly with tall forbs but rarely with thick carpets of bryophytes, in shallow standing water or on very wet ground
**Swamps and fens (p. 79)**

Flushes, mires with moving water (soligenous) and wet pastures generally dominated by rushes or sedges over a carpet of Sphagna or other bryophytes
**Soligenous mires and rush-pastures (p. 80)**

10  Vegetation dominated by ferns or forbs
         **11**

Vegetation dominated by bryophytes or by mixtures of bryophytes and small herbs in springs and wet soakways
         **12**

11  Vegetation dominated by ferns
**Fern communities (p. 85)**

Vegetation dominated by forbs
**Tall-herb and dwarf-herb vegetation (p. 87)**

12  Vegetation of springheads, wet soakways, and pools or very wet depressions in blanket and raised bogs
**Springs, soakways and bog pools (p. 89)**

Bryophyte-dominated vegetation of summits and high ridges, and of sheltered hollows, gullies and corrie walls at high altitudes where snow lies late
**Montane moss-heaths and snow-beds (p. 91)**

## Key to woodlands

Vegetation dominated by trees, including *Salix pentandra* and *S. cinerea*

1   Coniferous woodlands
         **2**

Deciduous woodlands
         **6**

2   Pine woods
         **3**

Planted woods of exotic conifers, such as Sitka spruce, Norway spruce or larch

Coniferous plantation

3    Pine woods with a heathy field layer

4

Pine woods with a grassy or very sparse and species-poor ground flora

5

4    Native pine woods or old and open plantations, with a heathy field layer of *Calluna vulgaris* and/or *Vaccinium* species, and commonly with a luxuriant layer of bryophytes

***Pinus sylvestris-Hylocomium splendens* woodland W18**

Native pine woods or old and open plantations, with a short heathy field layer of *Calluna vulgaris* and *Erica cinerea* conspicuously and thickly mingled with lichens

**Lichen-rich pine woods (see W18)**

5    Obviously native pine woods (trees uneven-aged and scattered rather than in straight lines) on peaty soils and with a field layer dominated by *Molinia caerulea*

**Pine woods with *Molinia* (see W18)**

Dense or more open woods of obviously planted trees, usually even-aged and in straight lines, and with ground vegetation of common grasses and mosses or a sparsely vegetated woodland floor littered with fallen needles

Pine plantation

6    Woodland with a field layer dominated by forbs, sedges, *Brachypodium sylvaticum*, *Deschampsia cespitosa* or *Phragmites australis* rather than other grasses or dwarf shrubs

7

Woodland with a field layer dominated by dwarf shrubs, grasses (except *Brachypodium sylvaticum*, *Deschampsia cespitosa* or *Phragmites australis*), bracken, bryophytes or *Luzula sylvatica*

13

7    Herb-rich woodland on damp or dry soils with *Mercurialis perennis*, *Brachypodium sylvaticum*, *Deschampsia cespitosa*, *Viola riviniana*, *Dryopteris filix-mas* and bryophytes such as *Eurhynchium praelongum*, *E. striatum*, *Thuidium tamariscinum* and *Mnium hornum*; the canopy commonly includes ash, elm and hazel

8

Herb-rich woodland on wet soils with at least one of *Filipendula ulmaria*, *Mentha aquatica*, *Galium palustre*, *Caltha palustris*, *Deschampsia cespitosa*, *Phragmites australis* and sedges common, and usually with much alder or willow in the canopy

9

8    Upland woodland with *Mercurialis perennis*, *Brachypodium sylvaticum*, *Deschampsia cespitosa*, *Viola riviniana*, *Oxalis acetosella* and other woodland

herbs, and abundant ferns and bryophytes; commonly under a canopy of ash, elm or hazel but birch and rowan are also common

**Fraxinus excelsior-Sorbus aucuparia-Mercurialis perennis woodland W9**

Lowland woodland similar to *Fraxinus-Sorbus-Mercurialis* woodland W9 (see above) but birch, rowan, *Oxalis acetosella* and ferns scarce or absent and bryophytes rather sparse. Bramble can be common. The lowland counterpart of *Fraxinus-Sorbus-Mercurialis* woodland which can occur on low ground in upland areas

*Fraxinus excelsior-Acer campestre-Mercurialis perennis* woodland W8

9    Swampy woodlands with a tall dense field layer of *Carex paniculata* under a canopy usually of birch, willow or alder; on moderately base-rich peat in floodplains, wet glens and basins; widespread but scarce in the lowlands, and rare at low altitudes in upland areas

*Alnus glutinosa-Carex paniculata* woodland W5

Wet or swampy woodlands with a field layer not dominated by *Carex paniculata*

**10**

10   Fen woodlands of willow or birch with *Phragmites australis* abundant or dominant in the field layer, either with woodland herbs or Sphagna; on very wet ground in floodplains, valley floors and basins; widespread but scarce in the lowlands, and rare at low altitudes in upland Great Britain

*Salix cinerea-Betula pubescens-Phragmites australis* woodland W2

Woodlands in which *Phragmites australis* is not abundant or dominant in the field layer

**11**

11   Swampy woodland with *Carex rostrata*, other species of *Carex*, *Menyanthes trifoliata*, *Potentilla palustris*, *Caltha palustris*, *Filipendula ulmaria*, *Calliergon cordifolium* and *C. giganteum* under a canopy usually of willows and birch

**Salix pentandra-Carex rostrata woodland W3**

Wet woodland without these swamp species

**12**

12   Wet woodland, commonly on flushed sloping ground, with *Filipendula ulmaria*, *Lysimachia nemorum*, *Deschampsia cespitosa*, *Athyrium filix-femina*, *Dryopteris dilatata* and mosses such as *Eurhynchium praelongum*, *Brachythecium rivulare* and *Calliergonella cuspidata*; canopy of alder, ash, hazel, birch or willows

**Alnus glutinosa-Fraxinus excelsior-Lysimachia nemorum woodland W7**

Wet *Salix cinerea* woodland with field layer including *Galium palustre*, *Juncus effusus* and *Mentha aquatica*, and with little or no *Lysimachia nemorum*, *Deschampsia cespitosa* and *Athyrium filix-femina*; lowland species such as *Hedera helix*, *Rubus fruticosus*, *Solanum dulcamara* and *Lycopus europaeus* can occur; bryophytes generally sparse; lowland woodland of wet soils on level ground or gentle slopes on floodplains, by rivers and beside ponds

*Salix cinerea-Galium palustre* woodland W1

13 Woodland with a field layer dominated by *Luzula sylvatica*

> **Woodland with a field layer of *Luzula sylvatica*; can be too species-poor to classify to NVC type but see *Quercus-Betula-Oxalis* woodland W11 and *Quercus-Betula-Dicranum* woodland W17**

Woodland with a field layer dominated by dwarf shrubs, grasses, bracken or bryophytes but not by *Luzula sylvatica*

**14**

14 Heathy woodlands with *Vaccinium myrtillus*, *Calluna vulgaris* and/or conspicuously extensive and diverse carpets of bryophytes over most available substrates; grassy forms, usually with much *Deschampsia flexuosa*, still have abundant and varied bryophytes; forbs not very common; commonly on steep rocky ground with thin acid soils; canopy usually of birch and/or oak

> ***Quercus petraea-Betula pubescens-Dicranum majus* woodland W17**

Grassy woodlands on slopes and level ground, usually without many rocks; although bryophytes can be common they do not form extensive mats over the ground, rocks and trees; small forbs can be very common

**15**

15 Woodlands on wet peaty soils, with a species-poor flora of *Molinia caerulea*, *Juncus effusus*, *Sphagnum fallax*, *S. palustre* and *Polytrichum commune*, and a canopy usually of birch, willow or alder

> ***Betula pubescens-Molinia caerulea* woodland W4**

Woodlands on well-drained slopes, with little or no *Molinia caerulea* and with a grassy field layer which can contain many small forbs; bracken can be very common

**16**

16 Woodlands with a dense field layer of *Deschampsia flexuosa*, commonly growing in rounded tufts so that the woodland floor looks as if it is covered with a green quilt, under a canopy of oak, birch or beech

**17**

Woodlands where the field layer comprises mixtures of *Holcus lanatus*, *H. mollis*, *Anthoxanthum odoratum*, *Agrostis capillaris* and *Deschampsia flexuosa*; usually with many woodland forbs, such as *Oxalis acetosella*, *Primula vulgaris*, *Viola riviniana*, *Potentilla erecta* and *Hyacinthoides non-scripta*, much bracken and in some places bramble

**18**

17 Species-poor woodlands on thin well-drained soils with a dense sward of *Deschampsia flexuosa* and in many places scattered bracken, under a canopy of oak and/or birch; not very mossy; the lowland counterpart of *Quercus-Betula-Dicranum* woodland W17; occurs on thin acid soils on some upland fringes

> *Quercus* species-*Betula* species-*Deschampsia flexuosa* woodland W16

Woodlands with a species-poor field layer of *Deschampsia flexuosa* under a canopy of planted beech; widespread in the lowlands and rather scarce in the uplands

> *Fagus sylvatica-Deschampsia flexuosa* woodland W15

18  Woodlands on well-drained soils, with a grassy field layer of *Holcus lanatus*, *Agrostis capillaris*, *Anthoxanthum odoratum*, forbs such as *Oxalis acetosella*, *Viola riviniana*, *Potentilla erecta*, *Primula vulgaris* and *Hyacinthoides non-scripta*, and commonly with much bracken

**Quercus petraea-Betula pubescens-Oxalis acetosella woodland W11**

Woodlands on well-drained soils, with a species-poor field layer usually dominated by *Holcus mollis*, bracken or bramble; *Anthoxanthum odoratum*, *Agrostis capillaris*, *Viola riviniana* and *Potentilla erecta* scarce or absent; the lowland counterpart of *Quercus-Betula-Oxalis* woodland W11 which can occur around upland fringes

*Quercus robur-Pteridium aquilinum-Rubus fruticosus* woodland W10

# Key to scrub

Vegetation dominated by *Crataegus monogyna*, *Prunus spinosa*, *Ulex europaeus*, *Cytisus scoparius*, *Juniperus communis* (except prostrate forms), *Salix aurita*, *S. arbuscula*, *S. lapponum*, *S. lanata*, *S. myrsinites*, *S. reticulata* or bramble

1  Scrub dominated by willows: *Salix aurita*, *S. arbuscula*, *S. lapponum*, *S. lanata*, *S. myrsinites* or *S. reticulata*

**2**

Scrub with a canopy of *Crataegus monogyna*, *Prunus spinosa*, *Ulex europaeus*, *Cytisus scoparius* or *Juniperus communis* ssp. *communis*, or underscrub of bramble

**5**

2  Scrub at moderate to high altitudes composed of montane willows (combinations of *Salix arbuscula*, *S. lapponum*, *S. lanata*, *S. myrsinites* and *S. reticulata*) and usually with a rich ground flora of tall herbs; usually on cliff ledges

**Salix lapponum-Luzula sylvatica scrub W20**

Scrub at low to moderate altitudes, dominated by *Salix aurita*

**3**

3  Willow scrub on wet peaty soils, with a species-poor flora of *Molinia caerulea*, *Juncus effusus*, *Sphagnum fallax*, *S. palustre* and *Polytrichum commune*

**Salix aurita scrub form of Betula pubescens-Molinia caerulea woodland W4**

Willow scrub with little or no *Molinia* and *Sphagnum*; forbs and sedges common

**4**

4  Swampy woodland with *Carex rostrata*, other species of *Carex*, *Menyanthes trifoliata*, *Potentilla palustris*, *Caltha palustris*, *Filipendula ulmaria*, *Calliergon cordifolium* and *C. giganteum*

**Salix aurita scrub form of Salix pentandra-Carex rostrata woodland W3**

Wet woodland, commonly on flushed sloping ground, with *Filipendula ulmaria*, *Lysimachia nemorum*, *Deschampsia cespitosa*, *Athyrium filix-femina*,

*Dryopteris dilatata* and mosses such as *Eurhynchium praelongum*, *Brachythecium rivulare* and *Calliergonella cuspidata*
**Salix aurita scrub form of Alnus glutinosa-Fraxinus excelsior-Lysimachia nemorum woodland W7**

5   Scrub dominated by *Juniperus communis* ssp. *communis* with a heathy or grassy field layer
**Juniperus communis ssp. communis-Oxalis acetosella woodland W19**

Scrub dominated by other species

**6**

6   Tall scrub dominated by *Crataegus monogyna* or *Prunus spinosa*

**7**

Scrub dominated by *Ulex europaeus*, *Cytisus scoparius* or *Rubus fruticosus*

**8**

7   Scrub dominated by *Crataegus monogyna*, commonly over a mixture of forbs, bramble and ivy; common in the lowlands and can occur at low altitudes in the uplands
*Crataegus monogyna-Hedera helix* scrub W21

Prickly and in many places impenetrable scrub dominated by *Prunus spinosa*, usually with sparse ground vegetation of forbs and in some places bramble or ivy; common in the lowlands and can occur at low altitudes in upland areas, for example near the sea in the west Highlands and Inner Hebrides
*Prunus spinosa-Rubus fruticosus* scrub W22

8   Scrub dominated by *Ulex europaeus* and/or *Cytisus scoparius*, with a grassy field layer or, in dense stands, with very little vegetation beneath the shrubs
**Ulex europaeus-Rubus fruticosus scrub W23**

Scrub dominated by *Rubus fruticosus*

**9**

9   *Rubus fruticosus* scrub with some *Holcus lanatus*, *Dactylis glomerata*, *Arrhenatherum elatius* and forbs on disturbed and abandoned ground at low altitudes in upland areas, especially around old buildings and at roadsides
*Rubus fruticosus-Holcus lanatus* underscrub W24

*Rubus fruticosus* scrub with much bracken and in many places *Rubus idaeus*, *Rosa canina* and some woodland species, such as *Hyacinthoides non-scripta* and *Teucrium scorodonia*
**Pteridium aquilinum-Rubus fruticosus underscrub W25**

## Key to dwarf-shrub heaths

Vegetation dominated by *Calluna vulgaris*, *Vaccinium* species, *Erica* species, *Arctostaphylos* species, *Empetrum nigrum*, *Ulex gallii*, prostrate *Juniperus communis* or *Dryas octopetala*

1   Heaths in which *Dryas octopetala* is abundant or dominant

**2**

Heaths dominated by other shrubs: *Calluna vulgaris*, *Vaccinium* species, *Ulex gallii*, *Empetrum nigrum*, *Erica* species, *Arctostaphylos* species or prostrate *Juniperus communis*

**3**

2    Heaths of *Dryas octopetala* and in some places also *Arctostaphylos uva-ursi* and *Empetrum nigrum*, with *Thymus polytrichus*, sedges such as *Carex flacca* and *C. panicea*, and a variety of small calcicolous forbs; at low altitudes on limestone and less commonly on other base-rich rocks (e.g. basalt) or on shell-sand

*Dryas octopetala-Carex flacca* **heath CG13**

Heaths with *Dryas* and a rich array of montane calcicoles and mesotrophic forbs, such as *Silene acaulis*, *Saxifraga aizoides*, *S. oppositifolia*, *Carex capillaris*, *Persicaria vivipara* and *Alchemilla alpina*, on ungrazed ledges at moderate to high altitudes, on mica-schist or similarly base-rich rocks

*Dryas octopetala-Silene acaulis* **ledge community CG14**

3    Wet heaths with a patchy variegated brown and ochre canopy of *Calluna vulgaris*, *Erica tetralix*, *Molinia caerulea*, *Trichophorum cespitosum* and, in south-west England and south Wales, *Ulex gallii* and *Agrostis curtisii*; *Sphagnum capillifolium*, *S. denticulatum* or *S. compactum* are common but *S. papillosum* is generally rare; *Eriophorum angustifolium* is common but *E. vaginatum* is absent or scarce; on shallow but wet peat; if *S. papillosum* and/ or *E. vaginatum* are common, see key to blanket, raised and valley mires (p. 70)

**4**

Dry or damp heaths dominated by ericoid shrubs, *Empetrum nigrum*, prostrate *Juniperus communis* or *Ulex gallii*, locally with much *Sphagnum capillifolium* but with little or no *Molinia caerulea*, *Erica tetralix* and *Trichophorum cespitosum*; on well-drained mineral soils or shallow humic soils over scree; in some places on dry peat but then lacking any bog species

**6**

4    Heaths with *Calluna vulgaris*, *Ulex gallii*, *Erica cinerea*, *E. tetralix*, *Molinia caerulea*, *Trichophorum cespitosum* and *Agrostis curtisii*; on wet peat or damp mineral soils up to about 500 m; only in south-west England and south Wales

*Ulex gallii-Agrostis curtisii* **heath H4**

Wet heaths in which *Ulex gallii* and *Agrostis curtisii* are scarce or absent

**5**

5    Wet heaths with an uneven patchy sward of *Calluna vulgaris*, *Erica tetralix*, *Molinia caerulea* and *Trichophorum cespitosum*, commonly with scattered shoots of *Narthecium ossifragum* and *Eriophorum angustifolium*; varies from tall swards co-dominated by *Calluna* and *Molinia* to shorter and more open swards with much *Erica cinerea* and *Racomitrium lanuginosum*; in the far north these shorter forms can have a noticeable white frosting of *Cladonia* lichens, and locally in the far north-west Highlands they can contain *Juniperus communis* ssp. *nana*; some stands contain scattered *Vaccinium myrtillus*, and others are dominated by *Trichophorum cespitosum* with very sparse dwarf shrubs and few other species

*Trichophorum cespitosum-Erica tetralix* **wet heath M15**

Wet heaths with much *Sphagnum compactum* under an even mixed sward of *Calluna* and *Erica tetralix* with sparse graminoids
**_Erica tetralix-Sphagnum compactum_ wet heath M16**

6    Heaths dominated by *Ulex gallii* with little or no *Calluna vulgaris*
**_Ulex_-dominated form of _Calluna vulgaris-Ulex gallii_ heath H8**

Dry heaths dominated by *Calluna vulgaris*, *Vaccinium* species, *Arctostaphylos* species or *Empetrum nigrum*; *Ulex gallii* may be co-dominant with *Calluna* but not the sole dominant

**7**

7    Dry heaths dominated by *Vaccinium myrtillus* or more rarely *V. vitis-idaea* or *Empetrum nigrum* with little or no *Calluna vulgaris*, appearing green from a distance

**8**

Heaths in which *Calluna vulgaris* is either dominant or at least co-dominant with other dwarf shrubs such as *Vaccinium myrtillus*, *Erica cinerea*, *Arctostaphylos uva-ursi*, *Juniperus communis* ssp. *nana* or *Ulex gallii*, appearing dark from a distance when the shrubs are not in flower, and with a rich purple hue in late summer when the heather is flowering

**12**

8    Sub-montane heaths of *Vaccinium myrtillus* or more rarely *V. vitis-idaea* or *Empetrum nigrum* ssp. *nigrum*, with species such as *Deschampsia flexuosa* and *Potentilla erecta*; can have a thick underlay of mosses (including *Sphagnum capillifolium*) but *Racomitrium lanuginosum* is not abundant, and montane species are rare

**9**

Montane heaths of short *Vaccinium myrtillus*, in many places with *V. vitis-idaea*, *V. uliginosum*, *Empetrum nigrum* ssp. *hermaphroditum*, *Diphasiastrum alpinum* and other montane species, with a ground layer of varied mixtures of large pleurocarpous mosses, *Racomitrium lanuginosum*, *Sphagnum capillifolium*, large leafy liverworts or of *Cladonia* lichens

**10**

9    *Vaccinium myrtillus*, *V. vitis-idaea* or *Empetrum nigrum* ssp. *nigrum* heaths with varying amounts of grasses and in some places *Alchemilla alpina*; commonly with a thick underlay of pleurocarpous mosses; *Cladonia* lichens may be common but montane species are absent
**_Vaccinium myrtillus-Deschampsia flexuosa_ heath H18**

Damp heaths dominated by *Vaccinium myrtillus*, in some places with *V. vitis-idaea* or *Empetrum nigrum* ssp. *nigrum*, and with much *Sphagnum capillifolium* in extensive mats of bryophytes under the dwarf shrubs; on steep shaded slopes at moderate altitudes in north-west England, Wales, and Scotland
**_Vaccinium myrtillus-Sphagnum capillifolium_ damp heath (see H21)**

10  Montane heaths of *Vaccinium* spp. and *Empetrum nigrum* ssp. *hermaphroditum* with a conspicuous white ground layer of *Cladonia arbuscula* and other lichens such as *Cetraria islandica*
**_Vaccinium myrtillus-Cladonia arbuscula_ heath H19**

Montane heaths of *Vaccinium myrtillus* and *Empetrum nigrum* ssp.
*hermaphroditum* with a mossy ground layer dominated by large
pleurocarpous mosses or *Racomitrium lanuginosum* together with varying
amounts of *Sphagnum capillifolium* and large liverworts, but at most a thin
scattering of lichens

**11**

11  Mossy ground layer dominated by *Racomitrium lanuginosum*, in some places
with *Sphagnum capillifolium*, large leafy liverworts and small quantities of
large pleurocarpous mosses
*Vaccinium myrtillus-Racomitrium lanuginosum* heath **H20**

Mossy ground layer dominated by large pleurocarpous mosses and *Sphagnum
capillifolium*; *Racomitrium lanuginosum* scarce
*Vaccinium myrtillus-Rubus chamaemorus* heath **H22**

12  *Calluna vulgaris* and *Arctostaphylos uva-ursi* co-dominant on dry stony soils
*Calluna vulgaris-Arctostaphylos uva-ursi* heath **H16**

*Calluna vulgaris* dominant or mixed with *Vaccinium myrtillus*, *Erica cinerea*
or *Ulex gallii*

**13**

13  Mixed heaths of *Calluna vulgaris* and *Ulex gallii*

**14**

Heaths of *Calluna vulgaris* and other dwarf shrubs but not *Ulex gallii*

**15**

14  Heaths with *Calluna vulgaris*, *Ulex gallii*, *Erica cinerea* and *Agrostis curtisii*;
the community is a blaze of purple and gold in late summer when the shrubs
are flowering; in south-west England and south Wales
*Ulex gallii-Agrostis curtisii* heath, *Agrostis curtisii-Erica cinerea* sub-
community **H4a**

Heaths with *Calluna vulgaris* and *Ulex gallii* but no *Agrostis curtisii*; the rich
golden flowers of *Ulex* are visible from afar and serve to distinguish these
heaths from pure heather moorland in the summer when the shrubs are
flowering; in the uplands of south-west England, throughout Wales, in
northern England, and very rare in southern Scotland
*Calluna vulgaris-Ulex gallii* heath **H8**

15  Sub-montane heaths with tall or at least erect *Calluna vulgaris* and no
montane species, appearing dense and dark from a distance except when the
heather is flowering in late summer

**16**

Montane heaths with *Calluna vulgaris* growing horizontally along the ground
(prostrate) or severely stunted, commonly with montane species, such as
*Cetraria islandica*, *Carex bigelowii* and *Diphasiastrum alpinum*; because the
stems of the heather are exposed to view these heaths have a pale reddish-
grey tinge quite different from the dark purple-brown of the taller, sub-
montane heaths

**21**

16  Mixed heaths of *Calluna vulgaris* and *Erica cinerea*; *Vaccinium myrtillus* scarce or absent

**17**

Mixed heaths of *Calluna vulgaris* and *Vaccinium myrtillus*, in some places with much *Empetrum nigrum* and/or *Vaccinium vitis-idaea*; if *Erica cinerea* occurs in quantity, *V. myrtillus* is also abundant

**18**

17  Maritime heaths with *Calluna vulgaris*, *Festuca ovina*, *Erica cinerea*, *Thymus polytrichus*, *Plantago lanceolata* and maritime species such as *Scilla verna*, *Armeria maritima* and *Plantago maritima*; on coastal slopes and cliffs
*Calluna vulgaris-Scilla verna* heath H7

*Calluna vulgaris* dominant or co-dominant with *Erica cinerea*; *Vaccinium myrtillus* scarce or absent; *Potentilla erecta* and *Carex binervis* common; the bright purple-pink flowers of *Erica cinerea* are conspicuous from midsummer onwards; commonly with abundant pleurocarpous mosses, in some places with grasses or a few small forbs, and in some other places with *Cladonia* lichens
**Calluna vulgaris-Erica cinerea heath H10**

18  Damp heaths with abundant *Sphagnum capillifolium* as well as other bryophytes under a canopy of *Calluna vulgaris* and *Vaccinium* species; on humic rankers over scree or very shallow damp peat; can occur from a few metres above sea-level to well above the tree-line

**19**

Sub-montane heaths either dominated by *Calluna vulgaris* or co-dominated by *Calluna* and *Vaccinium myrtillus*; in some places with *Vaccinium vitis-idaea* but with neither *Sphagnum capillifolium* nor montane species

**20**

19  Damp heaths with a canopy of *Calluna vulgaris* and smaller quantities of *Vaccinium myrtillus*, over a deep and usually extensive layer of mosses including *Sphagnum capillifolium*; *Racomitrium lanuginosum* can be common; in some places there are large mats and patches of leafy liverworts, such as the bright reddish-orange *Herbertus aduncus* ssp. *hutchinsiae*; where the liverworts are absent the *S. capillifolium* gives a deep red tinge to the moss layer; generally on steep and shaded slopes
**Calluna vulgaris-Vaccinium myrtillus-Sphagnum capillifolium heath H21**

Damp montane heaths with a short dense mixed canopy of *Calluna vulgaris* and *Vaccinium myrtillus* with much *Empetrum nigrum* ssp. *hermaphroditum* and a deep mossy layer of *Sphagnum capillifolium* mixed with *Rhytidiadelphus loreus*, *Hylocomium splendens*, *Dicranum scoparium* and *Pleurozium schreberi*; this mat of bryophytes has a distinctive golden-red colour from a distance; commonly dotted with *Rubus chamaemorus* and/or *Cornus suecica*; not confined to shaded slopes except in the south and east of the Highlands, where it occurs in sheltered hollows where snow obviously lies late
**Vaccinium myrtillus-Rubus chamaemorus heath H22**

20   *Calluna vulgaris* dominant, in some places with a little *Vaccinium myrtillus*
     and/or *Deschampsia flexuosa*, or *Molinia caerulea*; dense species-poor heaths
     with some *Pohlia nutans* but without a deep layer of mosses
                    ***Calluna vulgaris-Deschampsia flexuosa* heath H9**

     *Calluna vulgaris* abundant or dominant with much *Vaccinium myrtillus* in
     more varied heaths with a thick mat of large pleurocarpous mosses under the
     shrubs; in some places with abundant *Cladonia* lichens
                    ***Calluna vulgaris-Vaccinium myrtillus* heath H12**

21   Montane heaths dominated by *Calluna vulgaris*, in some places with
     *Empetrum nigrum* ssp. *hermaphroditum*, *Erica cinerea* or *Vaccinium* species
                                                                      **22**

     Mixed heaths of *Calluna vulgaris*, *Juniperus communis* ssp. *nana* and
     *Arctostaphylos alpinus*
                                                                      **25**

22   Prostrate or severely dwarfed heaths of *Calluna vulgaris* and in some places
     *Empetrum nigrum*, *Erica cinerea* or *Vaccinium* species, the dwarf shrubs less
     than 5 cm tall and surrounded by dense silvery mats of *Racomitrium
     lanuginosum* with, at most, a scattering of lichens
                                                                      **23**

     Prostrate or severely dwarfed heaths of *Calluna vulgaris* and in some places
     *Empetrum nigrum*, *Erica cinerea* or *Vaccinium* species, either with a dense,
     continuous and conspicuous creamy-white mat of lichens around the dwarf
     shrubs or with a species-poor canopy of *Calluna* with neither lichens nor
     *Racomitrium lanuginosum*
                                                                      **24**

23   Short, open and grassy heaths, usually with dwarfed clumps of *Calluna
     vulgaris* in a matrix of *Nardus stricta* with much *Racomitrium lanuginosum*
     and species such as *Diphasiastrum alpinum*, *Huperzia selago*, *Carex bigelowii*
     and *C. pilulifera*; heaths look distinctively speckled pale and dark from a
     distance
           ***Nardus stricta-Galium saxatile* grassland, *Racomitrium lanuginosum* sub-
                                                          community U5e**

     Heaths with a dense silvery-green mat of *Racomitrium lanuginosum* around
     the dwarf shrubs, and at most a scattering of lichens; *Nardus stricta* sparse or
     absent and *Calluna vulgaris* more continuous, giving a darker look to the
     vegetation; mainly in the western Highlands with outliers in Galloway and
     north Wales
                    ***Calluna vulgaris-Racomitrium lanuginosum* heath H14**

24   Heaths with a creamy-white mat of *Cladonia arbuscula* and other lichens
     around the dwarf shrubs; in the central, eastern and northern Highlands, the
     Southern Uplands and the Lake District
                    ***Calluna vulgaris-Cladonia arbuscula* heath H13**

     Prostrate heaths dominated by *Calluna vulgaris*, in some places with small
     amounts of *Empetrum nigrum* or *Vaccinium* species; can have a few other

species, such as *Carex pilulifera*, *Huperzia selago* or *Diphasiastrum alpinum*, but neither lichens nor *Racomitrium lanuginosum* are common

**Species-poor prostrate *Calluna vulgaris* heath (see H13 and H14)**

25  Mixed heaths of prostrate *Calluna vulgaris* and *Juniperus communis* ssp. *nana*, in some places with *Trichophorum cespitosum* and oceanic liverworts, such as *Pleurozia purpurea*; the shiny green clumps of juniper are conspicuous against the darker tones of the heather; on exposed stony ground at moderate to high altitudes in the western Highlands and the Hebrides, with fragmentary stands in the Lake District and north Wales

***Calluna vulgaris-Juniperus communis* ssp. *nana* heath H15**

Mixed prostrate heaths of *Calluna vulgaris*, *Arctostaphylos alpinus* and in many places *Loiseleuria procumbens* and *A. uva-ursi*; on stony ridges, moraines and gentle slopes at moderate to high altitudes in the north and north-west Highlands

***Calluna vulgaris-Arctostaphylos alpinus* heath H17**

# Key to blanket, raised and valley mires

Vegetation on deep wet peat which receives most of its water from rain rather than from lateral flushing (except for valley mires)

1  Bog vegetation on deep but rarely saturated peat with much *Eriophorum vaginatum* growing in tussocks with *Calluna vulgaris*, *Vaccinium* species and/or *Empetrum nigrum*, and with a bryophyte layer containing abundant *Sphagnum capillifolium* but few other Sphagna; in some places a virtually pure sward of tussocky *E. vaginatum*

**2**

Vegetation on deep and wet peat, usually with much *Sphagnum papillosum* as well as other Sphagna, and a mixed pale-coloured sward of species such as *Calluna vulgaris*, *Erica tetralix*, *Eriophorum vaginatum*, *E. angustifolium*, *Molinia caerulea*, *Trichophorum cespitosum*, *Myrica gale*, *Drosera* species and *Narthecium ossifragum*

**3**

2  Dark reddish-green swards with tufted evenly-mixed *Calluna vulgaris* and *Eriophorum vaginatum*, together with *Vaccinium* species, *Empetrum nigrum* (which can replace *Calluna* in stands at high altitudes) and a rich moss layer of *Sphagnum capillifolium* and pleurocarpous mosses such as *Hylocomium splendens*, *Rhytidiadelphus loreus*, *Pleurozium schreberi*, *Hypnum jutlandicum* and *Plagiothecium undulatum*; *Cladonia* lichens are very common in some northern and north-eastern stands

***Calluna vulgaris-Eriophorum vaginatum* blanket mire M19**

Pale swards dominated by dense tussocky *Eriophorum vaginatum*, commonly with a scattering of *Vaccinium* species, *Empetrum nigrum* and *Calluna vulgaris*; the moss flora is usually sparse but there can be some Sphagna, or much *Hypnum jutlandicum* or *Pleurozium schreberi*

***Eriophorum vaginatum* blanket and raised mire M20**

3  Valley mires on deep saturated peat in valley floors and hollows, in some places containing a sluggish stream; the vegetation consists of extensive mats

of *Sphagnum papillosum* with *S. denticulatum* or *S. fallax*, under a thin sward of *Narthecium ossifragum*, *Eriophorum angustifolium*, *Molinia caerulea*, *Erica tetralix* and *Calluna vulgaris*; without *Eriophorum vaginatum* or *Trichophorum cespitosum*; in the uplands, common only in south-west England, but with scattered occurrences as far north as the Pennines, Lake District and Scotland

**Narthecium ossifragum-Sphagnum papillosum valley mire M21**

Blanket, raised or less commonly valley mires with much *Eriophorum angustifolium*, *Trichophorum cespitosum* and dwarf shrubs; *Eriophorum vaginatum* usually present but can be sparse, especially in the far west Highlands and south-west England; abundant *Sphagnum papillosum* and *S. capillifolium* or *Racomitrium lanuginosum*

**4**

4   Mires with extensive carpets of *Sphagnum papillosum*, *S. capillifolium* and in many places *S. magellanicum*, beneath an open canopy of *Eriophorum* species, *Trichophorum cespitosum*, *Calluna vulgaris* and *Erica tetralix*; other dwarf shrubs may be present, such as *Empetrum nigrum* and, more distinctively and locally, *Vaccinium oxycoccos* and *Andromeda polifolia*; *Molinia caerulea* is rare; *Cladonia* lichens are locally common; the carpet of Sphagna can be almost continuous, and shallow pools and hollows tend to be filled with *Sphagnum cuspidatum*, but on drier bogs Sphagna can be more patchy; the characteristic vegetation of lowland raised mires but can form blanket mires in the uplands, especially in mid-Wales, northern England and Scotland

**Erica tetralix-Sphagnum papillosum raised and blanket mire M18**

Mires with much *Molinia caerulea*, *Trichophorum cespitosum*, *Eriophorum* species, *Myrica gale*, *Calluna vulgaris* and *Erica tetralix*; *Vaccinium oxycoccos* and *Andromeda polifolia* are rare or absent; Sphagna are usually no more abundant than vascular plants

**5**

5   Blanket bogs with a light ochre-coloured sward of *Eriophorum* species, *Trichophorum cespitosum*, *Molinia caerulea*, *Calluna vulgaris*, *Erica tetralix*, *Myrica gale*, *Sphagnum papillosum*, *S. capillifolium* and *Potentilla erecta*; in some stands Sphagna are scarce and there is much *Racomitrium lanuginosum* forming carpets or conspicuous raised hummocks; *Pleurozia purpurea* and *Cladonia* lichens are very common in some Scottish stands; small patches of *Sphagnum cuspidatum*, *S. tenellum* and *S. denticulatum* can be common; extensive in the west Highlands, more local in the central and eastern Highlands, Galloway, Lake District, Wales and south-west England

**Trichophorum cespitosum-Eriophorum vaginatum blanket mire M17**

Flushed channels in blanket bogs and wet heaths with a sparse open sward of *Erica tetralix*, *Molinia caerulea*, *Trichophorum cespitosum*, *Eriophorum angustifolium* and *Carex* species over a varied and patchy lower layer of *Sphagnum denticulatum*, *Campylopus atrovirens*, *Breutelia chrysocoma* and *Narthecium ossifragum*; *Sphagnum papillosum* and *Eriophorum vaginatum* scarce or absent; *Schoenus nigricans* is common in some stands at low altitudes in the western Highlands

**Trichophorum cespitosum-Erica tetralix wet heath, Carex panicea subcommunity M15a**

## Key to grasslands

Vegetation dominated by grasses other than *Phragmites australis* and *Phalaris arundinacea*

1    Wet or damp grasslands dominated by *Molinia caerulea*, either with a scattering of dwarf shrubs, an array of other grasses, or tall mesotrophic forbs

                                                  **2**

    Grasslands dominated by other species

                                                  **8**

2    Herb-rich *Molinia* grasslands with mesotrophic or basiphilous species, such as *Angelica sylvestris*, *Filipendula ulmaria*, *Geum rivale*, *Ranunculus acris*, *Succisa pratensis*, *Carex pulicaris*, *C. hostiana*, *Briza media*, *Cirsium palustre*, *C. dissectum*, *Lotus uliginosus*, *Mentha aquatica*, *Pulicaria dysenterica*, *Valeriana dioica*, *Crepis paludosa*, *Sanguisorba officinalis*, *Equisetum palustre* and *Calliergonella cuspidata*

                                                  **3**

    *Molinia* grasslands with dwarf shrubs or other grasses but few or none of the above mesotrophic or basiphilous species

                                                  **6**

3    *Molinia* grasslands, usually with two or more of *Carex pulicaris*, *C. hostiana*, *Briza media*, *Valeriana dioica*, *Cirsium dissectum*, *Crepis paludosa* and *Sanguisorba officinalis* common

                                                  **4**

    Herb-rich *Molinia* grasslands with little or no *Briza media*, *Carex pulicaris*, *C. hostiana*, *Valeriana dioica*, *Cirsium dissectum*, *Crepis paludosa* or *Sanguisorba officinalis*

                                                  **5**

4    Herb-rich *Molinia* grasslands lacking *Crepis paludosa* and usually with *Cirsium dissectum*; widespread in lowland England and Wales, and also occurs on low ground in upland areas of England, Wales and parts of Scotland

                        *Molinia caerulea-Cirsium dissectum* fen-meadow M24

    Herb-rich *Molinia* grasslands without *Cirsium dissectum* but with much *Crepis paludosa* and in many places with *Sanguisorba officinalis*; scarce in southern Scotland, northern England and north Wales

                              **Molinia caerulea-Crepis paludosa mire M26**

5    Herb-rich *Molinia* grasslands with one or more of *Angelica sylvestris*, *Filipendula ulmaria*, *Geum rivale*, *Cirsium palustre*, *Lotus uliginosus*, *Mentha aquatica* and *Calliergonella cuspidata* common; grasses other than *Molinia* scarce

         **Molinia caerulea-Potentilla erecta mire, Angelica sylvestris sub-community M25c**

    Moderately herb-rich *Molinia* grasslands with some *Succisa pratensis*, *Lotus uliginosus*, *Rumex acetosa* or *Ranunculus acris*, but little or no *Angelica sylvestris, Filipendula ulmaria, Geum rivale, Cirsium palustre, Mentha aquatica* or *Calliergonella cuspidata*; grasses other than *Molinia* are common

**Molinia caerulea-Potentilla erecta mire, Anthoxanthum odoratum sub-community M25b**

6 Very species-poor swards of *Molinia caerulea* with little more than a few shoots of *Potentilla erecta*; locally extensive on wet hillsides and in valley bottoms, especially in Wales

**Species-poor form of *Molinia caerulea-Potentilla erecta* mire M25**

*Molinia caerulea* grasslands with *Potentilla erecta* and either a scattering of dwarf shrubs or an array of other grasses

**7**

7 Boggy *Molinia-Potentilla* grasslands with shrubs such as *Calluna vulgaris, Erica tetralix* and *Myrica gale*, and commonly with other mire species such as *Narthecium ossifragum* and *Sphagnum capillifolium*; common throughout the western uplands

**Molinia caerulea-Potentilla erecta mire, Erica tetralix sub-community M25a**

Drier *Molinia-Potentilla* grasslands with much *Holcus lanatus, Anthoxanthum odoratum, Agrostis capillaris* and *Festuca* species; can have a few forbs, such as *Succisa pratensis, Rumex acetosa* and *Ranunculus acris*

**Molinia caerulea-Potentilla erecta mire, Anthoxanthum odoratum sub-community M25b**

8 Grasslands dominated by *Deschampsia cespitosa* or *D. flexuosa*

**9**

Grasslands dominated by other species

**13**

9 *Deschampsia flexuosa* grasslands

**10**

*Deschampsia cespitosa* grasslands

**11**

10 Sub-montane grasslands with tall swards of *Deschampsia flexuosa* mixed with *Calluna vulgaris, Galium saxatile* and *Potentilla erecta*

**Deschampsia flexuosa grassland U2**

Montane *Deschampsia flexuosa* grasslands on shaded slopes in high corries in the Cairngorms, with montane species such as *Juncus trifidus, Huperzia selago* and *Polytrichum alpinum*

**Salix herbacea-Racomitrium heterostichum snow-bed U12**

11 Wet lowland grasslands with a tall tussocky sward of *Deschampsia cespitosa, Holcus lanatus* and *Poa trivialis* mixed with other grasses and tall mesotrophic forbs; a grassland community which occurs locally at low altitudes in upland areas but its range does not overlap with that of the more montane *D. cespitosa* grasslands and it could hardly be confused with them

*Holcus lanatus-Deschampsia cespitosa* grassland MG9

Upland grasslands with a short tussocky sward of *Deschampsia cespitosa*; *Holcus lanatus* and *Poa trivialis* rare or absent; *Galium saxatile, Potentilla erecta* and *Festuca ovina*, or upland species, such as *Trollius europaeus* and *Geum rivale*, generally present

**12**

12 Species-poor grasslands with *Deschampsia cespitosa, Galium saxatile, Festuca ovina, Potentilla erecta* and pleurocarpous mosses on flushed shaded slopes and below cliffs

**Deschampsia cespitosa-Galium saxatile grassland, Anthoxanthum odoratum-Alchemilla alpina sub-community U13a**

Lush species-rich *Deschampsia cespitosa* grasslands with abundant tall mesotrophic forbs such as *Ranunculus acris, Trollius europaeus, Angelica sylvestris* and *Geum rivale*, and mosses such as *Rhytidiadelphus loreus* and *R. squarrosus*; on ungrazed slopes below cliffs and on ledges; also occurs as grazed stands of *Deschampsia cespitosa* with tall forbs growing as dwarfed basal rosettes

**Luzula sylvatica-Geum rivale tall-herb community, Agrostis capillaris-Rhytidiadelphus loreus sub-community U17c**

13 Grasslands dominated by *Nardus stricta*

**14**

Grasslands dominated by other species

**15**

14 Sub-montane grasslands with *Nardus stricta, Potentilla erecta, Galium saxatile* and common pleurocarpous mosses; can be quite species-rich with mesotrophic herbs such as *Ranunculus acris* and *Alchemilla glabra*; montane species rare or absent

**Nardus stricta-Galium saxatile grassland U5**

High-altitude *Nardus* grasslands with *Carex bigelowii* and other montane species such as *Cetraria islandica, Vaccinium uliginosum* and *Diphasiastrum alpinum*, and commonly with *Trichophorum cespitosum, Alchemilla alpina* or *Racomitrium lanuginosum*; in sheltered hollows, gullies and corries where snow lies late in spring

**Nardus stricta-Carex bigelowii grass-heath U7**

15 Grasslands dominated by *Sesleria caerulea*, with *Thymus polytrichus, Koeleria macrantha, Galium sterneri* and a range of other small calcicolous forbs; on limestone in northern England

**Sesleria caerulea-Galium sterneri grassland CG9**

Grasslands dominated by other species

**16**

16 Grasslands dominated by *Arrhenatherum elatius* and/or *Dactylis glomerata*

**17**

Grasslands dominated by other species

**19**

17 Open grasslands with large tussocks of *Arrhenatherum elatius* mixed with *Gymnocarpium robertianum, Geranium robertianum, Teucrium scorodonia* and *Festuca* species on limestone pavement and base-rich scree

**Gymnocarpium robertianum-Arrhenatherum elatius community OV38**

Denser grasslands with a thick sward of grasses and no *Gymnocarpium robertianum*

**18**

18 Rank species-rich grasslands of *Arrhenatherum elatius*, with woodland species such as *Heracleum sphondylium*, *Mercurialis perennis*, *Silene dioica* and *Dryopteris filix-mas*, together with plants of damp ground such as *Filipendula ulmaria*, *Valeriana officinalis*, *Geum rivale* and *Angelica sylvestris*
### *Arrhenatherum elatius-Filipendula ulmaria* tall-herb grassland MG2

Rank species-poor swards of *Arrhenatherum elatius* and/or *Dactylis glomerata*, with neither woodland species nor plants of damp ground; a lowland grassland type which can occur along field margins and roadsides at low altitudes in upland areas
*Arrhenatherum elatius* grassland MG1

19 Species-poor grasslands dominated by *Lolium perenne*, or co-dominated by *Lolium* and other grasses such as *Cynosurus cristatus* and *Festuca rubra*

**20**

Grasslands dominated by other species

**21**

20 Bright green swards of *Lolium perenne*, *Cynosurus cristatus*, *Holcus lanatus* and *Festuca rubra*, with other species including *Trifolium repens* and *Cerastium fontanum*; in improved permanent pastures, on roadside verges and on lawns
*Lolium perenne-Cynosurus cristatus* grassland MG6

Brilliant green and almost pure swards of *Lolium perenne*, with small quantities of other species such as *Phleum pratense*, *Poa trivialis* and *Dactylis glomerata*, and commonly with much *Trifolium repens*; improved grasslands and leys, mostly in enclosed fields and generally mown for hay or silage
*Lolium perenne* leys and related grasslands MG7

21 Grasslands with *Festuca ovina/vivipara*, *Agrostis capillaris*, *Thymus polytrichus* and a range of small forbs and bryophytes; the smell of the *Thymus* can be noticeable when walking in the vegetation, especially on warm still days

**22**

Grasslands with various mixtures of *Festuca* species, *Agrostis* species, *Holcus lanatus*, *Anthoxanthum odoratum* and *Cynosurus cristatus*, with little or no *Thymus polytrichus*; either herb-rich or species-poor but without small calcicolous forbs and bryophytes

**25**

22 Sparse, open grasslands with *Festuca ovina/vivipara*, *Agrostis capillaris*, *Thymus polytrichus*, *Minuartia verna*, *Thlaspi caerulescens* and usually *Linum catharticum*, *Rumex acetosa*, *R. acetosella* and *Campanula rotundifolia*; on spoil of old lead mines and metalliferous rocks
### *Festuca ovina-Minuartia verna* community OV37

Grasslands with a denser sward of *Festuca ovina/vivipara*, other grasses and *Thymus polytrichus*, and a range of forbs and sedges such as *Prunella vulgaris*, *Viola riviniana*, *Alchemilla alpina*, *Linum catharticum* and *Carex pulicaris*, but lacking metallophyte species such as *Thlaspi caerulescens* or *Minuartia verna*

**23**

23 Grasslands with *Festuca ovina, Helictotrichon pratense, Briza media, Koeleria macrantha, Thymus polytrichus, Helianthemum nummularium* and lowland forbs such as *Sanguisorba minor, Scabiosa columbaria* and *Carlina vulgaris*; the characteristic upland grassland species *Agrostis capillaris, Anthoxanthum odoratum* and *Potentilla erecta* are rare; widespread in the lowlands of north and west England and Wales; scarce in limestone uplands of Wales and northern England

> *Festuca ovina-Helictotrichon pratense* grassland, *Dicranum scoparium* sub-community CG2d (Note: this vegetation type has virtually the same species composition as the south-western *Festuca ovina-Carlina vulgaris* grassland, *Koeleria macrantha* sub-community CG1e)

Grasslands with *Festuca ovina/vivipara, Agrostis capillaris, Anthoxanthum odoratum, Thymus polytrichus* and *Potentilla erecta*; lowland forbs such as *Sanguisorba minor, Scabiosa columbaria* and *Carlina vulgaris* are generally absent; *Helictotrichon pratense, Briza media* and *Helianthemum nummularium* are generally rare but may be locally common, for example in the Breadalbanes where they grow in the company of northern and upland species

**24**

24 Grasslands with *Festuca ovina/vivipara, Agrostis capillaris, Anthoxanthum odoratum, Thymus polytrichus* and many small forbs such as *Plantago lanceolata, Prunella vulgaris, Viola riviniana* and *Trifolium repens*; in damper stands, *Carex pulicaris, C. panicea, Linum catharticum* and *Selaginella selaginoides* may be present; *Alchemilla alpina* is usually scarce or absent, and is never abundant

**Festuca ovina-Agrostis capillaris-Thymus polytrichus grassland CG10**

Grasslands with *Festuca ovina/vivipara, Agrostis capillaris, Anthoxanthum odoratum*, abundant *Alchemilla alpina*, and many small mesotrophic species such as *Thymus polytrichus, Selaginella selaginoides, Ranunculus acris* and *Carex pulicaris*

**Festuca ovina-Agrostis capillaris-Alchemilla alpina grassland CG11**

25 Grassland dominated by *Agrostis curtisii*, with *Potentilla erecta, Festuca ovina* and sprigs of *Calluna vulgaris*; in south-west England and south Wales

**Agrostis curtisii grassland U3**

Grassland dominated by *Festuca* species, *Agrostis* species, *Holcus lanatus, Anthoxanthum odoratum* or *Cynosurus cristatus* but not *Agrostis curtisii*

**26**

26 Lush, species-poor grasslands dominated by *Holcus lanatus* and/or *Festuca rubra*, with *Anthoxanthum odoratum, Poa pratensis, Dactylis glomerata, Trifolium repens, Plantago lanceolata*, and a sparse layer of mosses including *Scleropodium purum* and *Rhytidiadelphus squarrosus*; on neglected crofts at

low altitudes in the Hebrides and in meadows throughout the upland fringes; not described in the NVC

*Festuca rubra-Holcus lanatus-Anthoxanthum odoratum* provisional grassland community (Rodwell *et al.* 2000)

Grasslands not dominated by *Holcus lanatus* or *Festuca rubra*; sward with mixtures of *Festuca* species, *Agrostis* species, *Anthoxanthum odoratum* and in some places *H. lanatus* and *Cynosurus cristatus*; varying from species-poor to species-rich with mesotrophic species

**27**

27 Grasslands with *Festuca ovina/vivipara*, *Agrostis capillaris*, *A. canina* and *Anthoxanthum odoratum*; *Aira praecox* or small forbs such as *Galium saxatile* and *Potentilla erecta* can be very common in the sward; mostly species-poor and lacking mesotrophic herbs

**28**

Herb-rich grassland with a mixed sward of grasses and a rich array of mesotrophic forbs

**29**

28 Short and rather open swards of *Festuca ovina*, *Agrostis capillaris*, *Aira praecox*, *Rumex acetosella* and *Galium saxatile* on thin stony soils on dry sun-exposed slopes

**Festuca ovina-Agrostis capillaris-Rumex acetosella grassland, Galium saxatile-Potentilla erecta sub-community U1e**

Short to medium, more or less dense grasslands with *Festuca* species, *Agrostis* species, *Anthoxanthum odoratum*, *Potentilla erecta*, *Galium saxatile* and varied mixtures of pleurocarpous mosses or *Racomitrium lanuginosum*

**Festuca ovina-Agrostis capillaris-Galium saxatile grassland U4**

29 Swards of *Festuca ovina*, *F. rubra*, *Agrostis capillaris*, *A. canina* and *Anthoxanthum odoratum* with *Potentilla erecta*, *Galium saxatile*, a deep carpet of pleurocarpous mosses, and a rich flora of mesotrophic forbs such as *Ranunculus acris*, *Geum rivale*, *Filipendula ulmaria*, *Cirsium heterophyllum* and *Angelica sylvestris*

**Herb-rich form of *Festuca ovina-Agrostis capillaris-Galium saxatile* grassland U4**

Less mossy swards of *Festuca* species and *Agrostis* species with *Cynosurus cristatus*, *Holcus lanatus*, *Dactylis glomerata* and mesotrophic forbs such as *Trifolium pratense*, *Ranunculus acris* and *Rumex acetosa*; *Galium saxatile* scarce or absent

**30**

30 Wet herb-rich grasslands with much *Caltha palustris* and commonly with *Carex nigra* and forbs such as *Ranunculus* species, *Trifolium* species and *Filipendula ulmaria*, in a sward of *Holcus lanatus*, *Festuca rubra*, *Anthoxanthum odoratum*, *Cynosurus cristatus*, *Poa trivialis* and *Agrostis* species

**Cynosurus cristatus-Caltha palustris grassland MG8**

Drier grasslands with little or no *Caltha palustris*

**31**

31 Herb-rich grasslands with much *Geranium sylvaticum* and *Alchemilla* species, and in some places with *Geum rivale*, *Trollius europaeus* and *Cirsium heterophyllum*; in upland meadows and on riversides; locally common in northern England and scattered in Scotland

**Anthoxanthum odoratum-Geranium sylvaticum grassland MG3**

Grasslands with little or no *Geranium sylvaticum*, but much *Cynosurus cristatus*, *Plantago lanceolata*, *Centaurea nigra*, *Lotus corniculatus*, *Trifolium pratense*, *Rhinanthus minor* and *Ranunculus acris*; scattered through the uplands

**Cynosurus cristatus-Centaurea nigra grassland MG5**

## Key to rush- and sedge-heaths

Vegetation dominated by *Carex bigelowii*, *Juncus squarrosus* or *J. trifidus*

1 Swards dominated by *Carex bigelowii* with *Dicranum fuscescens*, *Racomitrium lanuginosum* or *Polytrichum alpinum* and other montane species, on high plateaux and ridges in the Highlands and rarely in North Wales

**Carex bigelowii-Polytrichum alpinum sedge-heath U8**

Swards dominated by *Juncus squarrosus* or *J. trifidus*

2

2 Swards dominated by *Juncus squarrosus*

3

Open swards dominated by *Juncus trifidus*

4

3 Swards with *Juncus squarrosus*, in some places damp with Sphagna, in other places drier with mixtures of *Festuca ovina*, *Nardus stricta* and other grasses; the vegetation can be dotted with *Vaccinium myrtillus* or *Calluna vulgaris*; less commonly it can be herb-rich with *Ranunculus acris*, *Thalictrum alpinum* and *Trollius europaeus*

**Juncus squarrosus-Festuca ovina grassland U6**

Blanket bog with much *Juncus squarrosus*, *Rhytidiadelphus loreus* and *Calluna vulgaris*, and also bog species such as *Erica tetralix*, *Narthecium ossifragum*, *Eriophorum* species and *Sphagnum papillosum*

**Trichophorum cespitosum-Eriophorum vaginatum blanket mire, Juncus squarrosus-Rhytidiadelphus loreus sub-community M17c**

4 Open swards of *Juncus trifidus*, *Carex bigelowii* and *Racomitrium lanuginosum*, together with other montane species and commonly with abundant lichens; on stony soils on high ridges and summits in the eastern and northern Highlands

**Juncus trifidus-Racomitrium lanuginosum rush-heath U9**

Very open swards of *Juncus trifidus*, *Festuca vivipara* and a few other species such as *Salix herbacea*, *Racomitrium lanuginosum*, *Ochrolechia frigida* and *Polytrichum alpinum*; on stony, wind-blasted but firm soils on exposed ridges in the western Highlands

**Carex bigelowii-Racomitrium lanuginosum moss-heath, species-poor form of the Silene acaulis sub-community U10c**

# Key to swamps and fens

Vegetation of standing water or very wet ground, dominated by sedges, *Phragmites australis*, *Phalaris arundinacea*, *Equisetum fluviatile* or *Eleocharis palustris* and commonly with tall forbs; mires dominated by *Filipendula ulmaria* and *Iris pseudacorus* are closely related to these swamps and fens, but are keyed out under tall-herb and dwarf-herb vegetation

1    Swamps dominated by *Eleocharis palustris*

**                                  *Eleocharis palustris* swamp S19**

    Swamps and fens dominated by other species

            **2**

2    Swamps dominated by *Equisetum fluviatile*

**                                  *Equisetum fluviatile* swamp S10**

    Swamps and fens dominated by other species

            **3**

3    Swamps and fens dominated by *Phragmites australis*, or in some places by *Eriophorum angustifolium*, with tall mesotrophic forbs

            **4**

    Swamps dominated by *Carex* species, *Phalaris arundinacea* or *Scirpus lacustris*

            **5**

4    Swamps with *Phragmites australis* either totally dominant or with a sparse array of lowland tall herbs, *Galium palustre*, *Mentha aquatica* or maritime species; widespread at low altitudes

                        *Phragmites australis* swamp S4

    Fens dominated by *Phragmites australis* or *Eriophorum angustifolium*, with a moderately rich assemblage of other species, including *Carex rostrata*, *Potentilla palustris*, *Menyanthes trifoliata*, *Angelica sylvestris*, *Lythrum salicaria* and *Lysimachia vulgaris*

**                *Carex rostrata-Potentilla palustris* tall-herb fen, *Lysimachia vulgaris* sub-community S27b**

5    Swamps or fens dominated by *Phalaris arundinacea*; widespread in the lowlands and scattered at low altitudes in upland areas

                        *Phalaris arundinacea* fen S28

    Swamps or fens dominated by other species

            **6**

6    Swamps dominated by *Schoenoplectus lacustris* in water more than 25 cm deep; throughout the lowlands but also, less commonly, in the uplands

                  *Schoenoplectus lacustris* ssp. *lacustris* swamp S8

    Swamps and fens dominated by *Carex* species

            **7**

7    Swamps dominated by *Carex paniculata*; mainly in the lowlands but scattered at low altitudes in upland areas

                        *Carex paniculata* swamp S3

Swamps not dominated by *Carex paniculata*

**8**

8 Swamps dominated by *Carex vesicaria*, either in pure swards or with a few other species (but *not* including *C. rostrata* or *Menyanthes trifoliata*)
**Carex vesicaria swamp, Carex vesicaria sub-community S11a and *Mentha aquatica* sub-community S11b**

Swamps and fens dominated by other *Carex* species, or if by *C. vesicaria* then with a well-developed understorey of forbs and other sedges, including *C. rostrata*

**9**

9 Swamps and fens dominated by *Carex rostrata*

**10**

Mixtures of *Carex rostrata*, *Potentilla palustris*, *Menyanthes trifoliata*, *Equisetum fluviatile* and *Eriophorum angustifolium*, in some places with much *C. vesicaria*, *C. aquatilis* or *Phragmites australis*
*Phragmites australis* swamp, *Menyanthes trifoliata* sub-community S4c
**Carex rostrata swamp, Menyanthes trifoliata-Equisetum fluviatile sub-community S9b**
**Equisetum fluviatile swamp, Carex rostrata sub-community S10b**
**Carex vesicaria swamp, Carex rostrata sub-community S11c**
**Carex rostrata-Potentilla palustris tall-herb fen, Carex rostrata-Equisetum fluviatile sub-community S27a**
(these NVC types converge to such an extent that it is often not possible to separate them)

10 *Carex rostrata* dominant in a thin and very species-poor sward in shallow water
**Carex rostrata swamp, Carex rostrata sub-community S9a**

*Carex rostrata* swamps and fens with much *Potentilla palustris* and *Menyanthes trifoliata*
**Carex rostrata swamp, Menyanthes trifoliata-Equisetum fluviatile sub-community S9b**
**Carex rostrata-Potentilla palustris tall-herb fen, Carex rostrata-Equisetum fluviatile sub-community S27a**
(these NVC types converge to such an extent that it is often not possible to separate them)

## Key to soligenous mires and rush-pastures

Vegetation of wet, flushed ground, usually dominated by sedges or rushes

1 Mires or wet pastures with a moss layer dominated by Sphagna or with a sparse sward of vascular plants on bare wet peat

**2**

Mires and wet pastures with bryophytes other than Sphagna dominant in the ground layer, although a few mesotrophic Sphagna may occur; the most common bryophytes are generally *Calliergonella cuspidata*, *Calliergon* species, *Rhizomnium punctatum*, *Bryum pseudotriquetrum*, *Campylium stellatum*, *Scorpidium scorpioides* and *Drepanocladus revolvens*

**14**

2    Springs, rills and soakways dominated by small forbs such as *Montia fontana*, *Chrysosplenium oppositifolium*, *Ranunculus omiophyllus*, *Hypericum elodes* and *Potamogeton polygonifolius*, usually with *Sphagnum denticulatum* and in some places other bryophytes

**3**

Mires dominated by sedges, rushes, *Molinia caerulea*, *Eriophorum angustifolium* or *Trichophorum cespitosum*, or with an open canopy of dwarf shrubs including *Erica tetralix*, *Myrica gale* and *Calluna vulgaris*

**5**

3    Vegetation consisting of dense mats of *Hypericum elodes*, *Potamogeton polygonifolius*, *Ranunculus flammula*, *Hydrocotyle vulgaris*, *Sphagnum denticulatum* and in some places *Carex panicea* and *C. viridula* ssp. *oedocarpa*; at low altitudes in the west
          ***Hypericum elodes-Potamogeton polygonifolius* soakway M29**

Vegetation with *Montia fontana* and/or either *Chrysosplenium oppositifolium* or *Ranunculus omiophyllus*

**4**

4    Vegetation with a thick mat of *Montia fontana* and/or *Chrysosplenium oppositifolium*, together with species such as *Agrostis canina*, *Epilobium palustre*, *Caltha palustris* and mosses such as *Philonotis fontana*, *Dicranella palustris* and *Sphagnum denticulatum*; in some places with a few montane species, such as *Saxifraga stellaris* and *Epilobium alsinifolium*; with a predominantly upland and northern distribution, usually above 400 m
          ***Philonotis fontana-Saxifraga stellaris* spring M32**

Vegetation with *Montia fontana*, *Ranunculus omiophyllus*, *R. flammula*, *Agrostis stolonifera*, *Juncus bulbosus* and *Sphagnum denticulatum* but no montane species; with a predominantly lowland and southern distribution, usually below 450 m
          ***Ranunculus omiophyllus-Montia fontana* rill M35**

5    Mires dominated by *Carex rostrata*

**6**

Mires dominated by other sedges, including *Eriophorum angustifolium*, or by rushes, or with a sparse array of dwarf shrubs such as *Erica tetralix*, *Calluna vulgaris* and *Myrica gale* together with *Molinia caerulea* and/or *Trichophorum cespitosum*

**8**

6    Species-poor *Carex rostrata* mires with a thin sward of *C. rostrata* over a layer of *Sphagnum fallax*, *S. palustre* and *Polytrichum commune*
          ***Carex rostrata-Sphagnum fallax* mire M4**

*Carex rostrata* mires with a ground layer of more mesotrophic Sphagna, such as *S. squarrosum*, *S. teres* or *S. warnstorfii*, and with some mesotrophic herbs in the sward

**7**

7    Moderately species-rich *Carex rostrata* mires with species such as *Potentilla palustris*, *C. nigra*, *Eriophorum angustifolium* and *Succisa pratensis* over a layer of *Sphagnum squarrosum* or *S. teres*; other mesotrophic herbs may also occur; widely distributed in north and west Great Britain

**Carex rostrata-Sphagnum squarrosum mire M5**

Moderately species-rich *Carex rostrata* mires with species such as *Viola palustris*, *Potentilla erecta*, *Selaginella selaginoides* and *Epilobium palustre* over a ground layer with *Sphagnum warnstorfii*, *S. teres*, *Rhizomnium pseudopunctatum* and *Aulacomnium palustre*; montane species such as *Thalictrum alpinum* and *Persicaria vivipara* may be present; rare in the central Highlands, northern England and north Wales

**Carex rostrata-Sphagnum warnstorfii mire M8**

8    Bog pools or flushed channels in bogs, usually with large amounts of *Eriophorum angustifolium*; *Erica tetralix*, *Calluna vulgaris*, *Myrica gale* and *Molinia caerulea* can occur

9

Mires or wet pastures dominated by sedges or rushes

10

9    Bog pools or flushed channels dominated by *Eriophorum angustifolium*, or bare spreads of peat on which *E. angustifolium* is the most abundant species

**Eriophorum angustifolium bog pool M3**

Flushed channels within blanket bog or wet heath, with mixtures of *Eriophorum angustifolium*, *Erica tetralix*, *Carex panicea*, *Trichophorum cespitosum*, *Molinia caerulea*, *Myrica gale*, *Sphagnum denticulatum*, *Campylopus atrovirens*, *Breutelia chrysocoma* and in many places *Drosera* species; *Molinia caerulea* and *Myrica gale* can be dominant; *Schoenus nigricans* can occur at low altitudes in the western Highlands

**Trichophorum cespitosum-Erica tetralix wet heath, Carex panicea sub-community M15a**

10   Mires dominated by sedges

11

Mires dominated by rushes

13

11   Mires dominated by *Carex panicea* and *Carex viridula* ssp. *oedocarpa* with *Potamogeton polygonifolius*; in peaty soakways

**Hypericum elodes-Potamogeton polygonifolius soakway M29 (species-poor form without *H. elodes*)**

Mires dominated by *Carex echinata*, *C. nigra* and other small sedges, in some places with *C. curta* and *Eriophorum angustifolium*

12

12   Mires with mixtures of *Carex echinata*, *C. nigra* and other small sedges, and with a ground layer of *Sphagnum fallax*, *S. denticulatum*, *S. palustre* and *Polytrichum commune*; common at low to moderate altitudes throughout the uplands

**Carex echinata-Sphagnum fallax/S. denticulatum mire, Carex echinata sub-community M6a and Carex nigra-Nardus stricta sub-community M6b**

Mires with small sedges, such as *Carex echinata, C. nigra, Eriophorum angustifolium* and commonly *C. curta*, with a ground layer of *Sphagnum denticulatum, S. papillosum, S. russowii* and in some stands *S. lindbergii*, and a scattering of montane species such as *C. bigelowii, Saxifraga stellaris* and *Festuca vivipara*; at moderate to high altitudes from the Lake District northwards

**Carex curta-Sphagnum russowii mire M7**

13   Mires dominated by *Juncus effusus* over a ground layer of *Sphagnum fallax, S. denticulatum, S. palustre* and *Polytrichum commune*

**Carex echinata-Sphagnum fallax/S. denticulatum mire, Juncus effusus sub-community M6c**

Mires dominated by *Juncus acutiflorus* over a ground layer of *Sphagnum fallax, S. denticulatum, S. palustre* and *Polytrichum commune*

**Carex echinata-Sphagnum fallax/S. denticulatum mire, Juncus acutiflorus sub-community M6d**

14   Wet pastures or extensive mires dominated by rushes mixed with grasses and tall forbs

**15**

Mires dominated by sedges with either tall or short forbs, often occurring as small flushes

**17**

15   Wet pastures or mires dominated by *Juncus acutiflorus*, with tall forbs, such as *Angelica sylvestris, Filipendula ulmaria, Ranunculus acris, Lychnis flos-cuculi, Mentha aquatica* and *Cirsium palustre*

**Juncus effusus/acutiflorus-Galium palustre rush-pasture, Juncus acutiflorus sub-community M23a**

Wet pastures or mires dominated by *Juncus effusus* with a range of grasses and tall forbs

**16**

16   Wet pastures with *Juncus effusus, Holcus lanatus, Agrostis canina, Galium palustre* and a few tall forbs such as *Cirsium palustre, Ranunculus flammula, Mentha aquatica, Angelica sylvestris, Viola palustris* and *Hydrocotyle vulgaris*

**Juncus effusus/acutiflorus-Galium palustre rush-pasture, Juncus effusus sub-community M23b**

Wet pastures with *Juncus effusus, Holcus lanatus, Agrostis stolonifera* and *Ranunculus repens* together with plants such as *Poa trivialis, Ranunculus acris, Trifolium repens* and in some places *Rumex obtusifolius* and *Cirsium arvense*; generally lacking fen species such as *Galium palustre, Cirsium palustre* and *Ranunculus flammula*; mainly lowland but extending into the upland fringes

**Holcus lanatus-Juncus effusus rush-pasture MG10**

17   Mires dominated by *Carex rostrata* with much *Eriophorum angustifolium, C. nigra, C. panicea, Menyanthes trifoliata, Potentilla palustris, Galium palustre* and many other mesotrophic herbs; with a dense underlay of mosses such as *Calliergonella cuspidata, Calliergon giganteum, Campylium stellatum*,

*Scorpidium scorpioides* and *Bryum pseudotriquetrum*; *Carex diandra* can be very common

**Carex rostrata-Calliergonella cuspidata/Calliergon giganteum mire M9**

Mires dominated by sedges other than *C. rostrata*

**18**

18 Mires dominated by *Carex saxatilis*, with montane species such as *Saxifraga stellaris*, *Persicaria vivipara* and *Thalictrum alpinum* and with a bryophyte layer including *Drepanocladus revolvens*, *Hylocomium splendens*, *Scapania undulata* and *Aneura pinguis*; at high altitudes in the Scottish Highlands, typically where snow lies late

**Carex saxatilis mire M12**

Mires dominated by sedges other than *Carex saxatilis*

**19**

19 Mires of small sedges such as *Carex nigra*, *C. echinata*, *C. panicea*, *C. viridula* ssp. *oedocarpa* and *C. hostiana*, mixed with tall mesotrophic forbs such as *Ranunculus acris*, *Crepis paludosa*, *Parnassia palustris*, *Geum rivale* and *Filipendula ulmaria*, and bryophytes such as *Rhizomnium punctatum* and *Calliergonella cuspidata*, or base-tolerant Sphagna (*Sphagnum contortum*, *S. teres*, *S. warnstorfii* and *S. squarrosum*)

Herb-rich small-sedge mire (see p. 423)

Mires of small sedges such as *Carex dioica*, *C. panicea* and *C. viridula* ssp. *oedocarpa*, with small forbs including *Potamogeton polygonifolius*, *Pinguicula vulgaris* and *Saxifraga* species, and mosses such as *Campylium stellatum*, *Scorpidium scorpioides*, *Drepanocladus revolvens* and *Blindia acuta*; tall forbs and base-tolerant Sphagna scarce or absent

**20**

20 Mires with an open sward of *Carex panicea* and *C. viridula* ssp. *oedocarpa* with much *Potamogeton polygonifolius* in peaty soakways

**Hypericum elodes-Potamogeton polygonifolius soakway M29 (species-poor form without H. elodes)**

Mires with *Carex panicea* and *C. viridula* ssp. *oedocarpa* but without *Potamogeton polygonifolius* and in many places with a rich flora of small calcicolous sedges and forbs; can have a noticeable sulphurous smell of decomposing vegetation

**21**

21 Rather sparse open mires of *Carex panicea*, *C. viridula* ssp. *oedocarpa*, *C. dioica*, *C. pulicaris*, *Pinguicula vulgaris* and *Selaginella selaginoides*, with tufts and clumps of mosses such as *Campylium stellatum*, *Drepanocladus revolvens*, *Scorpidium scorpioides* and *Blindia acuta*; *Saxifraga aizoides* scarce or absent, and other montane species usually not common; *Schoenus nigricans* is common in some stands; there can be a strong smell of decomposing vegetation

**Carex dioica-Pinguicula vulgaris mire M10**

Open and usually stony mires with *Carex panicea*, *C. viridula* ssp. *oedocarpa* and *Saxifraga aizoides*, typically with montane species such as *Juncus triglumis*, *Alchemilla alpina*, *Persicaria vivipara* and *Saxifraga stellaris*, and bryophytes including *Campylium stellatum*, *Drepanocladus revolvens*, *Blindia acuta* and *Aneura pinguis*

**22**

22 Flushes with *Koenigia islandica* evident during the spring to autumn months; *Saxifraga aizoides* common but can be absent; on basalt plateaux on Mull and Skye

### *Carex viridula* ssp. *oedocarpa-Koenigia islandica* flush M34

Mires without *Koenigia islandica* but mostly with *Saxifraga aizoides* except in north Wales

### *Carex viridula* ssp. *oedocarpa-Saxifraga aizoides* mire M11

## Key to fern communities

Vegetation dominated by ferns

1 Open communities on rocky substrates, including boulder fields, screes, limestone pavements and crevices in cliffs and walls, with abundant small to medium-sized ferns such as *Asplenium* species, *Gymnocarpium robertianum* or *Cryptogramma crispa*

**2**

Denser communities on soil or on larger cliff ledges, dominated by large ferns such as *Pteridium aquilinum*, *Oreopteris limbosperma* or *Dryopteris* species

**6**

2 Short open vegetation of rock crevices, dominated by *Asplenium* species

**3**

Taller communities dominated by other species

**4**

3 Assemblages of *Asplenium trichomanes*, *A. ruta-muraria*, *Festuca ovina*, *Thymus polytrichus* and the bryophytes *Homalothecium sericeum*, *Hypnum cupressiforme s.l.* and *Fissidens dubius*; in crevices of base-rich rocks and on mortared walls, in the lowlands and to moderate altitudes in the hills

### *Asplenium trichomanes-Asplenium ruta-muraria* community OV39

Assemblages of *Asplenium viride*, *A. ruta-muraria*, *A. trichomanes* and *Cystopteris fragilis*, with *Ctenidium molluscum*, *Fissidens dubius* and *Tortella tortuosa*; in shaded crevices and on rock ledges in the uplands

### *Asplenium viride-Cystopteris fragilis* community OV40

4 Vegetation dominated by *Gymnocarpium robertianum*; with tall tufts of *Arrhenatherum elatius*, sprawling plants of *Geranium robertianum*, and scattered *Teucrium scorodonia*; on limestone pavement and scree in southern Great Britain

### *Gymnocarpium robertianum-Arrhenatherum elatius* community OV38

Vegetation of scree or boulder fields with *Cryptogramma crispa* and no *Gymnocarpium robertianum*

**5**

5    Mixtures of *Cryptogramma crispa*, *Athyrium distentifolium*, *Dryopteris dilatata* and *D. oreades* growing in mats of humus with *Deschampsia cespitosa*, montane plants such as *Alchemilla alpina* and *Saxifraga stellaris*, and bryophytes including *Barbilophozia floerkei*, *Rhytidiadelphus loreus*, *Polytrichum alpinum* and *Kiaeria starkei*; in boulder fields on sheltered high slopes and in corries where snow lies late in spring

***Cryptogramma crispa-Athyrium distentifolium* snow-bed U18**

Vegetation with *Cryptogramma crispa*, *Deschampsia flexuosa*, *Festuca ovina*, *Galium saxatile*, *Campylopus flexuosus*, *Polytrichum formosum* and few or no montane species; on scree and boulders on sub-montane slopes

***Cryptogramma crispa-Deschampsia flexuosa* community U21**

6    Vegetation dominated by *Pteridium aquilinum*

**7**

Vegetation dominated by *Oreopteris limbosperma* or *Dryopteris* species

**8**

7    Stands of bracken with a species-poor ground flora which is either grassy, resembling a mossy *Festuca-Agrostis-Galium* grassland U4, heathy with *Vaccinium myrtillus* and/or *Calluna vulgaris*, or almost absent because of the presence of a deep mat of bracken litter

***Pteridium aquilinum-Galium saxatile* community U20**

Stands of bracken, many of which have *Rubus fruticosus* and a species-rich ground flora, either with woodland forbs such as *Hyacinthoides non-scripta*, *Primula vulgaris*, *Mercurialis perennis* and *Stachys sylvatica*, with *Teucrium scorodonia*, or with species of damp ground such as *Filipendula ulmaria*, *Geum rivale*, *Lysimachia nemorum*, *Carex panicea* and *C. pulicaris*

***Pteridium aquilinum-Rubus fruticosus* underscrub W25**

8    Fern communities dominated by *Dryopteris dilatata*, *D. filix-mas* or *D. affinis*, together with *Luzula sylvatica*, *Vaccinium myrtillus*, *Deschampsia cespitosa*, *Oxalis acetosella*, *Dicranum scoparium* and *D. majus*; on cliff ledges or steep slopes

***Luzula sylvatica-Vaccinium myrtillus* tall-herb community, *Dryopteris dilatata-Dicranum majus* sub-community U16a**

Fern communities dominated by *Oreopteris limbosperma* or *Dryopteris affinis* with little or no *Luzula sylvatica*

**9**

9    Fern communities dominated by *Oreopteris limbosperma* with a little *Blechnum spicant* and a grassy field layer with *Potentilla erecta*, *Galium saxatile* and *Oxalis acetosella*

***Oreopteris limbosperma-Blechnum spicant* community U19**

Fern communities dominated by *Dryopteris affinis* and with species such as *Galium saxatile*, *Potentilla erecta* and pleurocarpous mosses; on rocky banks and lower hill slopes

*Dryopteris affinis* provisional community (Rodwell *et al.* 2000)

# Key to tall-herb and dwarf-herb vegetation

Vegetation dominated by either short or tall herbs

1   Short vegetation made up of small forbs

          **2**

    Tall vegetation dominated by larger herbaceous species

          **6**

2   Open vegetation on dripping cliffs and rock faces, with *Saxifraga aizoides, S. oppositifolia, Selaginella selaginoides, Alchemilla glabra, A. alpina, Thalictrum alpinum* and a rich array of calcicolous bryophytes

          ***Saxifraga aizoides-Alchemilla glabra* banks U15**

    Open to dense vegetation on drier substrates, without abundant *Saxifraga* species

          **3**

3   Open, greyish, species-poor swards with much *Minuartia verna* and in many places with *Thlaspi caerulescens* and other small plants such as *Campanula rotundifolia, Linum catharticum, Rumex acetosa* and *R. acetosella*; on the spoil of lead mines and on veins of metalliferous rock

          ***Festuca ovina-Minuartia verna* community OV37**

    Short, herb-rich swards with a greater diversity of forbs and bryophytes and without metallophyte species such as *Minuartia verna* and *Thlaspi caerulescens*; in snow-beds and on high slopes and summits

          **4**

4   Rather sparse silvery-green mats of *Alchemilla alpina, Sibbaldia procumbens, Carex bigelowii, Polytrichum alpinum* and other small montane plants set in a mat of *Racomitrium* species, especially *R. fasciculare*; on high shaded slopes where snow lies late in the Scottish Highlands; there are similar-looking assemblages of *Alchemilla alpina* and *Potentilla erecta* (but lacking *Sibbaldia*) in comparable habitats in the Lake District and the western Highlands

          ***Alchemilla alpina-Sibbaldia procumbens* dwarf-herb community U14**

    Species-rich and usually greener-coloured swards of small plants including *Silene acaulis, Minuartia sedoides, Persicaria vivipara, Ranunculus acris, Thymus polytrichus* and *Alchemilla alpina* on high slopes and ridges

          **5**

5   Swards of *Silene acaulis, Minuartia sedoides, Thymus polytrichus, Alchemilla alpina, Persicaria vivipara, Festuca vivipara, Deschampsia cespitosa, Selaginella selaginoides* and other small plants, many of them calcicoles; with mosses such as *Racomitrium lanuginosum, Hylocomium splendens, Pleurozium schreberi* and *Hypnum lacunosum*; grasses and bryophytes are subordinate to the dwarf herbs; on high slopes and beneath cliffs, including places where snow lies moderately late

          ***Festuca ovina-Alchemilla alpina-Silene acaulis* dwarf-herb community CG12**

    Mossy swards of *Racomitrium lanuginosum* and *Carex bigelowii* with dwarf herbs such as *Silene acaulis, Minuartia sedoides, Armeria maritima, Ranunculus acris* and *Persicaria vivipara*, and also *Salix herbacea* and *Thymus polytrichus*; can be very herb-rich but looks more like a moss-heath with herbs than a dwarf-herb sward; on summits and high ridges which are blown clear of snow in winter

**Carex bigelowii-Racomitrium lanuginosum moss-heath, species-rich form of the Silene acaulis sub-community U10c**

6    Weedy vegetation dominated by *Urtica dioica*, mostly on disturbed ground

**7**

Vegetation of less-modified habitats dominated by *Iris pseudacorus*, *Caltha palustris* or varied mixtures of tall forbs, *Luzula sylvatica*, *Vaccinium myrtillus* and ferns

**8**

7    Dense beds of *Urtica dioica*, *Galium aparine*, *Poa trivialis*, *Arrhenatherum elatius* and in some places tall forbs such as *Anthriscus sylvestris* and *Heracleum sphondylium*; common throughout the lowlands; also in upland districts on disturbed ground around buildings, along roads and beside walls, and on the strandline

*Urtica dioica-Galium aparine* community OV24

Patchy stands of *Urtica dioica* dotted with *Cirsium arvense*, *C. vulgare* and grasses such as *Elymus repens*, *Holcus lanatus*, *Dactylis glomerata*, *Arrhenatherum elatius* and *Lolium perenne*; common throughout the lowlands; also occurs at low altitudes in upland areas on disturbed ground around buildings, on roadsides and on molehills or eutrophicated soil in pastures

*Urtica dioica-Cirsium arvense* community OV25

8    Dense swards of *Iris pseudacorus* with *Oenanthe crocata*, *Filipendula ulmaria*, *Poa trivialis* and species such as *Juncus effusus*, *J. acutiflorus*, *Rumex acetosa*, *Ranunculus acris*, *Galium* species, *Urtica dioica* and *Cirsium* species; some stands contain maritime species; along the shore, in wet hollows on raised beaches, and extending a few miles inland at low altitudes; mainly in the western Highlands and Hebrides

**Iris pseudacorus-Filipendula ulmaria mire M28**

Tall-herb vegetation not dominated by *Iris pseudacorus*

**9**

9    Herb-rich swards of *Caltha palustris* with *Ranunculus* species, *Trifolium* species, *Filipendula ulmaria*, *Carex nigra*, *Anthoxanthum odoratum*, *Holcus lanatus*, *Cynosurus cristatus* and *Festuca rubra*; in unimproved fields in Shetland, north-west Highlands, north Pennines and lowland England and Wales

**Cynosurus cristatus-Caltha palustris grassland MG8**

Vegetation not dominated by *Caltha palustris*; *Caltha* can occur in mixtures with other herbs, but is much less abundant than the other species

**10**

10    Species-poor assemblages of *Luzula sylvatica*, *Vaccinium myrtillus*, grasses and ferns on cliff ledges and lightly grazed slopes at moderate to high altitudes

**Luzula sylvatica-Vaccinium myrtillus tall-herb community U16**

Dense stands of *Filipendula ulmaria* in level to gently sloping low-altitude mires, or herb-rich assemblages on steep rocky slopes and dripping ledges with species such as *Alchemilla glabra*, *A. alpina*, *Saxifraga aizoides*, *Luzula sylvatica*, *Sedum rosea*, *Geum rivale* and many sedges and bryophytes

**11**

11  Tall mires or fens dominated by *Filipendula ulmaria*, commonly with *Angelica sylvestris*, *Valeriana officinalis*, *Galium palustre*, *Lychnis flos-cuculi*, *Cardamine pratensis* and *Ranunculus flammula*; on wet soils in low-altitude hollows and valleys, alongside streams, around open water and in wet woodland glades
**Filipendula ulmaria-Angelica sylvestris mire M27**

Tall-herb vegetation on cliff ledges, dripping rocks and steep slopes, with a rich suite of species including *Filipendula ulmaria*, *Alchemilla glabra*, *A. alpina*, *Saxifraga aizoides*, *Luzula sylvatica*, *Sedum rosea*, *Geum rivale* and many sedges and bryophytes

**12**

12  Open herbaceous vegetation on dripping rock-faces with *Saxifraga aizoides*, *S. oppositifolia*, *Selaginella selaginoides*, *Alchemilla glabra*, *A. alpina*, *Thalictrum alpinum* and many sedges and bryophytes
**Saxifraga aizoides-Alchemilla glabra banks U15**

Mixtures of *Luzula sylvatica*, *Deschampsia cespitosa* and tall forbs such as *Sedum rosea*, *Angelica sylvestris*, *Geum rivale*, *Trollius europaeus*, *Geranium sylvaticum* and *Alchemilla glabra*; on base-rich ledges and in some places on ungrazed slopes at moderate altitudes; can be extremely species-rich
**Luzula sylvatica-Geum rivale tall-herb community U17**

# Key to springs, soakways and bog pools

Bryophyte-dominated vegetation of wet habitats

1  Species-poor vegetation dominated by *Polytrichum commune*
*Polytrichum commune* community (see p. 424)

Vegetation not dominated by *P. commune*

**2**

2  Pools or very wet depressions in bogs, or soakways among various types of vegetation

**3**

Springheads where water emerges from the ground

**7**

3  Species-poor pools or very wet depressions in bogs; dominated by Sphagna

**4**

Soakways among a variety of vegetation types, dominated by bryophytes other than Sphagna or by herbs

**5**

4  Bog pools or very wet depressions with abundant *Sphagnum denticulatum* and, in many places, abundant *S. cuspidatum*; with a sparse layer of vascular plants, such as *Eriophorum angustifolium*, *Trichophorum cespitosum*, *Rhynchospora alba* and *Drosera* species; can contain *Menyanthes trifoliata*
**Sphagnum denticulatum bog pool community M1**

Bog pools or very wet depressions with abundant *Sphagnum cuspidatum* and/ or *S. fallax*, and usually lacking *S. denticulatum*; vascular plants include *Erica tetralix* and a few small species such as *Rhynchospora alba*, *Vaccinium oxycoccos* and *Andromeda polifolia*

**Sphagnum cuspidatum/S. fallax bog pool community M2**

5    Soakways with *Campylopus atrovirens* and a sparse sward of *Narthecium ossifragum*, *Trichophorum cespitosum*, *Carex panicea*, *Nardus stricta* and in some places *Erica tetralix*; within blanket bogs, wet heaths or grasslands at fairly low to moderately high altitudes; *Campylopus shawii* is locally common in the Outer Hebrides and on Skye

**Trichophorum cespitosum-Erica tetralix wet heath, Carex panicea subcommunity M15a**

More herb-rich soakways without *Erica tetralix*, *Trichophorum cespitosum* and other bog species, but with small herbs such as *Montia fontana*, *Potamogeton polygonifolius* and *Ranunculus flammula*, sedges and mosses such as *Philonotis fontana*

**6**

6    Soakways with *Hypericum elodes*, *Potamogeton polygonifolius*, *Molinia caerulea*, *Hydrocotyle vulgaris*, *Sphagnum denticulatum* and in many places *Anagallis tenella* and various sedges; can lack *H. elodes* and consist mainly of *P. polygonifolius*, *Carex panicea*, *C. nigra* and *C. viridula* ssp. *oedocarpa*

**Hypericum elodes-Potamogeton polygonifolius soakway M29**

Soakways with *Ranunculus omiophyllus*, *R. flammula*, *Montia fontana*, *Sphagnum denticulatum* and in many places *Agrostis stolonifera*, *Juncus bulbosus*, *J. articulatus* and *Potamogeton polygonifolius*

**Ranunculus omiophyllus-Montia fontana rill M35**

7    Springheads dominated by *Anthelia julacea* growing in a tight silvery-green mat or cushion studded with *Sphagnum denticulatum*, *Saxifraga stellaris*, *Oligotrichum hercynicum* and other small bryophytes and herbs; generally at high altitudes in places where snow lies late

**Anthelia julacea-Sphagnum denticulatum spring M31**

Springs in which *Anthelia julacea* is not dominant

**8**

8    Springs dominated by *Pohlia wahlenbergii* var. *glacialis*, with a few other species such as *Saxifraga stellaris*, *Deschampsia cespitosa* ssp. *alpina*, *Scapania undulata* and *Pohlia ludwigii*; conspicuous pale apple-green patches at high altitudes in corries and on shaded slopes where snow lies late in the Scottish Highlands

**Pohlia wahlenbergii var. glacialis spring M33**

Springs in which *Pohlia wahlenbergii* var. *glacialis* is not dominant, although the odd tuft may occur in springs at high altitudes

**9**

9    Springs dominated by *Palustriella commutata* or *Cratoneuron filicinum*, with *Bryum pseudotriquetrum*, *Festuca rubra* and small forbs, sedges and grasses; on hills with base-rich rocks

**10**

Springs dominated by species other than *Palustriella commutata* or
*Cratoneuron filicinum*

**11**

10 *Palustriella commutata* or *Cratoneuron filicinum* springs with *Festuca rubra*,
*Bryum pseudotriquetrum*, *Cardamine pratensis*, *Carex panicea*, *C. flacca* and a
few other species; widespread in the uplands
*Palustriella commutata-Festuca rubra* **spring M37**

*Palustriella commutata* or *Cratoneuron filicinum* springs with *Carex nigra*,
*Selaginella selaginoides* and a rich flora of small forbs, grasses and sedges,
including montane plants such as *Persicaria vivipara*, *Juncus triglumis* and
*Epilobium anagallidifolium*; at moderate to high altitudes on limestone in
northern England and on base-rich rocks in Scotland
*Palustriella commutata-Carex nigra* **spring M38**

11 Springs with *Montia fontana*, *Ranunculus omiophyllus*, *R. flammula*,
*Sphagnum denticulatum* and a few small forbs, sedges and grasses; at low to
moderate altitudes in the uplands of south-west England, Wales and north-
west England
*Ranunculus omiophyllus-Montia fontana* **rill M35**

Springs dominated by bryophytes such as *Philonotis fontana*, *Dicranella
palustris*, *Sphagnum denticulatum*, *Warnstorfia fluitans*, *W. exannulata*, *Bryum
pseudotriquetrum* and *Scapania undulata*; the brightly coloured patches of
bryophytes are dotted with small herbs such as *Montia fontana*, *Saxifraga
stellaris*, *Chrysosplenium oppositifolium*, *Agrostis canina* and *Epilobium*
species; at moderate to high altitudes from Wales northwards
*Philonotis fontana-Saxifraga stellaris* **spring M32**

## Key to montane moss-heaths and snow-beds

Bryophyte-dominated vegetation of damp or dry substrates at high altitudes

1 Moss-heaths consisting of extensive silvery carpets of *Racomitrium
lanuginosum* or rarely *R. ericoides* on exposed summits and high ridges, or
predominantly bare patches of stony and wind-blasted ground on very
exposed summits, cols and ridges

**2**

Moss-heaths or liverwort mats not dominated by *Racomitrium lanuginosum*
and mostly on steep shaded slopes, in hollows or in gullies at high altitudes
where snow lies late

**6**

2 Heaths with much *Nardus stricta*, usually with scattered small plants of other
vascular species, such as *Diphasiastrum alpinum*, *Huperzia selago* and
*Calluna vulgaris*
*Nardus stricta-Galium saxatile* **grassland,** *Racomitrium lanuginosum* **sub-
community U5e**

*Racomitrium* heaths or predominantly bare wind-blasted stony surfaces with
very little or no *Nardus stricta* or *Calluna vulgaris*

**3**

3    Predominantly bare, stony, wind-blasted ground with a sparse sprinkling of
     plants, almost too open to be called vegetation at all

                                                                              **4**

     Vegetation with a thick mat of *Racomitrium lanuginosum* or, less commonly,
     *R. ericoides* over stony summits or ridges or on scree

                                                                              **5**

4    Sparse open vegetation on wind-blasted summits and ridges at high altitudes
     with species such as *Juncus trifidus, Festuca vivipara, Alchemilla alpina, Salix
     herbacea, Racomitrium lanuginosum* and, in many places, the lichen
     *Ochrolechia frigida*
              **Carex bigelowii-Racomitrium lanuginosum moss heath, species-poor form of
                                   the *Silene acaulis* sub-community U10c**

     Spreads of fine gravel at moderate to high altitudes with a scattering of
     species such as *Festuca ovina/vivipara, Agrostis canina, Vaccinium myrtillus,
     V. vitis-idaea, Saxifraga stellaris, Campanula rotundifolia, Thymus polytrichus,
     Viola riviniana, Potentilla erecta, Oligotrichum hercynicum* and *Racomitrium
     ellipticum*

                                                         Fell-field (see p. 422)

5    *Racomitrium* heaths with a speckling of *Carex bigelowii, Vaccinium myrtillus,
     Diphasiastrum alpinum, Galium saxatile, Festuca vivipara* and other small
     species; *Racomitrium lanuginosum* is generally the dominant species but on
     some summits around Ben Nevis and perhaps elsewhere there are *R. ericoides*
     heaths which are otherwise identical; (*R. ericoides* also grows in pure stands
     on gravel, along tracks and on river shingle at low altitudes; this community
     was not sampled by the NVC and has no connection with montane moss-
     heaths, although it looks superficially similar)
              **Carex bigelowii-Racomitrium lanuginosum moss-heath U10**

     Species-poor mats of *Racomitrium lanuginosum* on dry, stable block scree
              *Racomitrium lanuginosum* scree community (see p. 420)

6    Moss-heaths dominated by *Rhytidiadelphus loreus*, commonly with
     *Hylocomium splendens*, and with a sprinkling of small vascular species such
     as *Deschampsia cespitosa, Carex bigelowii* and *Vaccinium myrtillus*
     **Deschampsia cespitosa-Galium saxatile grassland, *Rhytidiadelphus loreus* sub-
                                                             community U13b**

     Moss-heaths and snow-beds dominated by bryophytes other than
     *Rhytidiadelphus loreus*

                                                                              **7**

7    Snow-beds dominated by *Racomitrium heterostichum, R. fasciculare* or dimin-
     utive liverworts

                                                                              **8**

     Snow-beds dominated by *Kiaeria* species, *Polytrichum sexangulare* or *Pohlia
     ludwigii*

                                                                              **9**

8    Patches of *Racomitrium* species, especially *R. fasciculare*, scattered with
     *Alchemilla alpina*, *Sibbaldia procumbens* and other montane species, such as
     *Luzula spicata*
          ***Alchemilla alpina-Sibbaldia procumbens* dwarf-herb community U14**

     Snow-beds consisting either of patches of *Racomitrium heterostichum* with
     small vascular species such as *Luzula spicata*, *Silene acaulis*, *Saxifraga
     stellaris* and *Persicaria vivipara*, or of very low, crusty, blackish-grey patches
     of tiny liverworts such as *Gymnomitrion concinnatum*, *G. corallioides*,
     *Marsupella* species, *Nardia breidleri* and *Anthelia juratzkana*, speckled with
     *Salix herbacea*, *Gnaphalium supinum*, *Oligotrichum hercynicum* and other
     small species
          ***Salix herbacea-Racomitrium heterostichum* snow-bed U12**

9    Snow-beds consisting of mats of *Kiaeria starkei*, *K. blyttii*, *K. falcata* or
     *Polytrichum sexangulare*, with a few other montane species such as *Huperzia
     selago*, *Saxifraga stellaris*, *Anthelia juratzkana* and *Conostomum tetragonum*
          ***Polytrichum sexangulare-Kiaeria starkei* snow-bed U11**

     Snow-beds dominated by *Pohlia ludwigii* with *Polytrichum sexangulare*,
     *Deschampsia flexuosa* and *Nardia scalaris*
          *Pohlia ludwigii* snow-bed (Rothero 1991) (see ***Salix-Racomitrium* snow-bed U12**)

# Accounts of vegetation types

Detailed accounts are provided of the 99 NVC communities which occur mainly or exclusively in upland Great Britain, or which are principally lowland but are sufficiently common in the uplands to warrant inclusion. These accounts are ordered by NVC type, beginning with woodland and scrub communities (with NVC code letter W), followed by mires (M), heaths (H), mesotrophic grasslands (MG), calcicolous grasslands (CG), calcifuge grasslands and montane communities (U), swamps and fens (S), and vegetation of open habitats (OV). These accounts are followed by shorter descriptions of other types of upland vegetation and habitat which are not described in the NVC.

# W3 *Salix pentandra-Carex rostrata* woodland

Typical view of W3 *Salix-Carex rostrata* woodland thinning out into fen, swamp and open water.

W3 is also shown in the picture for M26.

## Description

This is wet woodland with a low, grey canopy of willows. *Salix cinerea* and *S. pentandra* are the two most common species, although *S. pentandra* is absent from some stands; there can also be some *Betula pubescens* or *Alnus glutinosa*. In some places the willows are old, gnarled, twisted bushes, grey with lichens and hanging with epiphytic bryophytes. They stand knee-deep in a lush, swampy sward of *Carex rostrata* and tall forbs such as *Angelica sylvestris*, *Filipendula ulmaria*, *Potentilla palustris* and *Geum rivale*. There is commonly a sprinkling of *Caltha palustris*, its golden flowers lighting up the drab woodland floor in spring. Other small plants grow here too, including *Cardamine pratensis* with its delicate lilac flowers, *Galium palustre*, *Menyanthes trifoliata* and *Valeriana dioica*. Over the wet peat there is a thin weft of mosses such as *Calliergonella cuspidata*, *Calliergon cordifolium*, *C. giganteum*, *Eurhynchium praelongum* and *Climacium dendroides*.

## Differentiation from other communities

*Salix-Carex* woodland resembles *Alnus-Fraxinus-Lysimachia* woodland W7, which is also wet and herb-rich. It most closely resembles the *Carex-Cirsium* sub-community W7b, which is the form of the community on the wettest ground and which can occur under a canopy of willows, especially in the western Highlands and the Hebrides. *Salix-Carex* woodland is wetter and more swampy, with a distinctive field layer of tall, lush poor-fen herbs, and there is more *Carex rostrata*, *Equisetum fluviatile*, *Calliergon giganteum* and *C. cordifolium* and less *Lysimachia nemorum* and *Carex remota* than in *Alnus-Fraxinus-Lysimachia* woodland.

*Salix-Carex* woodland is similar to the lowland *Salix cinerea-Galium palustre* woodland W1, which also has a field layer including poor-fen herbs. However, *Salix-Carex* woodland is more species-rich, with more *Filipendula ulmaria*, *Caltha palustris*, *Angelica sylvestris*, *Carex rostrata* and *Geum rivale*. These herbs can also grow in *Salix-Galium* woodland, but usually as scattered individuals rather than as lush mixed swards. *Juncus effusus* is more common in the *Salix-Galium* community than in *Salix-Carex* woodland.

### Ecology

This is woodland of wet basins, ungrazed mires and the edges of fens and open water. The soils are fen peats kept wet by base-rich water. It is most common over base-rich rocks such as limestone and basalt, usually below about 400 m.

### Conservation interest

This scarce type of woodland is valuable for nature conservation. The rare *Lysimachia thyrsiflora*, *Corallorhiza trifida* and *Pyrola rotundifolia* have been recorded in *Salix-Carex* woodland. The community forms an important part of the succession from open

*Salix-Carex* woodland is scarce but widely scattered at low altitudes from north Wales northwards.

There seems to be nothing quite like this form of woodland outside Great Britain and Ireland, although montane willow scrub in Scandinavia can have a similar ground flora.

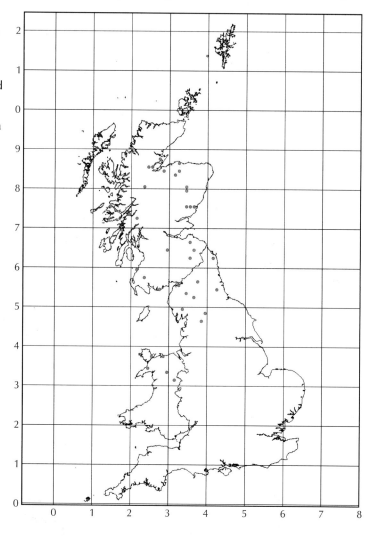

water to dry woodland, and commonly occurs in mosaics with various uncommon types of fen and swamp.

## Management

The original natural vegetation of some of the richer *Carex rostrata* mires and fens in the uplands might have been woodland of this type; where there is a source of willow seed nearby it would be desirable to fence some of these mires against grazing animals to see whether trees would colonise. Many existing woods are still threatened by grazing, although some stands are so wet that they cannot be more than lightly grazed except in the driest of summer weather or when the ground is frozen solid in winter.

# W4 *Betula pubescens-Molinia caerulea* woodland

W4 *Betula-Molinia* woodland (foreground) on wet, peaty soil on gentle slope, with grassy W11 *Quercus-Betula-Oxalis* woodland (non-stippled) on drier slope behind. W17 *Quercus-Betula-Dicranum* woodland (stippled) on thin acidic soils on rocky slope in distance.

W4 is also shown in the pictures for M21 and M26.

## Description

This is wet woodland with a green, grassy, ground flora. *Betula pubescens, Alnus glutinosa, Salix aurita* and *S. cinerea*, some of them old, distorted and covered with lichens and bryophytes, stand over a lush field layer of *Molinia caerulea*. *Quercus petraea* can grow here too. The mosses *Sphagnum palustre, S. fallax* and *Polytrichum commune* grow in rich-green carpets over the soft, wet ground. In the western Highlands and the Inner Hebrides there are scrub forms of *Betula-Molinia* woodland with a canopy of *S. aurita* or *S. cinerea* or both, but no larger trees.

There are three sub-communities, which correspond with variation in the wetness of the soil. The *Dryopteris dilatata-Rubus fruticosus* sub-community W4a is the driest type, with *Dryopteris dilatata, Rubus fruticosus* and *Lonicera periclymenum*. The *Juncus effusus* sub-community W4b has a moderately rich assortment of vascular species, including *Juncus effusus, J. acutiflorus, Holcus mollis, H. lanatus, Deschampsia cespitosa, Potentilla erecta, Viola palustris* and *Cirsium palustre*. The *Sphagnum* sub-community W4c occurs on the wettest soils. It has sheets of *Sphagnum fallax* and *S. palustre*, and rather less *S. fimbriatum, S. papillosum* and *S. squarrosum*. A few bog species such as *Erica tetralix, Calluna vulgaris, Eriophorum angustifolium* and *E. vaginatum* may also grow here. There is also a species-poor form of *Betula-Molinia* woodland with little more than *Molinia* and *Potentilla erecta* under the trees. This does not fit any of the sub-communities described in the NVC.

## Differentiation from other communities

In no other woodland type are *Molinia*, *Sphagnum* species and *Polytrichum commune* as common as in the *Betula-Molinia* community. *Quercus-Betula-Oxalis* W11 and *Quercus-Betula-Dicranum* W17 woodlands are drier and generally have little *Sphagnum*, although *S. quinquefarium* is locally common in some stands of the *Quercus-Betula-Dicranum* community. *Salix-Carex* woodland W3 and *Alnus-Fraxinus-Lysimachia* woodland W7 are more herb-rich, with little or no *Molinia* and *Sphagnum*. Willow scrub forms of *Betula-Molinia* woodland have more *Salix aurita*, *Molinia* and *Sphagnum* and fewer herbs than the lowland *Salix cinerea-Galium palustre* woodland W1.

## Ecology

The community is common on flushed slopes, in poorly drained gullies and valley bottoms, and in hollows with impeded drainage. It commonly forms small patches among other types of woodland, especially *Quercus-Betula-Oxalis* and *Quercus-Betula-Dicranum* woodlands. The species-poor form can occur in large stands. *Betula-Molinia* woodland is the counterpart of *Alnus-Fraxinus-Lysimachia* woodland on acid soils, and the soils are wet, moderately acid peats. On adjacent open ground it can grade

*Betula-Molinia* woodland is common throughout most of upland Great Britain, especially in the wet climate of north Wales, the Lake District and western Scotland. It is also widespread but scarce in the lowlands. The *Juncus* sub-community is one of the most common types of woodland in the uplands.

There are similar wet woodlands in mainland Europe.

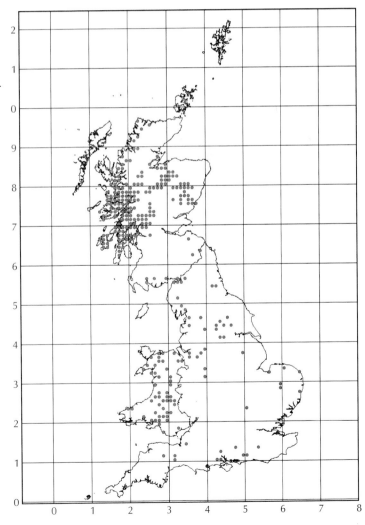

into mires, heaths and grasslands, especially *Molinia-Potentilla* mire M25 and *Trichophorum-Erica* wet heath M15.

### Conservation interest

At one time many swampy glens in the uplands must have been filled with impenetrable thickets of this kind of wet woodland. In particular, north Wales, the Lake District, the western Highlands and the Inner Hebrides come to mind. The remaining stands probably show something of the original, more natural vegetation of these places. Few rare vascular species have been recorded in *Betula-Molinia* woodland, but the epiphytic flora of bryophytes and lichens can be rich and, in the west, can include some important oceanic species. The most common habitat for the scarce subterranean parasitic liverwort *Cryptothallus mirabilis* is under *Sphagnum fallax*, *S. palustre* and *S. fimbriatum* in boggy birch woodland (Hill *et al.* 1991). The rare *S. riparium* has also been recorded in this community.

### Management

The *Dryopteris-Rubus* sub-community persists only where grazing is light or absent; the other two sub-communities are usually at least moderately grazed. In many heavily grazed stands there are no young trees. This may eventually lead to a failure to regenerate and ultimately to the loss of the trees altogether. *Molinia-Potentilla* mire and *Trichophorum-Erica* wet heath can be grazed derivatives of *Betula-Molinia* woodland. Draining can dry out the soils and eventually lead to a change to a drier type of woodland, such as *Quercus-Betula-Oxalis* or *Quercus-Betula-Dicranum*.

# W7 *Alnus glutinosa-Fraxinus excelsior-Lysimachia nemorum* woodland

## Synonyms
Alder stand types 7Aa, 7Ab, 7Bc, 7D and 7Eb (Peterken 1981).

Typical view of alder-dominated W7 *Alnus-Fraxinus-Lysimachia* woodland, opening out into M23 *Juncus-Galium* rush-pasture (foreground and middle right of picture).

## Description
This is wet, herb-rich woodland. The canopy can vary dramatically from dwarfed *Salix aurita* barely more than one or two metres tall, through hazel coppice and birch scrub, to tall ash and alder trees. *Alnus glutinosa, Betula pubescens, Fraxinus excelsior, Quercus petraea, Sorbus aucuparia, Salix aurita, S. cinerea, S. caprea* and *Corylus avellana* have all been recorded here either as the sole species in the canopy or in mixtures. Many stands are moribund with a single layer of old trees and little or no regeneration. There is a green, grassy, herb-rich field layer of species such as *Lysimachia nemorum, Filipendula ulmaria, Athyrium filix-femina, Holcus mollis* and *Poa trivialis*. The field layer can be short or lush, depending on grazing levels. There is a tangled mat of bryophytes over the woodland floor, made up of species such as *Eurhynchium praelongum, Plagiomnium undulatum, Calliergonella cuspidata, Brachythecium rutabulum* and *B. rivulare*. In some places, especially in the western Highlands, there are scrub forms of *Alnus-Fraxinus-Lysimachia* woodland with a low canopy of *S. aurita* or *S. cinerea* or both.

There are three sub-communities. The *Urtica dioica* sub-community W7a is the more lowland type, and is characterised by carpets of *Chrysosplenium oppositifolium* and *Ranunculus repens*, commonly beneath a lush growth of *Urtica dioica, Galium aparine* and *Angelica sylvestris*. *Phalaris arundinacea, Oenanthe crocata* and *Iris pseudacorus* are common in some western stands. The *Carex remota-Cirsium palustre* sub-community W7b occurs on the most permanently waterlogged soils, and has a rich mixture of herbs, including *Cirsium palustre*, *Valeriana officinalis* and *Crepis paludosa*. *Carex remota* is common, and *C. laevigata* and *C. pendula* can also occur. The *Deschampsia cespitosa* sub-community W7c occurs on drier soils. *Deschampsia cespitosa* is especially common here, together with *Anthoxanthum odoratum*, *Agrostis capillaris*, *Oxalis acetosella*, *Dryopteris dilatata* and *D. affinis*.

### Differentiation from other communities

The most similar types of upland woodland are *Betula-Molinia* W4, *Quercus-Betula-Oxalis* W11 and *Fraxinus-Sorbus-Mercurialis* W9. Neither of the first two communities has as many herbs as *Alnus-Fraxinus-Lysimachia* woodland. *Betula-Molinia* woodland has more *Molinia caerulea*, Sphagna and *Polytrichum commune*, and *Quercus-Betula-Oxalis* woodland is drier with more grasses and bracken. *Fraxinus-Sorbus-Mercurialis* woodlands are herb-rich but occupy drier soils and have more *Oxalis acetosella*, *Viola riviniana*, *Mercurialis perennis* and *Eurhynchium striatum*. Willow scrub forms of *Alnus-Fraxinus-Lysimachia* woodland with a canopy of *Salix cinerea* or *S. aurita* can resemble the lowland *Salix cinerea-Galium palustre* woodland W1, but their ground flora usually includes *Filipendula ulmaria*, *Lysimachia nemorum*, *Athyrium filix-femina*, *Dryopteris dilatata* and the mosses *Eurhynchium praelongum*, *Brachythecium rivulare* and

The community is widespread and common at low altitudes throughout the British uplands with outliers in the lowlands of southern England. It is scarce in the far north of Scotland. Scrubby stands with a canopy of *Salix aurita*, *S. cinerea* or *Corylus avellana* are particularly common in the Inner Hebrides.

There are similar woods in mainland Europe.

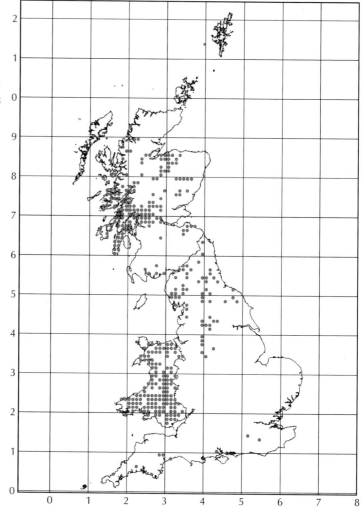

*Calliergonella cuspidata*. Because of their ground flora, these types of willow scrub are better classed as forms of *Alnus-Fraxinus-Lysimachia* woodland than as either *Salix-Galium* woodland or willow scrub forms of the *Betula-Molinia* community.

## Ecology
These are woodlands of flushed slopes, valleys and streamsides throughout the upland fringes. The underlying rocks may be either acid or basic, but the soils are wet brown earths with a neutral pH. The community occurs either as pure stands or as patches within other types of woodland, especially *Quercus-Betula-Oxalis* woodland and *Fraxinus-Sorbus-Mercurialis* woodland.

## Conservation interest
Woodlands resembling these must have been the original natural vegetation of many flushed slopes with moderately basic soils in the uplands. The flora of *Juncus-Galium* rush-pasture M23, *Filipendula-Angelica* mire M27 and *Iris-Filipendula* mire M28 has much in common with that of the ground flora of *Alnus-Fraxinus-Lysimachia* woodland, and many stands of these mire types appear to be derived from this type of woodland. Although this is not a habitat of rare vascular plants, the epiphytic flora of lichens and bryophytes can be rich. The wet ground makes the atmosphere under the trees very humid, and many exacting oceanic bryophyte species and lichens of the *Lobarion pulmonariae* alliance grow in this community in western Scotland. Calcicole species are common on ash and hazel, which have base-rich bark.

## Management
Many stands are not currently regenerating because of the intensity of grazing. In established woods, alder will regenerate from seed only on disturbed ground. In Wales and perhaps elsewhere some *Alnus-Fraxinus-Lysimachia* woodland is moribund because the management is such that there is no bare disturbed ground where trees can regenerate. Some stands have been managed as coppice. This can halt the succession to drier woodland, possibly because the growth of the trees is restricted by continual cropping and consequently the uptake of water from the soil is less than if the trees were allowed to grow to their full size.

# W9 *Fraxinus excelsior-Sorbus aucuparia-Mercurialis perennis* woodland

## Synonyms

*Fraxinus-Brachypodium sylvaticum* nodum, mixed deciduous woodland and *Betula-*herb nodum (basiphilous facies) (McVean and Ratcliffe 1962); *Fraxinus excelsior-Brachypodium sylvaticum* Association, *Corylus avellana-Oxalis acetosella* Association *p.p.* and *Betula pubescens-Cirsium heterophyllum* Association (Birks 1973); *Querco-Ulmetum glabrae* (Birse and Robertson 1976); Ash-wych elm stand type 1Ab and 1D *p.p.*, Hazel-ash stand type 3C *p.p.*, Alder stand type 7D *p.p.* and Birch stand type 12B *p.p.* (Peterken 1981).

Typical view of ash-dominated W9 *Fraxinus-Sorbus-Mercurialis* woodland on a low-altitude hillside.

## Description

These are herb-rich woodlands on fairly dry to moist soils. Beneath a canopy of *Fraxinus excelsior, Ulmus glabra, Sorbus aucuparia, Betula pubescens, Corylus avellana* and in some places *Quercus petraea* there is a lush field layer of *Mercurialis perennis, Sanicula europaea, Urtica dioica, Brachypodium sylvaticum, Oxalis acetosella, Viola riviniana, Primula vulgaris, Filipendula ulmaria* and *Deschampsia cespitosa*. The ferns *Athyrium filix-femina, Dryopteris filix-mas, D. affinis* and *D. dilatata* are common. There are many bryophytes too, including *Eurhynchium praelongum, E. striatum, Plagiomnium undulatum* and *Thuidium tamariscinum*. In north-western Scottish stands, where the climate is cool and humid and there is little atmospheric pollution, the trees can be almost totally covered with epiphytes, including lichens of the *Lobarion pulmonariae* and *Graphidion scriptae* alliances and many oceanic bryophytes. Some stands, especially in western Scotland, have a canopy dominated by hazel, with few or no taller trees.

There are two sub-communities. The Typical sub-community W9a is the more wide-spread type, and has *Geum urbanum*, *Circaea lutetiana* and *Dryopteris dilatata* in the field layer. The *Crepis paludosa* sub-community W9b has a more restricted and decid-edly upland distribution. Here *G. urbanum* is generally replaced by *G. rivale*, and this, together with other characteristic species such as *Crepis paludosa*, *Deschampsia cespitosa*, *Cirsium heterophyllum* and *Geranium sylvaticum*, shows the influence of cooler, damper conditions.

## Differentiation from other communities

*Quercus-Betula-Oxalis* woodland W11 is grassier than *Fraxinus-Sorbus-Mercurialis* woodland, with less ash, more oak and birch, and fewer tall herbs such as *Mercurialis perennis* and *Sanicula europaea*. *Alnus-Fraxinus-Lysimachia* woodland W7 is wetter than *Fraxinus-Sorbus-Mercurialis* woodland, and has more *Lysimachia nemorum*, *Cirsium palustre*, *Carex remota*, *Chrysosplenium oppositifolium* and wetland mosses such as *Brachythecium rivulare*, *Calliergonella cuspidata* and *Rhizomnium punctatum*. *Fraxinus excelsior-Acer campestre-Mercurialis perennis* woodland W8 is the lowland counterpart of this community. Its flora is very variable and can be quite similar to that of the Typical sub-community of *Fraxinus-Sorbus-Mercurialis* woodland but there is usually less *Oxalis acetosella* and fewer bryophytes, lichens and ferns.

## Ecology

The soils under *Fraxinus-Sorbus-Mercurialis* woodland are moist or rather dry fertile brown earths derived from base-rich rocks, such as Moine schist, limestone and basalt. Most stands are on steep slopes or in ravines. The uneven, irregular topography of these habitats gives rise to a complicated mixture of soil conditions and hence a diverse ground flora, with *Fraxinus-Sorbus-Mercurialis* woodland commonly forming mosaics with other types of woodland, such as the *Alnus-Fraxinus-Lysimachia* and *Quercus-Betula-Oxalis* communities. *Fraxinus-Sorbus-Mercurialis* vegetation is the characteristic woodland of limestone pavements in the uplands.

## Conservation interest

*Fraxinus-Sorbus-Mercurialis* woodland is a species-rich element in the upland land-scape, and is an important habitat for some uncommon plants including *Trollius europaeus*, *Cirsium heterophyllum* and *Geranium sylvaticum*. The rare *Actaea spicata*, *Bromus benekenii*, *Crepis mollis*, *Gagea lutea* and *Polygonatum verticillatum* have also been recorded in this type of vegetation. Ash, elm and hazel have base-rich bark, and commonly have rich epiphytic floras, including scarce bryophytes and fine examples of the *Lobarion pulmonariae* lichen assemblage. All the British species of *Lobaria* and *Sticta* grow in this community, together with a number of other large, attractive foliose lichens such as *Pseudocyphellaria crocata*, *P. intricata*, *P. norvegica*, *Nephroma laevigatum* and *N. parile*. Small crustose lichens are much less conspicuous but are equally important members of the flora; the recently discovered endemic *Graphis alboscripta* is one of these (Gilbert 2000). The rare fungus *Hypocreopsis rhododendri* grows on hazels and some other shrub species in *Fraxinus-Sorbus-Mercurialis* wood-lands in the west Highlands and the Inner Hebrides.

## Management

A ground flora characteristic of *Quercus-Betula-Oxalis* woodland can occur below a canopy characteristic of *Fraxinus-Sorbus-Mercuralis* woodland. This occurs where heavy grazing eliminates the grazing-sensitive tall herbs of *Fraxinus-Sorbus-Mercuralis* wood-land. A reduction in the intensity of grazing in some *Quercus-Betula-Oxalis* woodland

## W9 *Fraxinus excelsior-Sorbus aucuparia-Mercurialis perennis* woodland

The community is widely distributed through the British uplands from Dartmoor to Sutherland, but has not been recorded on the Outer Hebrides, Orkney or Shetland. The wetter *Crepis* sub-community is distinctly northern.

There is related vegetation in Ireland, Norway and Sweden.

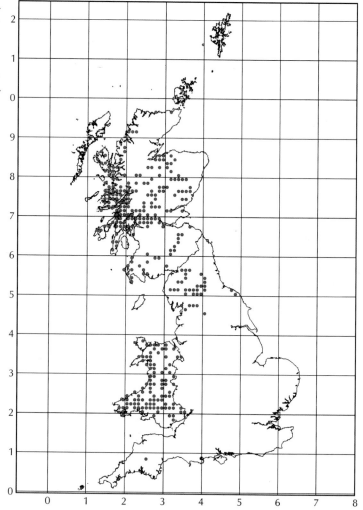

might allow more tall herbs to flourish and the vegetation to revert to *Fraxinus-Sorbus-Mercurialis* woodland. Some stands of the *Luzula-Geum* tall-herb community U17, *Sesleria-Galium* grassland CG9, *Festuca-Agrostis-Thymus* grassland CG10, herb-rich *Festuca-Agrostis-Galium* grassland U4 and *Pteridium-Rubus* underscrub W25, and the more herb-rich examples of *Calluna-Erica* heath H10d, might also revert to *Fraxinus-Sorbus-Mercurialis* woodland in the absence of grazing and given a source of seed. Hazel coppicing has taken place in some *Fraxinus-Sorbus-Mercurialis* woodlands, but is not generally a desirable form of management here, especially where there are rich bryophyte and lichen floras on trees and rocks. This is because coppicing results in an open canopy that is not favourable to those bryophytes and lichens which thrive in humid microclimates.

# W11 *Quercus petraea-Betula pubescens-Oxalis acetosella* woodland

## Synonyms

*Betuletum Oxaleto-Vaccinietum p.p.* and *Betula*-herb nodum (McVean and Ratcliffe 1962); *Betula pubescens-Vaccinium myrtillus* Association *p.p.* and *Corylus avellana-Oxalis acetosella* Association *p.p.* (Birks 1973); *Lonicero-Quercetum* (Birse and Robertson 1976); Hazel-ash woodland *p.p.*, Oak-lime woodland *p.p.* and Birch-oak woodland *p.p.* (Peterken 1981).

W11 *Quercus–Betula–Oxalis* woodland (non-stippled) on well-drained slope, with more mossy/heathy W17 *Quercus–Betula–Dicranum* woodland (stippled) on thinner acid soils on rocky slope in distance. W4 *Betula-Molinia* woodland in foreground on wet, peaty soil.

## Description

These are dry woodlands with a grassy ground flora. The canopy is made up of *Betula pubescens*, *B. pendula*, *Quercus petraea*, *Q. robur* and *Sorbus aucuparia*. There can be a shrub layer in which *Corylus avellana* is usually the most common species; some stands are dominated by hazel, with few or no larger trees. *Ilex aquifolium* can be common too, especially in the west of Scotland. The birches and oaks vary from tall and straight to low and twisted. They can have dense growths of bryophytes and lichens on their trunks and lower branches. In some stands the oaks and hazels are coppiced. The field layer is generally made up of *Agrostis capillaris*, *Holcus mollis*, *Anthoxanthum odoratum* and other grasses, together with small forbs such as *Oxalis acetosella*, *Potentilla erecta* and *Viola riviniana*. *Pteridium aquilinum* can be very common. Dwarf shrubs are rare. There are a great many bryophytes, especially *Thuidium tamariscinum*, *Hylocomium splendens*, *Rhytidiadelphus squarrosus*, *Scleropodium purum*, *Polytrichum formosum* and other large mosses.

Four sub-communities are described in the NVC. In the *Dryopteris dilatata* sub-community W11a at least one of *Dryopteris dilatata*, *Rubus fruticosus* and *Lonicera periclymenum* predominates in a rather tall, untidy field layer which is either lightly grazed or ungrazed; *Hyacinthoides non-scripta* is common here. *Blechnum spicant* does indeed grow in the *Blechnum spicant* sub-community W11b, but just as characteristic here are *Primula vulgaris*, *H. non-scripta* and the moss *Hylocomium brevirostre*. Rocks and the trunks and branches of trees can be covered with rich assemblages of bryophytes and lichens. The *Anemone nemorosa* sub-community W11c has a noticeably northern and more Scandinavian appearance, with much *Rhytidiadelphus triquetrus*, and, scattered among the grasses and mosses, *Anemone nemorosa*, *Luzula pilosa* and *Trientalis europaea*. The *Stellaria holostea-Hypericum pulchrum* sub-community W11d is the least distinctive sub-community. Among the grasses are a few forbs, including *Stellaria holostea*, *Hypericum pulchrum*, *Veronica chamaedrys* and *V. officinalis*; *H. non-scripta* is locally common.

Throughout upland Great Britain there are many stands of *Quercus-Betula-Oxalis* woodland that are very species-poor, with a field layer of *Agrostis capillaris*, *Anthoxanthum odoratum* and *Holcus* species. Some of these stands are of young trees colonising what was once grassland; others are older woods that have suffered from heavy grazing. These impoverished, grass-dominated stands do not fit into any of the recognised sub-communities. There are also stands of oak or birch woodland with a ground flora of little more than *Luzula sylvatica* which are not described in the NVC, although they vary enough to have affinities with several NVC types. Stands with neither dwarf shrubs nor extensive carpets of mosses are best regarded as forms of *Quercus-Betula-Oxalis* woodland.

## Differentiation from other communities

*Quercus-Betula-Oxalis* woodland has fewer dwarf shrubs and bryophytes, and more grasses, than *Quercus-Betula-Dicranum* woodland W17. It is more grassy and less herb-rich than either *Alnus-Fraxinus-Lysimachia* W7 or *Fraxinus-Sorbus-Mercurialis* W9 woodland. It is drier than *Betula-Molinia* woodland W4, with less *Molinia caerulea*, Sphagna and *Polytrichum commune*. The lowland *Quercus robur-Pteridium aquilinum-Rubus fruticosus* woodland W10 may closely resemble *Quercus-Betula-Oxalis* woodland, but can generally be distinguished by the scarcity of *Agrostis capillaris*, *Anthoxanthum odoratum*, *Deschampsia flexuosa*, *Oxalis acetosella*, *Potentilla erecta* and *Viola riviniana*.

## Ecology

This is woodland of free-draining acid soils on the upland margins, generally below 400 m. Some woods are comprised of pure stands of *Quercus-Betula-Oxalis* woodland, but the community more commonly occurs in mosaics with other types of woodland. It gives way to *Quercus-Betula-Dicranum* woodland on steep rocky slopes, and to *Betula-Molinia* woodland in wet hollows and gullies and on level waterlogged ground.

## Conservation interest

Most stands of *Quercus-Betula-Oxalis* woodland are not species-rich; few rare plants have been recorded here. However, *Quercus-Betula-Oxalis* woodlands are attractive, especially in the spring when the fragrant golden-green leaves of birch and oak unfold, and bluebells and primroses are flowering. There is some interest in most of the sub-communities. The *Blechnum* sub-community in the north-west can have rich assemblages of bryophytes and lichens, including the liverworts *Plagiochila spinulosa*, *P. punctata*, *P. killarniensis*, *Drepanolejeunea hamatifolia* and *Harpalejeunea molleri*, and the lichens *Lobaria pulmonaria*, *L. virens*, *L. scrobiculata*, *L. amplissima*, *Sticta* species, *Pseudocyphellaria* species, *Degelia* species, *Parmelia laevigata*, *Menegazzia terebrata* and various *Usnea* species. The *Anemone* sub-community in the north-east can contain good populations of *Trientalis europaea*. Some lightly-grazed examples of the *Dryopteris*

The community occurs throughout the British uplands from south-west England to north-west Scotland. It has a western distribution in England and Wales but not in Scotland, where it extends far to the east. The sub-communities have distinctive distributions: the *Dryopteris* sub-community is south-western, the *Blechnum* sub-community is north-western, the *Anemone* sub-community is north-eastern, and the *Stellaria-Hypericum* sub-community and the species-poor grassy form of the community are widespread. The form of *Quercus-Betula-Oxalis* woodland that is dominated by *Luzula sylvatica* occurs in ungrazed or lightly grazed situations, mainly from north Wales northwards.

There are similar grassy birch woods in the western parts of mainland Europe, but there is probably nothing like the *Blechnum* sub-community outside Great Britain and Ireland.

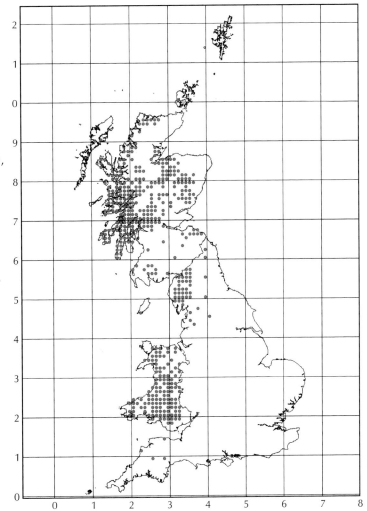

sub-community have a lush field layer, with bramble and honeysuckle scrambling over the ground and reaching up into the trees. Honeysuckle can grow well in these woods and can form vast towering thickets reaching high into the canopy. This might give a glimpse of the more natural vegetation of the past, when many of the woods in the south and west may have had this type of structure.

Like all semi-natural woodlands, these upland oak and birch woods are important breeding habitats for birds. There can be large populations of species that avoid treeless ground. It is a refreshing experience in spring or early summer to come down from the hill into woodland and hear the birds singing. Willow Warbler *Phylloscopus trochilus*, Wood Warbler *P. sibilatrix*, Chiffchaff *P. collybita*, Redstart *Phoenicurus phoenicurus* and Spotted Flycatcher *Muscicapa striata* are among the most common of these birds. Other breeding species include Buzzard *Buteo buteo*, Sparrowhawk *Accipter nisus* and Kestrel *Falco tinnunculus*. In Wales these woods are famed for their populations of Pied Flycatchers *Ficedula hypoleuca*, and are also important for Red Kites *Milvus milvus*.

## Management
Grassy woodlands such as these are usually the result of moderately heavy grazing; most examples of the herb-rich or heathy *Fraxinus-Sorbus-Mercurialis* and *Quercus-Betula-*

*Dicranum* woodlands are less heavily grazed. The *Dryopteris* sub-community seems to be the most susceptible to grazing, and the form with a dense field layer of *Luzula sylvatica* occurs only where there is little or no grazing. Continued heavy grazing prevents regeneration of the canopy, and may contribute to the eventual loss of the trees and the replacement of woodland by *Festuca-Agrostis-Galium* grassland or other acid swards. Many examples of *Quercus-Betula-Oxalis* woodland have been managed as coppice but this is not generally a desirable form of management here, especially where the trees and rocks are clothed with rich assemblages of bryophytes and lichens.

# W17 *Quercus petraea-Betula pubescens-Dicranum majus* woodland

## Synonyms
*Betuletum Oxaleto-Vaccinetum* (McVean and Ratcliffe 1962); *Betula pubescens-Vaccinium myrtillus* Association *p.p.* (Birks 1973); Birch-oak woodland and Birch woodland (Peterken 1981).

W17 *Quercus-Betula-Dicranum* woodland (stippled) on thin acidic soils on rocky slope in distance, with grassy W11 *Quercus-Betula-Oxalis* woodland (non-stippled) on well-drained slope in middle distance and W4 *Betula-Molinia* woodland in foreground on wet, peaty soil.

## Description
This community includes heathy and mossy woodlands with a canopy of *Quercus petraea* and *Betula pubescens* dotted with *Sorbus aucuparia* and, in some places, *Ilex aquifolium*. If there is a shrub layer at all it is just a thin understorey of *Corylus avellana* and young specimens of the canopy trees. Dwarf shrubs predominate in the field layer: dark, shaggy *Calluna vulgaris* or dense, green mats of *Vaccinium myrtillus*. More distinctively, there are great quantities of bryophytes, growing in variegated mats and patches over the ground, covering stumps and boulders, and muffling the bases of trees. The most common species are the mosses *Hylocomium splendens*, *Rhytidiadelphus loreus*, *Pleurozium schreberi*, *Plagiothecium undulatum*, *Thuidium tamariscinum*, *Polytrichum formosum*, *Dicranum majus* and *D. scoparium*.

There are four sub-communities. The *Isothecium myosuroides-Diplophyllum albicans* sub-community W17a occurs on the steepest and most rocky ground, including cliff ledges and the sides of deep ravines. It has the richest bryophyte flora, including oceanic liverworts such as *Scapania gracilis* and *Plagiochila spinulosa*. The oceanic filmy ferns *Hymenophyllum wilsonii* and *H. tunbrigense* also occur here. The Typical sub-community W17b is very heathy, with a thick field layer of heather or bilberry or both, but otherwise has no special distinguishing species. The *Anthoxanthum odoratum-Agrostis capillaris* sub-community W17c is grassier than the others, and there can be few or even no dwarf shrubs, but there is still a rich array of bryophytes growing in extensive carpets. In the *Rhytidiadelphus triquetrus* sub-community W17d, the crowded thick shoots of *Rhytidiadelphus triquetrus* with their broad, pale-green leaves are commonly mixed with the more slender *Scleropodium purum* among *Calluna*. *Juniperus communis* ssp. *communis* can form a shrub layer in this sub-community.

There are also stands of birch and oak woodland where the field layer is an almost pure sward of *Luzula sylvatica* interspersed with a few dwarf shrubs and entwined with mosses. These clearly have an affinity with *Quercus-Betula-Dicranum* woodland.

### Differentiation from other communities

*Quercus-Betula-Dicranum* woodland is separated from other types of deciduous woodland by its mossy and heathy ground flora. Dwarf shrubs can be scarce in the *Anthoxanthum-Agrostis* sub-community and some forms of the *Isothecium-Diplophyllum* sub-community, but in such places the vegetation can be distinguished from *Quercus-Betula-Oxalis* woodland W11 by the luxuriance and diversity of mosses and liverworts. The more lowland *Vaccinium myrtillus-Dryopteris dilatata* sub-community of *Quercus* species-*Betula* species-*Deschampsia flexuosa* woodland W16b has much *Vaccinium* but is less mossy than *Quercus-Betula-Dicranum* woodland. *Pinus-Hylocomium* woodland W18 has a heathy and mossy ground flora similar to that in this community, but has a canopy of pine instead of deciduous trees.

### Ecology

This is the characteristic upland woodland of steep, rocky hillsides on thin, moist, free-draining acid mineral soils. The underlying rocks are usually hard and acid, and include granite, Pre-Cambrian sandstone and gneiss, Dalradian schist, Ordovician and Silurian slate and shale, and Carboniferous Millstone Grit. Most stands are below 450 m, but *Quercus-Betula-Dicranum* woodland extends up to almost 700 m on a few sheltered slopes in the western Highlands. It generally replaces *Quercus-Betula-Oxalis* woodland on thinner and more acid soils, on steeper, rockier slopes, and at higher altitudes.

### Conservation interest

Some stands of *Quercus-Betula-Dicranum* woodland on steep rocky slopes and in ravines may never have been felled, and may be survivors of the original natural woodland. Many examples of the community are rather species-poor and without scarce plants, but the western *Isothecium-Diplophyllum* sub-community is one of the most important British habitats for oceanic bryophytes and lichens. A number of exacting, humidity-demanding species grow here, including the oceanic liverworts *Plagiochila punctata*, *P. killarniensis*, *P. exigua*, *P. atlantica*, *Adelanthus decipiens*, *Lepidozia cupressina*, *Herbertus aduncus* ssp. *hutchinsiae*, *Leptoscyphus cuneifolius*, *Lejeunea* species, *Radula* species, *Frullania* species, *Aphanolejeunea microscopica*, *Drepanolejeunea hamatifolia* and *Harpalejeunea molleri*. The large foliose lichen *Lobaria pulmonaria* can grow in huge shaggy patches on oak trunks, and in some places *L. virens*, *L. scrobiculata* and *L. amplissima* are common too. Other scarce lichens include *Parmelia laevigata*, *Menegazzia terebrata* and various *Usnea* species. These woods are as important for upland birds as *Quercus-Betula-Oxalis* woodland.

### Management

Much heathy *Quercus-Betula-Dicranum* woodland has been converted to the more grassy *Quercus-Betula-Oxalis* type by grazing; there are many places in the uplands where the two communities are separated by a fence or a wall, the only difference being

*Quercus-Betula-Dicranum* woodland occurs throughout the British uplands from Dartmoor to Sutherland. It is particularly common in north Wales and in the Scottish Highlands. The sub-communities have varying distributions: the Typical and *Anthoxanthum-Agrostis* sub-communities are widespread, the *Isothecium-Diplophyllum* sub-community is western, and the *Rhytidiadelphus* sub-community is north-eastern. Mossy oak-birch woodland with a ground flora dominated by *Luzula sylvatica* has been recorded in lightly grazed places from north Wales northwards.

There are woodlands resembling this community elsewhere in Europe. Woodland similar to the *Isothecium-Diplophyllum* sub-community occurs in the west of Ireland, where the bryophyte flora can be just as rich as in the western Highlands (e.g. Ratcliffe 1968, Mitchell and Averis 1988).

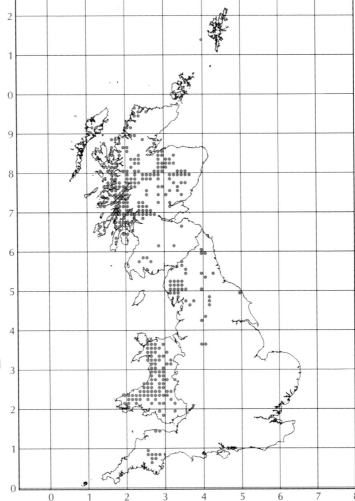

the grazing intensity on either side. Much *Quercus-Betula-Oxalis* woodland might revert to *Quercus-Betula-Dicranum* woodland if it was not grazed so intensively. In the absence of grazing, some stands of *Calluna-Vaccinium* heath H12 might also develop into *Quercus-Betula-Dicranum* woodland, provided that there was a suitable seed source. Reducing the grazing in existing stands of *Quercus-Betula-Dicranum* woodland would let the dwarf shrubs grow taller and allow tree regeneration. In some recently-established ungrazed enclosures, honeysuckle and bramble have grown up in heathy stands of this community, even though these species are generally absent from the floristic tables for *Quercus-Betula-Dicranum* woodland and other heathy vegetation types in the NVC.

There has been some concern that if these woods are not grazed at all, the field layer of *Vaccinium myrtillus* and *Deschampsia flexuosa* will grow tall and dense and shade out bryophytes on the bases of trees and on rocks. It is widely believed that this is a potential problem throughout Great Britain, although it seems to have been demonstrated only in north Wales (Edwards and Birks 1986). What seems to have happened in Wales is that the long history of heavy grazing over 500 or 600 years has made the woodland ground flora so short that the bryophytes have become confined to the lower parts of trees and rocks where the climate is most humid and equable. As a result, they are at risk from over-shading when the field layer is allowed to grow tall. In the more consistently wet climate of western Scotland, the bryophytes grow higher up the rocks and the trunks of trees, and would probably not be over-shaded by a taller ground layer (Averis and Coppins 1998).

The ground flora of *Quercus-Betula-Dicranum* woodland is able to persist under planted conifers, especially in larch plantations but also under other trees where the planting is not dense. There is the potential to reinstate semi-natural woodland in these situations, by gradually removing the conifers and replanting with native species, at least over small areas.

Coppicing has been widely practised in these woodlands, but is not generally a desirable form of management, because the open canopy does not favour the rich bryophyte and lichen floras on trees and rocks so characteristic of this type of woodland.

# W18 *Pinus sylvestris-Hylocomium splendens* woodland

## Synonyms
*Pinetum Hylocomieto-Vaccinietum* (McVean and Ratcliffe 1962); *Erica cinerea-Pinus sylvestris* Plantation (Birse and Robertson 1976); Pinewood plot types 1–9 (Peterken 1981).

Typical view of W18 *Pinus-Hylocomium* woodland, thinning out into open heath at lower left of picture.

## Description
These are woods with a canopy of *Pinus sylvestris*: tall, spreading, stately trees with red-brown or grey-brown bark and dense, dark-blue-green needles. Pine trees have clean stems, usually without a great growth of bryophytes, but can be well covered with small grey lichens such as *Hypogymnia physodes*, *Bryoria fuscescens* and *Platismatia glauca*. Many woods are almost pure pine, although there can be a little *Betula pubescens*, *B. pendula*, *Sorbus aucuparia*, *Populus tremula* and *Ilex aquifolium*. In some stands there is a shrub layer of *Juniperus communis* ssp. *communis*, especially in the eastern Highlands. *Calluna vulgaris*, *Vaccinium myrtillus* and *V. vitis-idaea* are the most conspicuous species in the field layer, and are accompanied by *Deschampsia flexuosa*. There are big hummocks and cushions of mosses, mainly large common species such as *Pleurozium schreberi*, *Rhytidiadelphus loreus*, *Hylocomium splendens*, *Plagiothecium undulatum* and *Dicranum scoparium*.

There are five sub-communities: the first three are characteristic of dry situations and the last two occur on damper ground. The *Erica cinerea-Goodyera repens* sub-community W18a occurs in both semi-natural and plantation pinewoods. *Erica cinerea* is common, though it rarely grows thickly, and *Vaccinium myrtillus* and *V. vitis-idaea* are rare or absent. The pinewood orchid *Goodyera repens* grows here, its slender spikes of cream-white flowers enlivening the dark woodland floor. The *Vaccinium myrtillus-Vaccinium vitis-idaea* sub-community W18b is the typical form of dry pinewood. There

115

is a dense green layer of *Vaccinium* species under the trees but other than this there are no special distinguishing species. The *Luzula pilosa* sub-community W18c is another dry woodland, characterised by *Luzula pilosa, Galium saxatile* and *Oxalis acetosella*. The *Sphagnum capillifolium/quinquefarium-Erica tetralix* sub-community W18d is the typical form of damp pinewood. Very characteristic here are the wine-red and bright-green cushions of *Sphagnum capillifolium* and *S. quinquefarium*, and there can also be some *Erica tetralix* and *Molinia caerulea* among the *Calluna* and *Vaccinium* species. The *Scapania gracilis* sub-community W18e resembles the *Sphagnum-Erica* sub-community, but has patches of the liverworts *Scapania gracilis* and *Anastrepta orcadensis* among the carpets of Sphagna and other mosses.

*Molinia* can grow rather sparsely in the *Sphagnum-Erica* sub-community, but is dominant in the field layer of some other forms of pine woodland; such stands do not fit anywhere in the NVC (Rodwell *et al.* 2000). Locally, in north-eastern Scotland, there are open pine plantations in which the ground vegetation is conspicuously frosted with large, whitish-cream lichens of the genus *Cladonia* (Watson and Birse 1990; Dargie 1994; Gilbert 2000). This vegetation, which resembles some of the more continental European pine woodlands, cannot be classed within *Pinus-Hylocomium* woodland as currently circumscribed, and may even warrant a separate NVC community (Rodwell *et al.* 1998).

### Differentiation from other communities

Only in the *Quercus-Betula-Dicranum* woodland W17 do the field and ground layers resemble those of *Pinus-Hylocomium* woodland, but this has a canopy of birch or oak with at most only a little pine. Apart from the more heathy examples of *Juniperus-Oxalis* woodland W19a, which are dominated by juniper with little or no pine, there are no other types of British woodland that can be confused with *Pinus-Hylocomium* woodland. The field and ground layers of this type of woodland have much in common with *Calluna-Vaccinium* H12, *Vaccinium-Deschampsia* H18 and *Calluna-Vaccinium-Sphagnum* H21 heaths. In open pine woodland the spatial and floristic boundaries between *Pinus-Hylocomium* woodland and these heaths can be very diffuse.

### Ecology

This is a community of impoverished, podsolised, acid, free-draining soils. It commonly occurs on sandy or gravelly fluvioglacial deposits or directly over base-poor metamorphic or sedimentary rocks, such as quartzite or sandstone. The soils may be dry or damp; most of the wetter stands are in the west. The maximum altitudinal limit at present is 640 m on Creag Fhiaclach in the western Cairngorms (Nethersole-Thompson and Watson 1981). Elsewhere in the Cairngorms, the upper limit of the woods appears to have been lowered by land use and management (Pearsall 1968), and the general maximum altitude attained by *Pinus-Hylocomium* woodland is around 500–550 m. In the western Highlands, it is mostly below 300 m. *Pinus-Hylocomium* woodland accounts for some of the largest expanses of native woodland in Scotland, but there are also many very small stands of the community scattered among heathland or forming mosaics with birch woodland.

### Conservation interest

The value of this community for nature conservation is well known. The total extent of native pinewood in Scotland (including stands that have regenerated recently from native stock of local origin) is estimated to be approximately 25,000 ha (MacKenzie 1999). Over 15,000 ha are protected within candidate SACs, including a very large proportion of the total area of ancient pine woodland. Many stands of *Pinus-Hylocomium*

*Pinus-Hylocomium* woodland is locally common in the Scottish Highlands. It is absent from the extreme north, the extreme south-west and the Hebrides. The *Erica-Goodyera*, *Vaccinium* and *Luzula* sub-communities are recorded mainly in the east, and the *Sphagnum-Erica* and *Scapania* sub-communities are western. The largest and best-known stands are in the eastern Highlands, in the straths and glens of the Spey and the Dee either side of the Cairngorms, where the *Vaccinium* sub-community predominates. There are some large stands further west too: in Glen Affric and Glen Cannich in Inverness-shire, and around Shieldaig and Beinn Eighe in Wester Ross. In the south-west, there are some fine stands in the Black Wood of Rannoch and near Tyndrum. A few planted stands of pine woodland south of the High-lands have also been classed as *Pinus-Hylocomium* woodland (Rodwell 1991a).

Pine woodland similar to this community occurs in western Norway.

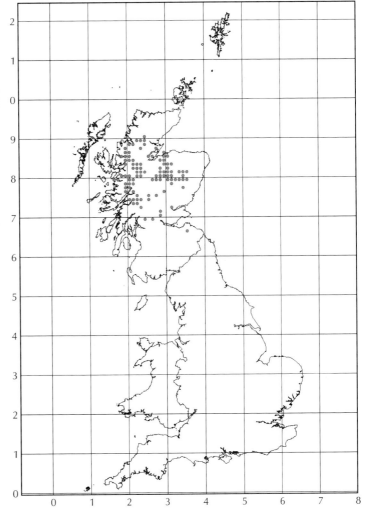

woodland are within National Nature Reserves or in the hands of independent conserva-tion bodies, such as the Royal Society for the Protection of Birds.

The habitat has a distinctive bird fauna which includes Scottish Crossbill *Loxia scotica*, Crested Tit *Parus cristatus* (nesting in dead stumps) and Capercaillie *Tetrao urogallus*. Other breeding birds include Kestrels *Falco tinnunculus*, Buzzards *Buteo buteo*, Greenfinches *Carduelis chloris*, Redstarts *Phoenicurus phoenicurus*, Treecreepers *Certhia familiaris*, Siskins *Carduelis spinus*, Chaffinches *Fringilla coelebs*, Wood War-blers *Phylloscopus sibilatrix* and Goldcrests *Regulus regulus*. Fieldfares *Turdus pilaris* and Redwings *Turdus iliacus* breed in a few places. Ospreys *Pandion haliaetus* and Gos-hawks *Accipiter gentilis* also nest in pine woodland, and Golden Eagles *Aquila chrysaetos* occasionally have their tree eyries in Scots pine within these woods. Pinewoods are a stronghold for Red Squirrels *Sciurus vulgaris* in Scotland and are also home to Pine Martens *Martes martes*.

*Pinus-Hylocomium* woodland provides an important habitat for the scarce plant spe-cies *Goodyera repens*, *Linnaea borealis* and *Moneses uniflora*. Most of the known popula-tions of the rare moss *Dicranum subporodictyon* are on rocks within or close to this type of woodland in the western Highlands, and the rare moss *Buxbaumia viridis* grows on pine logs in this community. The lichen flora is distinctive and interesting.

The range of pinewood specialists includes *Bryoria capillaris, B. furcellata, Calicium parvum, Hypocenomyce friesii, Lecanora cadubriae, Lecidea turgidula, Ochrolechia microstictoides* and *Protoparmelia ochrococca*, as well as rare species such as *Cladonia carneola, C. cenotea, C. sulphurina* and *C. botrytes. Cetraria juniperina*, which grows on the trunks of pine and juniper, has not been seen since the nineteenth century but could well be refound (Gilbert 2000). New discoveries are still being made, and it is quite possible that many, if not most, of the species known in the pinewoods of western Scandinavia may also grow in Scotland (Brian Coppins pers. comm.). The pinewoods of Glen Strathfarrar and Glen Affric have the richest floras of characteristic pinewood lichens (Coppins 1990), perhaps because they are in the central Highlands and have both eastern and western species.

## Management

Most stands of *Pinus-Hylocomium* woodland are grazed by deer or livestock, and sustained heavy grazing can prevent tree regeneration. The intensity of grazing has been reduced at some sites to allow pine regeneration and to help the age structure of the woods become more diverse. *Pinus-Hylocomium* woodland commonly occurs in association with various forms of dry and wet heath, and pine trees can become established in the dwarf-shrub vegetation if grazing animals are removed (e.g. Wormell 2000).

# W19 *Juniperus communis* ssp. *communis-Oxalis acetosella* woodland

## Synonyms

*Juniperus-Thelypteris* nodum (McVean and Ratcliffe 1962); *Juniperus communis*-fern scrub A1 (Birks and Ratcliffe 1980).

Typical view of W19 *Juniperus-Oxalis* woodland: scattered patches of juniper with occasional birch, among heath and grassland.

## Description

These shaggy, uneven, dark-glaucous-green stands of scrub are dominated by *Juniperus communis* ssp. *communis* with its twisting spires of branches and its astonishing diversity of form, ranging from flat, spreading bushes to tall, columnar specimens. Accompanying the juniper there can be scattered trees of *Betula pendula*, *B. pubescens* or *Sorbus aucuparia*. *Rosa canina*, *Sambucus nigra*, *Prunus spinosa*, *P. padus* and other shrubs can make a small contribution to the canopy.

There are two sub-communities: one with a heathy ground flora and one with a grassy, herb-rich ground flora. The *Vaccinium vitis-idaea-Deschampsia flexuosa* sub-community W19a takes in heathy stands on acid soils. Under the juniper bushes there is a thick layer of dwarf shrubs, comprising *Calluna vulgaris*, *Vaccinium myrtillus* and *V. vitis-idaea*, interleaved with *Deschampsia flexuosa*, *Agrostis capillaris* and *A. canina*. The ground is carpeted with a deep layer of mosses: generally large common species such as *Hylocomium splendens*, *Thuidium tamariscinum*, *Plagiothecium undulatum*, *Rhytidiadelphus loreus* and *Pleurozium schreberi*. In contrast, the field layer of the *Viola riviniana-Anemone nemorosa* sub-community W19b includes *Festuca ovina*, *F. rubra*, *Holcus lanatus*, *Anthoxanthum odoratum*, *Urtica dioica*, *Poa trivialis* and *Veronica chamaedrys*, as well as *Viola riviniana* and *Anemone nemorosa*. There can be lush patches of ferns, including *Gymnocarpium dryopteris*, *Phegopteris connectilis* and *Oreopteris limbosperma*, and *Pyrola media* is locally common. Under these plants is a loose weft of mosses including *Eurhynchium praelongum*, *Brachythecium rutabulum*, *Hypnum cupressiforme s.l.* and *Plagiomnium undulatum*.

### Differentiation from other communities

*Juniperus-Oxalis* scrub is distinguished by the dominance of juniper in the canopy. Many stands of the community contain scattered birch or pine, and woodlands containing mixtures of birch and juniper, or pine and juniper, are widespread; these are transitional between *Juniperus-Oxalis* woodland and *Quercus-Betula-Oxalis* W11, *Quercus-Betula-Dicranum* W17 or *Pinus-Hylocomium* W18 woodlands. Typical stands of these three communities have at most only an open shrub layer of juniper that is clearly subordinate to birch or pine.

### Ecology

*Juniperus-Oxalis* woodlands occur at moderate to high altitudes between 200 m and 650 m on a variety of rocks from mica-schist and limestone to granite. The soils are free-draining and moist with a fast turnover of nutrients. There is a slight preference for damp hollows and for north-facing slopes where the soils are moister (McVean and Ratcliffe 1962).

This is a northern and boreal community of northern England and eastern Scotland, with a few outliers in western Scotland. It is most common and extensive in the eastern Highlands (e.g. Tidswell 1988) and in parts of the Lammermuir Hills, the Lake District and Upper Teesdale.

There is related vegetation in western Norway.

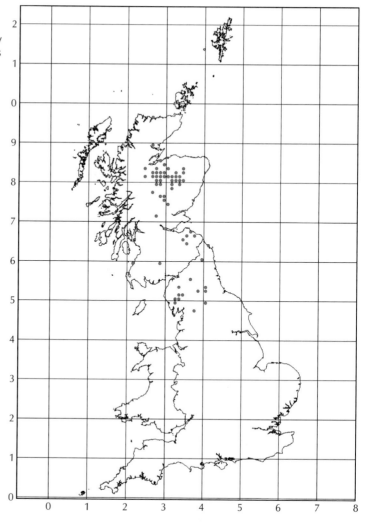

Many stands of *Juniperus-Oxalis* woodland may be relict understoreys of woods where birch and pine regeneration has been prevented by grazing (Tidswell 1988; Rodwell 1991a). In many places there are scattered birches among the juniper, for example in parts of the Morrone Birkwood in eastern Scotland and on Birk Fell in the Lake District. In the more natural woodlands of Norway there is juniper scrub at the upper altitudinal limit of pine and birch woodland and in clearings; the field layer can be dominated by grasses, forbs and ferns in mixtures similar to those in the *Viola-Anemone* sub-community. The only good British example of juniper replacing pine at the altitudinal limit of woodland is on Creag Fhiaclach in the Cairngorms at about 600 m. Here it may be natural climax scrub rather than a remnant of more complex woodland. Other stands of juniper scrub have been recorded at around this altitude in the eastern Highlands and they too could be natural.

### Conservation interest
*Juniperus-Oxalis* woodland is an uncommon vegetation type with a restricted British distribution. The few examples that appear to be natural climax vegetation are of particular interest for nature conservation. Scarce plant species recorded in the community include *Linnaea borealis*, *Potentilla crantzii*, *Orthilia secunda*, *Pyrola media*, the bryophytes *Cynodontium strumiferum*, *Tetralophozia setiformis*, *Lophozia longidens* and *Anastrophyllum saxicola*, and the lichen *Cetraria pinastri*.

### Management
With only light grazing, trees can be expected to colonise juniper scrub, as at Morrone in eastern Scotland. Juniper itself appears to benefit from alternating periods of intense and light grazing; more grazing creates the open conditions that are necessary for the establishment of seedlings, but too much grazing can eliminate seedlings and prevent regeneration. Under inappropriate management, such as burning or heavy grazing, stands can degenerate and eventually be replaced by heath or grassland. Studies have shown that some stands are quite even-aged and appear to represent 'bursts' of juniper regeneration during a period when there was some disturbance of the ground (McPhail and Taylor 1997). Juniper is severely damaged or destroyed by fire. However, if adjacent stands of heathlands are burned, this will create open areas where juniper seedlings can establish. Thus juniper scrub has a place in dynamic boreal woodland ecosystems, forming part of the seral succession from heathland to woodland after fire.

# W20 *Salix lapponum-Luzula sylvatica* scrub

## Synonyms
*Salix lapponum-Luzula sylvatica* nodum (McVean and Ratcliffe 1962); *Salix lapponum-Luzula sylvatica* scrub A2 (Birks and Ratcliffe 1980).

Typical view of W20 *Salix-Luzula* scrub (dark stippled): small patches of willow scrub on steep, ungrazed cliff ledges, associated with U16 *Luzula-Vaccinium* and U17 *Luzula-Geum* tall-herb vegetation (light stippled).

## Description
This rare and fragmented vegetation is one of the last refuges of scarce montane willows. The willows grow together in a low, scrubby, silvery canopy. *Salix lapponum* and *S. arbuscula* are the most common species, but *S. lanata*, *S. myrsinites*, *S. phylicifolia* or *S. reticulata* can also occur, either in mixtures or as pure stands. In the larger examples there can be a bewildering array of hybrids. The plants grow low and twisted, many clinging precariously to steep, shelving rock ledges streaming with water, their roots clamped to the rocks. Under the willows a great profusion of other plants clothes the ledges and spills over the cliff faces.

Although no sub-communities of *Salix-Luzula* scrub are described in the NVC, there are at least three distinct forms of montane willow scrub (e.g. Averis and Averis 1999a). One is a community of more acid soils, with a green carpet of *Vaccinium myrtillus*, *V. uliginosum*, *Deschampsia cespitosa*, *Luzula sylvatica*, *Dryopteris dilatata* and *Blechnum spicant*; it resembles a *Luzula-Vaccinium* tall-herb community U16 growing under a canopy of willow. The second is a low scrub of *S. myrsinites* growing over limestone outcrops at low to moderate altitudes; this has been recorded at Inchnadamph in Sutherland, at Rassal Ashwood in Wester Ross and on Meall Mór on the south side of Glen Coe in Argyll. No other willows occur. Some examples have a rich understorey of tall herbs and ferns; in others the sward is open and sparse. The third type is the most common, and resembles *Luzula-Geum* tall-herb vegetation U17 but with willows. It can be an almost unbelievably rich assemblage of tall herbs and ferns, the most common of which are *Alchemilla glabra*, *Angelica sylvestris*, *Geum rivale*, *Luzula sylvatica*, *Sedum rosea*, *Filipendula ulmaria*, *Ranunculus acris* and, in stands south of the Great Glen, *Geranium sylvaticum*. The herbs flourish out of the reach of sheep and deer and are bright with flowers in summer. There are usually great mats of mosses and liverworts under the shrubs and herbs; the most common species are big pleurocarpous mosses such as *Hylocomium splendens*, *Rhytidiadelphus loreus*, *Pleurozium schreberi* and *Thuidium tamariscinum*, and a few widespread calcicoles such as *Neckera crispa* and *Ctenidium molluscum*.

## Differentiation from other communities

*Salix lapponum*, *S. arbuscula* and, less commonly, the other montane willows can grow in stands of the *Luzula-Vaccinium* and *Luzula-Geum* tall-herb communities, as well as in montane *Dryas-Silene* heath CG14, but only in *Salix-Luzula* scrub do they form a continuous canopy. Montane willows can also occur very sparsely in *Carex-Pinguicula* mire M10 and *Carex-Saxifraga* mire M11, but here the ground flora consists of sedges, small forbs and bryophytes, with fewer tall herbs.

## Ecology

*Salix-Luzula* scrub occurs on steep, rocky ground which is either inaccessible or difficult for grazing animals to reach. Most examples are on rock ledges, crags, or rock outcrops in ravines; there are a few stands in boulder fields. This is a climax community occurring at high altitude (generally above 600 m) in a harsh climate. Most stands are on wet base-rich soils over Moine or Dalradian schist, but a few examples occur on more acid substrates (Rodwell 1991a), and *Salix myrsinites* scrub is mainly in well-drained situations on limestone. The willows can grow on slopes of all aspects, but they flourish particularly well on slopes facing north or east, where the soils are cooler and wetter and where snow lies late in spring. Although extant stands are almost all on inaccessible rocky ground, such as cliff ledges, ravine sides and boulder fields, these are unlikely to be the only natural habitats of *Salix-Luzula* scrub in Great Britain. Willow scrub would almost certainly once have spread more widely over flushed slopes where there are now herb-rich grasslands and mires.

## Conservation interest

This montane scrub is an interesting floristic link between the upland vegetation of Great Britain and that of the larger mountainous regions of Scandinavia. It is the rarest of all the upland woodland and scrub communities, with a total extent in Great Britain of only a few hectares. It is now confined to fragmentary stands on rocky slopes and ledges, although it was almost certainly much more common at one time around the upper limit of woodland on hill slopes and in corries wherever there are moderately basic rocks, as it still is in Norway and Sweden. Almost all of the known examples are protected within candidate SACs.

All the montane willows are rare plants, and the rich array of vascular plants in the field layer can include other scarce species such as *Polystichum lonchitis* and *Carex atrata*. *Salix arbuscula*, *S. myrsinites*, *S. lapponum* and *S. lanata* are all northern, boreal plants which reach their southern limit in Great Britain. Willow scrub is also an important habitat for invertebrates and birds. Willow leaves are eaten by the larvae of sawflies, many of which are themselves rare. In Norway, Great Snipe *Gallinago media*, Bluethroat *Luscinia svecica*, Wood Sandpiper *Tringa glareola* and Lapland Bunting *Calcarius lapponicus* breed in willow scrub. The last three have bred occasionally in Scotland (although not in this habitat) and might be tempted to do so more frequently if there is more montane willow scrub.

## W20 *Salix lapponum-Luzula sylvatica* scrub

The distribution of *Salix-Luzula* scrub is centred on the base-rich band of Dalradian rocks that runs across the southern Highlands from Ben Lui in the west to Caenlochan in the east. Even here it is rare. The largest stand remaining in Scotland is in Coire Sharrock in Glen Clova in Angus, where it covers about half a hectare of a steep, rocky slope. There is also montane willow scrub on the south side of Glen Coe in Argyll, in the hills around Glens Affric, Cannich and Strathfarrar in Inverness-shire, in the Cairngorms and the Drumochter hills, around Beinn Dearg at the head of Loch Broom, around Rassal in Wester Ross, and on Ben More Assynt and Ben Hope in Sutherland. Fragmentary outliers occur in the Moffat Hills in the Southern Uplands (not shown on the map). *Salix lapponum* grows on the cliffs of Helvellyn in the Lake District, although not in this type of scrub.

There is similar vegetation in Scandinavia and related forms of willow scrub occur in the Alps.

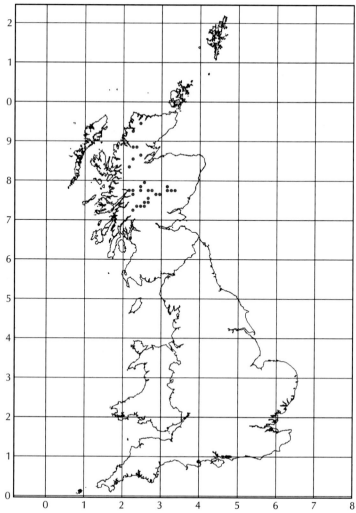

## Management

Of all the rare types of vegetation in the British uplands, this is one of the most difficult to manage and conserve. The willows that define the community are themselves rare, especially *Salix lanata* and *S. reticulata*. They are all dioecious species, with male and female flowers on separate plants. Many of the surviving populations are small and consist of plants of only one sex (e.g. Mardon 1990; Averis and Averis 1999a). Pollen is transported between plants by wind and by bumble-bees (Lusby and Wright 1996), and this process will be hindered or even completely prevented when the male and female plants are so rare and grow so far apart. Even where the sexes occur together, there is commonly a mixture of species, and since willows hybridise readily, offspring are just as likely to be hybrids as pure members of a single species. It is likely that the plants rarely set seed, and even if they do, seeds will germinate only on bare soil where there is little competition from other plants. In the Highlands most suitable sites are grazed, and the young willow plants are eaten by sheep, deer, small mammals, slugs and snails. *Salix arbuscula* grows in some *Carex-Pinguicula* and *Carex-Saxifraga* mires, where the ground is wet and unstable and the vegetation perhaps not grazed as intensively as the surrounding land; these communities may therefore provide a more congenial habitat for regeneration.

*Salix-Luzula* scrub is now almost confined to cliff ledges of soft, crumbling rocks where it is easy for stands to be lost as the cliffs erode away. The willows and tall herbs of the understorey cannot tolerate heavy grazing, so it is necessary to fence exclosures in order to reinstate this type of vegetation. This is being undertaken in Glen Doll in the Caenlochan National Nature Reserve and on Ben Lawers with some success, although it is hard to erect and maintain fences in this habitat and in the severe climate at high altitudes. Planting may also be necessary, especially to create populations of both sexes. Indeed, most willows root readily from cuttings. The ultimate aim would be to recreate the natural altitudinal sequence of vegetation, with willow scrub extending down into herb-rich birch woodland as it does in Scandinavia today.

# W23 *Ulex europaeus-Rubus fruticosus* scrub

## Synonyms
*Pteridium aquilinum-Ulex europaeus* Association (Birse and Robertson 1976).

Typical habitats of W23 *Ulex-Rubus* scrub:

1   Scattered patches among low-altitude heath and grassland
2   Along banks of streams and rivers
3   Colonising river shingles

W23 is also shown in the picture for U1.

## Description
This scrub is dominated by *Ulex europaeus* with its dark-green, spiny shoots and golden, coconut-scented flowers. In some stands *U. europaeus* is replaced by *Cytisus scoparius*. Beneath the gorse and broom there is usually just a sparse and species-poor flora of plants such as *Rubus fruticosus*, *R. idaeus*, *Agrostis capillaris* and *Pteridium aquilinum*. *R. fruticosus* is most common in lowland stands of *Ulex-Rubus* scrub and is absent from many stands in the uplands.

There are three sub-communities. In the *Anthoxanthum odoratum* sub-community W23a, the vegetation beneath the gorse is rather grassy and has several species in common with *Festuca-Agrostis-Galium* grassland U4, including *Anthoxanthum odoratum*, *Agrostis capillaris*, *Holcus lanatus*, *Galium saxatile*, *Potentilla erecta* and the moss *Rhytidiadelphus squarrosus*. The *Rumex acetosella* sub-community W23b shares many species with the *Anthoxanthum* sub-community but has less *Anthoxanthum* and *P. erecta*. It also has a weedier look, with *Rumex acetosella*, *Hypochaeris radicata*, *Senecio jacobaea* and *Plantago lanceolata*. The *Teucrium scorodonia* sub-community W23c is distinguished by *Teucrium scorodonia*, and there can also be a bushy undergrowth of *Hedera helix*. Some forms of *Ulex-Rubus* scrub have little or no vegetation below a dense bushy canopy of *Ulex* or *Cytisus*, and cannot be assigned to any of the sub-communities (Rodwell 1991a). For such a common type of vegetation, *Ulex-Rubus* scrub has been

*Ulex-Rubus* scrub is common on the fringes of the uplands from south-west England to northern Scotland. It is also widespread in lowland Great Britain though this is not shown on the map.

Similar vegetation occurs in Ireland and in western mainland Europe.

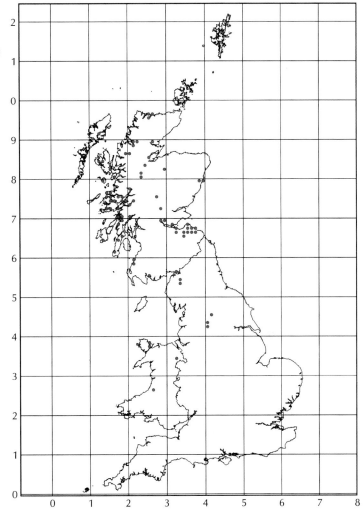

greatly under-sampled in the NVC. Hence it is not surprising that one can find stands that do not fit well into any sub-community: for example, stands with a rank grassy or weedy field layer containing much *Arrhenatherum elatius*, *Dactylis glomerata*, *Urtica dioica* and *Galium aparine*.

### Differentiation from other communities

Scattered bushes of *Ulex europaeus* or *Cytisus scoparius* can occur in a wide range of vegetation types (in particular, grasslands, heaths and bracken underscrub) but *Ulex-Rubus* scrub is the only type of British vegetation in which these species are dominant.

### Ecology

*Ulex-Rubus* scrub occurs on acid, freely draining soils on gentle to very steep, rocky slopes at low altitudes. Its highest localities are at about 300–350 m on south-facing slopes as far apart as the eastern Highlands and south-west England. The vegetation is mainly secondary, developing after woodland clearance or on abandoned pasture. Progression to woodland may be held in check by reintroduction of stock or by burning. The natural habitats of *Ulex-Rubus* scrub are likely to be steep, rocky slopes on thin soils that cannot support a continuous canopy of tall trees, unstable habitats such as riverside shingle banks, and temporarily disturbed ground after fires in woodland and heath.

## Conservation interest

*Ulex-Rubus* scrub is not notable for rare species, but it is an important component of varied mosaics of vegetation, especially in south-west England and Wales, and contributes to the structural diversity of moorland vegetation. It can be an important habitat for birds, for example Linnet *Carduelis cannabina*, Yellowhammer *Emberiza citrinella*, Whinchat *Saxicola rubetra*, Stonechat *Saxicola torquata* and Whitethroat *Sylvia communis*.

## Management

*Ulex europaeus* can invade grasslands where the intensity of grazing has been reduced, and once established can tolerate a considerable amount of grazing. Many patches of *Ulex-Rubus* scrub appear to have spread out from steep, rocky and unstable ground into surrounding grassland after stock grazing has been discontinued. In this way it can threaten the existence of short, species-rich grasslands that depend on heavy grazing. *U. europaeus* is resistant to fire and appears to thrive when burnt, sprouting up from the base of the stem. Even so, it is often burnt off in spring in an attempt to stop it colonising grassland. In many places there are ungrazed tracts of gorse scrub where woodland regeneration is limited by periodic fires, a lack of seed sources, and time. Where gorse is not rejuvenated by fire the bushes can grow tall and leggy. The canopy opens up and more light reaches the ground. As a result, more continuous vegetation, such as grassland, can develop under the gorse. Presumably the bushes eventually die and if there is a period with a low intensity of grazing the cycle could start again.

# W25 *Pteridium aquilinum-Rubus fruticosus* underscrub

Typical view of W25 *Pteridium-Rubus* underscrub (in foreground and stippled): patches of bracken-dominated vegetation scattered among grassland, woodland and mires at low altitude.

W25 is also shown in the picture for MG5.

## Description

These are stands of *Pteridium aquilinum* in which the bracken is entangled with *Rubus fruticosus* and in some places *Rosa canina* or *Rubus idaeus*, forming a dense and rather prickly mixture. In south-west England *Hedera helix* is common in this community.

There are two sub-communities: one herb-rich, the other species-poor. The *Hyacinthoides non-scripta* sub-community W25a is the more herb-rich of the two, and is characterised by woodland species including *Hyacinthoides non-scripta*, *Urtica dioica*, *Heracleum sphondylium*, *Silene dioica*, *Glechoma hederacea* and *Angelica sylvestris*, and tall grasses such as *Holcus mollis* and *Dactylis glomerata*. The less species-rich *Teucrium scorodonia* sub-community W25b is distinguished by *Teucrium scorodonia* and *Holcus lanatus*, and there can be some *Digitalis purpurea*, *Agrostis capillaris* and *Anthoxanthum odoratum*. Herb-rich stands of bracken in which *Rubus* species are rare or absent, but with an understorey including mesotrophic forbs such as *Filipendula ulmaria*, *Geum rivale*, *Lysimachia nemorum*, *Cirsium heterophyllum* and *Alchemilla glabra*, are quite common on damp soils over basalt, limestone and other base-rich rocks in north Wales, the west Highlands and the Inner Hebrides (e.g. A M Averis 2001b; Averis and Averis 1995a, 1995b, 1999b, 2000a, 2000c). They are related to the *Hyacinthoides* sub-community, but are not currently described in the NVC.

## Differentiation from other communities

*Pteridium-Rubus* underscrub is most likely to be confused with the other bracken-dominated NVC type: the *Pteridium-Galium* community U20. The two communities look superficially similar and it is only at close range that the characteristic bramble and other woodland plants of *Pteridium-Rubus* underscrub are visible beneath the fern. The most likely confusion is with the *Anthoxanthum* sub-community of the *Pteridium-Galium* community U20a, which can be moderately herb-rich, but this is grassier, with more *Galium saxatile* and *Potentilla erecta*, less bramble and fewer mesotrophic herbs.

### Ecology

In the uplands, *Pteridium-Rubus* underscrub is a community of the lower hill slopes, occurring below about 300 m on deep, free-draining, fertile soils, many of which are derived from basic parent rocks such as limestone or basalt. This is well within the zone of woodland, and indeed the flora of the *Hyacinthoides* sub-community can be similar to that of *Fraxinus-Sorbus-Mercurialis* woodland W9. The damp herb-rich form of the community can resemble a bracken-dominated form of the field layer of *Alnus-Fraxinus-Lysimachia* woodland W7.

### Conservation interest

The dense stands of bracken which constitute *Pteridium-Rubus* underscrub and the *Pteridium-Galium* community are usually thought of as a conservation problem rather than as an asset. However, in *Pteridium-Rubus* underscrub, the flora under the fern fronds is relatively rich, especially in the damper stands, and compensates somewhat

*Pteridium-Rubus* underscrub is widespread in lowland Great Britain and at low altitudes in upland areas though this is not shown on the map. Within the uplands it is most common in the Inner Hebrides, the western Highlands, the Southern Uplands, the Lake District, Wales and south-west England. It has also been recorded in Jersey, and may occur in mainland western Europe.

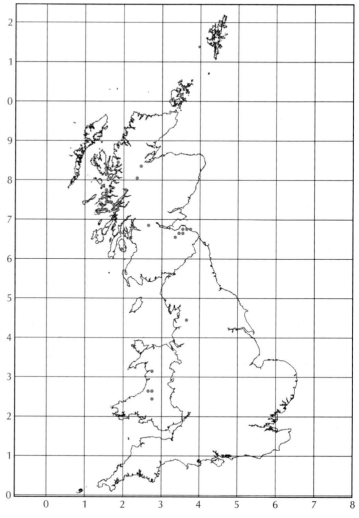

for the invasive canopy of bracken. In certain places, this community contains some of the more diverse, herb-rich and interesting vegetation among vast tracts of species-poor vegetation on peat and acid soils. Many of the herb-rich bracken stands in the Inner Hebrides have a flora that compares well with some of the species-rich grasslands in the same area. Like *Pteridium-Galium* vegetation, stands of *Pteridium-Rubus* underscrub can be an important breeding habitat for passerines, such as Stonechat *Saxicola torquata*, Whinchat *Saxicola rubetra* and Meadow Pipit *Anthus pratensis*.

## Management
*Pteridium-Rubus* underscrub typically develops in response to management treatments such as woodland clearance, burning and grazing. Many stands have probably been derived from woodland, but the dense canopy of bracken can hinder the re-establishment of trees.

# M1 *Sphagnum denticulatum* bog pool community

## Synonyms
*Trichophoreto-Eriophoretum* (pools) (McVean and Ratcliffe 1962); *Eriophorum angustifolium-Sphagnum cuspidatum* association (Birks 1973); *Sphagnum auriculatum* bog pool community (Rodwell 1991b).

M1 *Sphagnum denticulatum* bog pool community in wet depressions among extensive M17 *Trichophorum-Eriophorum* blanket mire on deep, waterlogged peat at low altitude.

M1 is also shown in the picture for M21.

## Description
These are shallow peaty pools, wet hollows and soakways in blanket bogs or topogenous mires, filled with a half-submerged, half-floating mass of *Sphagnum denticulatum*. They form colourful and conspicuous red-gold patches over the mire surface. *S. denticulatum* is commonly accompanied by *S. cuspidatum*. Pricking up through the *Sphagnum* is a sparse scatter of plants, including *Narthecium ossifragum* with its bright-yellow, star-shaped flowers, the carnivorous *Drosera* species with their red-fringed, sticky leaves, *Menyanthes trifoliata*, *Juncus bulbosus*, *Eriophorum angustifolium* and, less commonly, *Rhynchospora alba* with its sharp, cream-coloured flowers.

## Differentiation from other communities
*Sphagnum cuspidatum/fallax* bog pools M2 resemble this community, but have little or no *Sphagnum denticulatum* and not much *Menyanthes trifoliata*. *Sphagnum denticulatum* bog pools have more *Sphagnum* species, *Carex* species and small aquatic herbs, and less *Eriophorum angustifolium*, than the more species-poor *Eriophorum angustifolium* bog pool M3. *Sphagnum denticulatum* bog pools are unlikely to be confused with other types of vegetation. Stands are usually quite distinct from the surrounding bog communities with their more varied flora of dwarf shrubs and graminoids such as *Eriophorum vaginatum* and *Molinia caerulea*. In both *Sphagnum*-dominated bog

pool communities the peat is much wetter and *Sphagnum* species are usually totally dominant. The boundaries may be rather blurred in shallow and grazed valley mires, but wetter areas with bog pool vegetation are usually recognisable.

## Ecology

The *Sphagnum denticulatum* community is typical of bog pools in the oceanic west of Great Britain. The bog vegetation within which it occurs is mostly *Trichophorum-Eriophorum* mire M17 in the north and west, and *Narthecium-Sphagnum* valley mire M21 in the south-west, but it also occurs within stands of *Erica-Sphagnum papillosum* blanket mire M18 and *Calluna-Eriophorum* blanket mire M19, especially in the west Highlands and the Hebrides. Stands are usually on level or very gently sloping ground, at altitudes below 300 m. The substrate is thick, wet, acid peat with a pH between 3 and 5. In parts of north-western Scotland the community can cover large areas, especially where the bog pools are large, elongated and close together. However, it is more common to find stands covering areas of only a few square metres.

This community occurs throughout the west of Great Britain between Sutherland and Cornwall, with outliers in the New Forest where it is associated with valley mires. This westerly distribution may to some extent be an artificial pattern, reflecting more thorough mire drainage in the east rather than a dependence on an oceanic climate, but there is insufficient evidence to support either premise (Rodwell 1991b).

There is similar vegetation in the oceanic blanket bogs of western Ireland, where *Rhynchospora alba* and *Erica tetralix* are common (Horsfield et al. 1991), and *Schoenus nigricans* may also occur. Similar communities also occur in western Scandinavia.

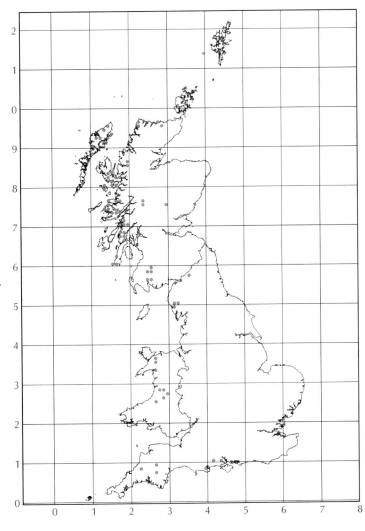

**Conservation interest**

These bog pools provide a habitat for a number of rare species, including *Hammarbya paludosa*, *Rhynchospora fusca*, *Carex limosa*, *C. magellanica*, *Scheuchzeria palustris*, *Sphagnum pulchrum* and *S. majus*. They are also valuable feeding sites for upland birds such as Golden Plover *Pluvialis apricaria*, Greenshank *Tringa nebularia* and Dunlin *Calidris alpina*, which nest on bogs and feed their chicks on insects. Bog pools can also be important breeding sites for frogs. Upland populations of frogs are becoming increasingly important as more of their lowland habitats are lost to development and drainage.

**Management**

The community is easily damaged by drainage and burning, although it is rather more resistant to burning than the drier parts of the surrounding bogs. Grazing and nutrient enrichment of topogenous valley mires by sheep, cattle and especially ponies may also be a problem in southern England and south Wales. Some blanket mires and associated bog pools have been damaged by peat extraction, although *Sphagnum denticulatum* bog pools can develop as secondary vegetation in peat cuttings.

# M2 *Sphagnum cuspidatum/fallax* bog pool community

## Synonyms
*Sphagnum cuspidatum/recurvum* bog pool (Rodwell 1991b).

M2 *Sphagnum cuspidatum/fallax* bog pool community in wet depressions among M18 *Erica-Sphagnum* raised and blanket mire on deep, waterlogged peat at low altitude.

## Description
Like the *Sphagnum denticulatum* bog pool M1 this is a community of pools, hollows, seepage lines and soakways among bogs, but here the most common species are the soft, feathery, mid-green *Sphagnum cuspidatum* and the neat, grass-green *S. fallax*. The vascular flora consists of a sparse array of dwarf shrubs including *Erica tetralix*, and small herbs.

There are two sub-communities. The *Rhynchospora alba* sub-community M2a has a moss layer of *Sphagnum cuspidatum* pierced by *Rhynchospora alba*, *Drosera* species, *Narthecium ossifragum* and, in some places, *Andromeda polifolia*. The *Sphagnum fallax* sub-community M2b has as much *S. fallax* as *S. cuspidatum*, if not more, and a heathier vascular flora comprising plants such as *Vaccinium oxycoccos*, *Calluna vulgaris* and *Eriophorum vaginatum*. In north-west England and western Scotland, where the community has not been adequately sampled, the more obvious split within the community is between species-poor stands with little other than a layer of *Sphagnum cuspidatum* or *S. fallax* or both, and richer stands with an open sward of shrubs and herbs including typical species from both of the two sub-communities.

## Differentiation from other communities
This community is very similar to the *Sphagnum denticulatum* bog pool but *S. fallax* is much more common, and *S. denticulatum* is usually absent. Occurrences of *Andromeda polifolia* in bog pools are mostly in the *Sphagnum cuspidatum/fallax* community. The

differentiation of *Sphagnum*-dominated bog pools from associated types of mire is usually straightforward (see comments on p. 132).

## Ecology

This community forms pools on the surface of wet, base-poor blanket mires or raised mires. In the lowlands it is mostly associated with raised mires but in the uplands it can occur in *Erica-Sphagnum papillosum* mire M18, *Trichophorum-Eriophorum* mire M17 or *Calluna-Eriophorum* mire M19. The pools are usually sharply demarcated from the surrounding mire surface. *Sphagnum cuspidatum/fallax* pools also occur within *Narthecium-Sphagnum* valley bogs M21 in south-west England, and here they tend to merge into the surrounding mire vegetation. Although most stands are at low altitudes, *Sphagnum cuspidatum/fallax* vegetation can occur in blanket mires above 500 m.

## Conservation interest

*Sphagnum cuspidatum/fallax* bog pools are an important habitat for *Carex pauciflora, C. magellanica, C. limosa, Andromeda polifolia, Sphagnum pulchrum* and *S. balticum*. The rare *Scheuchzeria palustris* and *Rhynchospora fusca* have also been recorded here. Like *Sphagnum denticulatum* bog pools, this community is valuable for insects, as a feeding

Bog pools of this type are widespread throughout blanket and valley mires in the uplands, from Cornwall north to the Highlands. It is not as western in its distribution as the *Sphagnum denticulatum* bog pool; it occurs further east in parts of the country where the climate is drier and where winters are cold.

There is similar vegetation in Ireland, and in mainland Europe from Scandinavia to northern France.

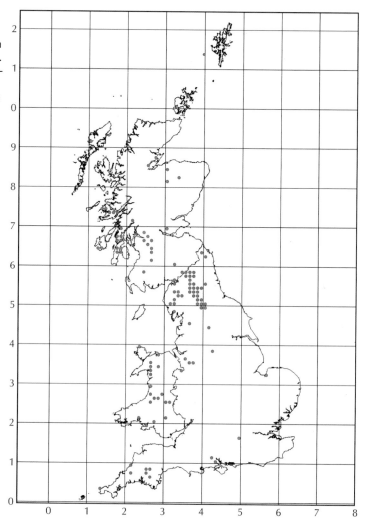

area for upland birds, and as a breeding ground for frogs. *Rhynchospora alba* is a food plant of the Large Heath butterfly *Coenonympha tullia*.

## Management

Many stands of *Sphagnum cuspidatum/fallax* vegetation have been lost when bogs have been drained, converted to agricultural land, or destroyed by the commercial extraction of peat. The community is most common on unmodified peatlands, although it can develop in peat cuttings and artificial pools. It is best conserved by protecting bogs from drainage and burning.

# M3 *Eriophorum angustifolium* bog pool community

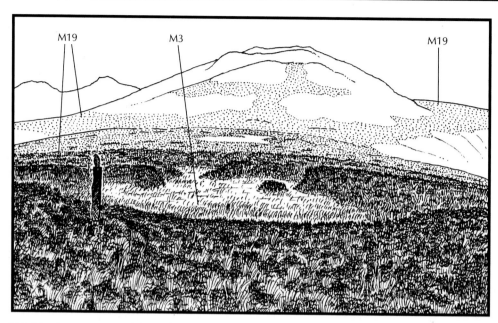

M3 *Eriophorum angustifolium* bog pool community (in centre) in wet depressions among hagged M19 *Calluna-Eriophorum* blanket mire (foreground and stippled) on peaty watershed at moderate altitude.

## Description

Most stands of this community consist of dark-red-green, species-poor swards of *Eriophorum angustifolium* growing on wet, redistributed peat in disturbed areas of blanket bog. *E. angustifolium* can grow in dense swards here, but equally commonly it occurs as sparse shoots scattered over the dark expanses of peat. Beneath the cotton grass there is rarely a continuous layer of plants; it is more usual to see spreads of bare peat with scattered plants of *Carex echinata*, *Molinia caerulea*, *Calluna vulgaris*, and the mosses *Polytrichum commune*, *Sphagnum papillosum*, *S. fallax* and *Campylopus flexuosus*. Less commonly, there is a more extensive layer of Sphagna and *P. commune*. In some places there are also a few plants of *Menyanthes trifoliata*, *Narthecium ossifragum*, *Erica tetralix*, and the bog liverworts *Odontoschisma sphagni* and *Gymnocolea inflata*.

## Differentiation from other communities

This community is hard to confuse with any other vegetation type. *Eriophorum angustifolium* can be very common in blanket mire, but in *Eriophorum angustifolium* bog pools it is strongly dominant, the flora is very impoverished, and there are rarely any dwarf shrubs. There is more *E. angustifolium* and less *Sphagnum* in most stands of *Eriophorum angustifolium* vegetation than in the *Sphagnum denticulatum* and *Sphagnum cuspidatum/fallax* bog pool communities. *E. angustifolium* can be very common in some stands of *Carex-Potentilla* tall-herb fen S27, but that vegetation is much more herb-rich.

## Ecology

*Eriophorum angustifolium* bog pools occur typically on wet, exposed, acid peat in blanket bogs, where *E. angustifolium* is usually one of the first colonising species. The community can persist where the water-level fluctuates, as well as occurring in permanently flooded pools or dried-up hollows. Although it can occur in situations such as flushed channels, it is generally most extensive where the original bog vegetation has been lost, for example on peat cuttings or where there has been damage by vehicles or by trampling animals. The community can cover substantial areas of flat, exposed peat between hags in eroding bogs. It can also occur in natural hollows where other forms of bog pool vegetation may have been impoverished by burning. It is possible that vegetation resembling the original mire communities may re-establish once the peat has been stabilised by the *Eriophorum*, but this has not been studied (Rodwell 1991b). The matrix of bog vegetation within which *Eriophorum angustifolium* bog pools occur is usually *Calluna-Eriophorum* mire M19 or *Eriophorum vaginatum* mire M20, but may also be *Trichophorum-Eriophorum* mire M17 or *Trichophorum-Erica* wet heath M15. As a component of montane blanket mire *Eriophorum angustifolium* bog pools reach altitudes of over 900 m.

**Conservation interest**

*Eriophorum angustifolium* bog pools are generally of less value for nature conservation than other forms of bog pool vegetation. One benefit of the community is that it can help

The distribution of the community follows that of blanket bogs and wet heaths. It is widespread in the north and west, especially among eroding *Calluna-Eriophorum* mires.

The community also occurs on eroding blanket bog in Ireland (Hobbs and Averis 1991b), and there is similar vegetation elsewhere in northern and western Europe.

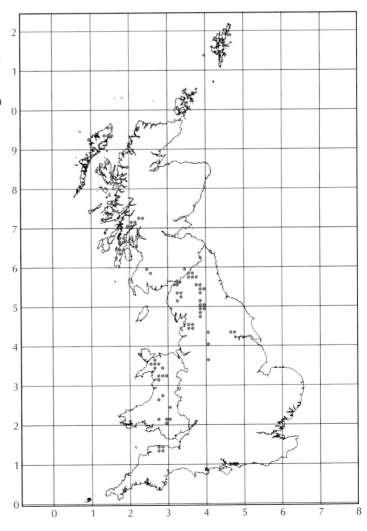

to stabilise bare peat that might otherwise be washed away; in such places it may be the first stage in the regeneration of more diverse bog vegetation.

**Management**

In many cases this community is a sign of poor management, or at least of management that is detrimental to blanket bog vegetation. For example, *Eriophorum angustifolium* communities are common in Strath Dionard in Sutherland, where the bog vegetation has been destroyed and the peat compressed by the use of tracked vehicles over many years. There are many stands on eroding peat within degraded blanket bogs in the southern Pennines and in mid- and south Wales. Many factors may have contributed to erosion in these bogs, including atmospheric pollution, heavy grazing and severe fires. Elsewhere, such as on the bogs of North Harris, the community occupies periodically inundated channels in blanket bogs where it is presumably more natural.

# M4 *Carex rostrata-Sphagnum fallax* mire

## Synonyms
*Carex lasiocarpa-Menyanthes trifoliata* Association *p.p.* and *Carex rostrata-Carex limosa* nodum *p.p.* (Birks 1973); Sub-montane *Carex rostrata-Sphagnum recurvum* mire H3c (Birks and Ratcliffe 1980); *Carex rostrata-Sphagnum recurvum* mire (Rodwell 1991b).

Typical habitats of M4 *Carex rostrata-Sphagnum fallax* mire:

1   Wet peaty hollow
2   Along edges of stream
3   Around edge of pool
4   Hillside flushes

## Description
This is a tall, even, grey-green sward, with the leaves of *Carex rostrata* swaying in the slightest breeze over a closely packed, bright-green, spongy carpet of *Sphagnum palustre*, *S. fallax*, *S. denticulatum*, *S. papillosum* and *Polytrichum commune*. There is little else to enrich the sward except for the odd plant of species such as *Potentilla erecta*, *Viola riviniana*, *Agrostis canina* and *Carex nigra*.

## Differentiation from other communities
This community represents the oligotrophic end of a series of *Carex rostrata* mires. Two other types of *Carex rostrata* mire have a moss layer consisting of Sphagna: *Carex-Sphagnum squarrosum* mire M5 and *Carex-Sphagnum warnstorfii* mire M8. These two communities are neither as acid nor as species-poor as *Carex rostrata-Sphagnum fallax* mire. Their *Sphagnum* carpet is made up of mildly base-tolerant species, such as *S. squarrosum*, *S. warnstorfii*, *S. teres* and *S. contortum*. They are dotted with *Potentilla palustris* and *Aulacomnium palustre*, as well as more obviously mesotrophic plants including *Parnassia palustris*, *Lychnis flos-cuculi*, *Filipendula ulmaria*, *Selaginella selaginoides*, *Thalictrum alpinum* and *Carex pulicaris*. The moss layer of *Carex rostrata-*

## M4 Carex rostrata-Sphagnum fallax mire

*Carex rostrata-Sphagnum fallax* mires are distributed throughout the British uplands from south-west England to northern Scotland. Although most stands are small, large spreads can occur where conditions are suitable, for example in shallow, ill-drained basins among blanket bog and soligenous mires on the upland plateaux of mid- and north Wales.

Related vegetation occurs widely in Scandinavia and elsewhere in western Europe.

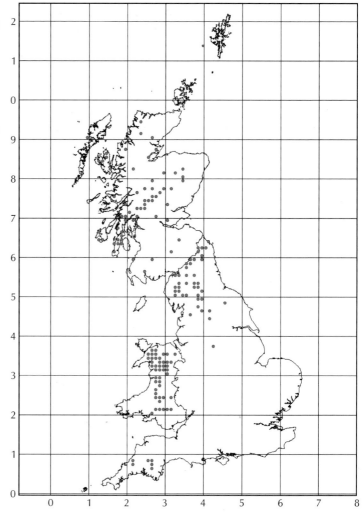

*Sphagnum fallax* mire is similar to that of *Carex echinata-Sphagnum* mire M6, which is split into four sub-communities according to the dominant sedges or rushes. Perhaps the vegetation currently described as *Carex rostrata-Sphagnum fallax* mire could equally well be classed as a *Carex rostrata* sub-community of *Carex echinata-Sphagnum* mire.

### Ecology

*Carex rostrata-Sphagnum fallax* mires occur on wet peat where there is some lateral water movement or in hollows with stagnant water. Stands are typically hidden among bogs and grassland in shallow depressions where water drains diffusely through poor acid soils. They are also common around lochs where water seeps slowly down slopes. Where they occur in mosaics with other mire communities they commonly occupy the wettest ground close to the source of irrigation. The peats are acid with a surface pH of around 4.0. Most examples are at moderate altitudes, but the community can occur above 600 m.

### Conservation interest

*Carex rostrata-Sphagnum fallax* mire is not particularly notable for rare species, although some stands include *Carex chordorrhiza* or *Lysimachia thyrsiflora*. C.

*lasiocarpa*, *C. limosa* and *C. magellanica* can also grow in this vegetation. In the hills and moorlands mires of this type can be important feeding places for upland birds, such as Snipe *Gallinago gallinago*, Curlew *Numenius arquata*, Golden Plover *Pluvialis apricaria* and Red Grouse *Lagopus lagopus*.

## Management

Burning and grazing can damage *Carex rostrata-Sphagnum fallax* mire, although the wet ground of many stands affords some protection against this. These mires need a high water-table, and if they are drained the soils dry out and become less suitable for *Carex rostrata* and Sphagna. There are also many places where the vegetation has been damaged by the use of all-terrain vehicles in the hills. In the more resilient grasslands and heaths it can be hard to tell where these vehicles have been driven, but one pass through a stand of *Carex rostrata-Sphagnum fallax* mire is enough to leave a characteristic set of wheel-tracks full of water.

# M5 *Carex rostrata-Sphagnum squarrosum* mire

**Synonyms**
*Carex rostrata-Aulacomnium palustre* Association (Birks 1973).

Typical habitats of M5 *Carex rostrata-Sphagnum squarrosum* mire:

1 Wet peaty hollow
2 Along edges of stream
3 Around edge of pool
4 Hillside flushes

## Description

This is a wet mire with an extensive or patchy layer of the mesotrophic or base-tolerant *Sphagnum* species *S. squarrosum*, *S. fallax* and *S. teres* below a thin, grey-green canopy of sedges. The mats of Sphagna are variegated with the yellow-green shoots of *Aulacomnium palustre*, the translucent, ochre-green leaves of *Rhizomnium punctatum*, and the golden, branched shoots of *Calliergonella cuspidata* and *Campylium stellatum*. *Carex rostrata* is usually the most common sedge, but *C. nigra*, *C. curta* and *C. echinata* may also occur. There is typically a rich array of tall forbs such as *Mentha aquatica*, *Lychnis flos-cuculi*, *Myosotis scorpioides*, *Galium palustre*, *Potentilla palustris* and *Filipendula ulmaria*, as well as *Menyanthes trifoliata*, *Equisetum fluviatile*, *Hydrocotyle vulgaris* and *Potamogeton polygonifolius*. These mires can be attractive in summer when the forbs are in flower and the bright colours stand out from the muted greens and browns of the surrounding vegetation.

## Differentiation from other communities

*Carex-Sphagnum squarrosum* mire superficially resembles the other mire communities dominated by *Carex rostrata*. The best distinguishing species are the mosses *Sphagnum squarrosum* and *S. teres*. In *Carex rostrata-Sphagnum fallax* mire M4, the moss layer consists mainly of *S. fallax*, *S. palustre* and *Polytrichum commune*. *Carex-Sphagnum warnstorfii* mire M8 has less *S. squarrosum*, *Potentilla palustris*, *Eriophorum angustifolium* and *Succisa pratensis*, and more *S. warnstorfii* than *Carex rostrata-Sphagnum squarrosum* mire. In *Carex-Calliergonella* mire M9 Sphagna are rare, and the most common bryophytes are *Calliergonella cuspidata*, *Calliergon* species and other mesotrophic mosses. Vegetation similar to *Carex-Sphagnum squarrosum* mire but with little or no *C. rostrata* is included among the small-sedge mires described on p. 424.

## Ecology

This is a community of loch-sides, pools and fens where there is mild base enrichment, either from the underlying rock or from irrigating water. It occurs where the pH is a little higher than in the *Carex rostrata-Sphagnum fallax* mire, but where conditions are not as base-rich as they are in the *Carex-Calliergonella* mire. The substrate is a soft, wet peat that protects the community from fire and grazing animals. *Carex-Sphagnum squarrosum* mire usually occurs in mosaics with other mire communities, where local variations in wetness and nutrient status are reflected in complex patterns of vegetation. It can occur from a few tens of metres above sea-level to almost 900 m, well into the montane zone.

This community is widespread but scarce in both upland and lowland parts of northern and western Great Britain.

Related vegetation types occur locally in Scandinavia and parts of central Europe.

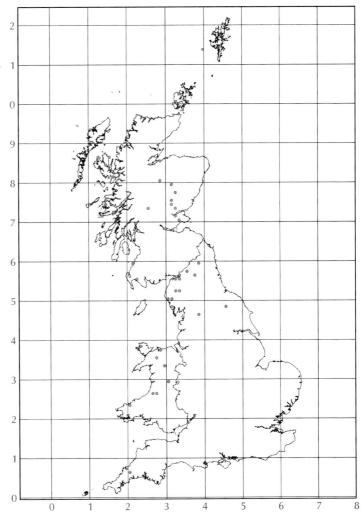

## Conservation interest

*Carex-Sphagnum squarrosum* mire is a valuable part of the series of oligotrophic and mesotrophic mire communities that occur in the British uplands. It frequently occurs as a component of mosaics of different mire types in which relationships between the environment and the various types of vegetation are clearly displayed. *Carex-Sphagnum squarrosum* mire is a species-rich assemblage and the characteristic forbs and sedges are usually able to flower and set seed; this is important to maintain the genetic diversity of the plants. The uncommon *Carex limosa, Dactylorhiza purpurella* and *D. incarnata* occur in some stands of this type of mire.

## Management

In its normal waterlogged state *Carex-Sphagnum squarrosum* mire is protected from fire and grazing. However, stands can dry out in hot summer weather and are then accessible to grazing animals. Grazing is presumably at least sufficient to prevent willows becoming established. The community may be near-natural at higher altitudes, but many stands are within the sub-montane zone and might revert to swampy *Salix-Carex* woodland W3 in the absence of grazing. In common with all other mire communities, *Carex-Sphagnum squarrosum* mire can be damaged by drainage and by eutrophication.

# M6 *Carex echinata-Sphagnum fallax/denticulatum* mire

## Synonyms
*Carex*-moss mire H3b, *Juncus effusus-Sphagnum recurvum* mire H2a and *Juncus acutiflorus-Sphagnum recurvum* mire H2b (Birks and Ratcliffe 1980); *Carex echinata-Sphagnum recurvum/auriculatum* mire (Rodwell 1991b).

Typical habitats of M6 *Carex echinata-Sphagnum* mire:

1　Hillside flushes
2　Wet level ground in valley floor
3　In ditch
4　Along edges of stream
5　Around edges of pool

M6 is also shown in the pictures for M20, M21 and U4–U6.

## Description
These are soligenous mires with a sward of sedges or rushes over a dense, green and golden layer of the mosses *Sphagnum fallax*, *S. denticulatum*, *S. palustre* and *Polytrichum commune*; *S. papillosum* and *Rhytidiadelphus squarrosus* can be common here too. The sward is usually interleaved with *Agrostis canina* and *Molinia caerulea*. The only common forbs are *Viola palustris* and *Potentilla erecta*, though there may also be a little *Galium saxatile* or *Cirsium palustre*.

There are four sub-communities, defined by the sedge or rush species that predominate in the sward. The *Carex echinata* sub-community M6a is a dull-green assemblage of *Carex echinata* together with other sedges (mainly *C. panicea* and *C. nigra*), in some places with a few fen plants such as *Potentilla palustris* and *Menyanthes trifoliata*, and the sundew *Drosera rotundifolia*. The *Carex nigra-Nardus stricta* sub-community M6b is a paler sward of *C. nigra*, *C. panicea*, *Eriophorum angustifolium*, *Juncus squarrosus* and *Nardus stricta*. *Festuca ovina* and *Anthoxanthum odoratum* are common here and the vegetation can resemble flushed grassland. There is also another type of vegetation – usually a topogenous mire – which is distinctive but appears to fit within this sub-community. This is an even sward of pure or nearly pure *C. nigra* growing through an extensive carpet of Sphagna and *Polytrichum commune*. Rushes dominate the other two sub-communities: the *Juncus effusus* sub-community M6c and the *Juncus acutiflorus* sub-community M6d are characterised by tall, deep-green swards of one or other species. Although these are all acid, species-poor mires, the *Juncus acutiflorus* sub-community can include *Myosotis secunda*, *Ranunculus acris* and other forbs that are scarce in the other sub-communities.

### Differentiation from other communities

The *Carex echinata* and *Carex-Nardus* sub-communities of *Carex echinata-Sphagnum* mire can be distinguished from other sub-montane small-sedge mires because they have a carpet of Sphagna rather than more mesotrophic mosses. They are also relatively species-poor, without the array of small forbs characteristic of the richer mires. The montane counterpart of these acid flushes is *Carex-Sphagnum russowii* mire M7. This can look very similar to *Carex echinata-Sphagnum* mire but has less *Molinia* and *Sphagnum palustre*, and usually contains some montane species such as *Carex bigelowii*, *Sphagnum russowii* or *Saxifraga stellaris*. *Carex rostrata-Sphagnum fallax* mire M4 has a moss layer similar to that in *Carex echinata-Sphagnum* mire, but the graminoid sward there is of *C. rostrata* rather than rushes or short sedges.

The *Juncus effusus* and *Juncus acutiflorus* sub-communities of *Carex echinata-Sphagnum* mire can only be confused with *Juncus-Galium* rush-pasture M23, which has a richer flora of herbs, and mosses such as *Calliergonella cuspidata*, *Brachythecium rutabulum* and *B. rivulare* rather than Sphagna and *Polytrichum commune*.

### Ecology

These mires occur in wet hollows, seepage lines, flushes, shallow gullies cutting down hillsides, and along the margins of streams within expanses of blanket mire, dwarf-shrub heath or acid grassland. They also occur around slow-flowing springs at the heads of rivers. For example, the River Tweed in the Scottish borders rises in a large, wet, spongy hollow filled with this type of mire vegetation. Similarly in Wales, the Wye and the Severn rise among these mires on the shallow dome of Pumlumon. *Carex echinata-Sphagnum* mires also cover level, ill-drained valley floors, and are common in neglected and abandoned pastures on the upland margins.

The soils beneath *Carex echinata-Sphagnum* flushes are deep, wet and usually peaty. The irrigating water is acid with a pH between 4.4 and 5.7 (McVean and Ratcliffe 1962). The supply of plant nutrients is greater than it is in the stagnant *Sphagnum denticulatum* M1, *Sphagnum cuspidatum/fallax* M2 and *Eriophorum angustifolium* M3 bog pool communities, but less than it is in the base-rich small-sedge mires *Carex-Pinguicula* M10, *Carex-Saxifraga* M11 and *Carex saxatilis* M12. *Carex echinata-Sphagnum* mire is mainly a community of the sub-montane zone up to about 400 m, although in northern England, in Wales and on Dartmoor, there are stands of the *Juncus effusus* sub-community above 550 m along the sheltered gullies of winding, sluggish streams.

### Conservation interest

*Carex echinata-Sphagnum* mires do not have a rich flora and are not the home of many rare plant species. They do, however, contribute to the diversity of the vegetation of the upland margins. They can be the only places with wet, soft ground amid great tracts of dry heather moorland and short unimproved grassland. As such, they are valuable habitats for insects and spiders, and provide feeding grounds for a range of upland birds whose chicks require invertebrate food. Waders such as Curlew *Numenius arquata*,

*Carex echinata-Sphagnum* mires are common throughout the uplands from Cornwall north to Shetland. They are the most widespread soligenous mires in the British uplands. The two sedge-dominated sub-communities are ubiquitous. The two rush-dominated sub-communities are widespread in Wales, northern England and southern Scotland but are less common in the northern Highlands and the Outer Hebrides. For example, the *Juncus acutiflorus* sub-community is very common in the south-west Highlands but is scarce further north. *Carex echinata-Sphagnum* mire has also been recorded in the lowlands, for example on Anglesey and in a few places in Cheshire, Dorset and south-eastern England.

There is similar vegetation in western Europe, including Iceland, but the *Juncus acutiflorus* and *Juncus effusus* sub-communities are almost confined to Great Britain and Ireland (White and Doyle 1982).

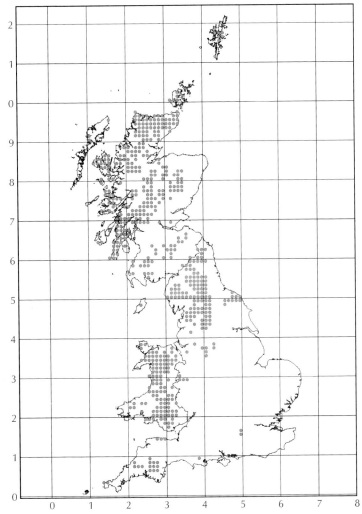

Snipe *Gallinago gallinago* and Redshank *Tringa totanus*, and ducks such as Mallard *Anas platyrhynchos* and Teal *Anas crecca*, often conceal their nests among the tufts of rushes and sedges. In winter the water-table is higher than in summer and because the water is moving it does not freeze as readily as still water. At this time of year *Carex echinata-Sphagnum* flushes are used by Snipe wintering on the hills; these mires are among the few places they can get their beaks into when most of the ground is frozen hard.

## Management

Almost all examples of *Carex echinata-Sphagnum* mire are grazed by sheep or deer. In some places they are heavily grazed and the tufts of sedges and rushes are browsed into short tussocks. In such situations, *Polytrichum commune* can prevail at the expense of Sphagna, encouraged by poaching and by nutrients in dung and urine. In heavily grazed flushes, shrubs and trees are generally absent and there are few forbs among the swards of sedges and rushes. In the absence of grazing many of these mires would be colonised by willows and birch, and would eventually develop into carr woodland. If the mires are drained, Sphagna are replaced by *P. commune* and other mosses, and the rushes can thicken up into a tall, dense sward. The combination of drainage and heavy grazing can convert *Carex echinata-Sphagnum* mire into *Juncus-Galium* rush-pasture or *Holcus-Juncus* rush-pasture MG10, and ultimately to species-poor, acid *Nardus-Galium* grassland U5 or *Juncus-Festuca* grassland U6. *Carex echinata-Sphagnum* flushes should never be drained on grouse moors, as the invertebrates they support provide important food for grouse and wader chicks.

# M7 *Carex curta-Sphagnum russowii* mire

## Synonyms
*Sphagneto-Caricetum alpinum* and *Carex aquatilis-rariflora* nodum (McVean and Ratcliffe 1962); Montane *Carex echinata-Sphagnum recurvum* mire H3h (Birks and Ratcliffe 1980).

Typical habitats of M7 *Carex-Sphagnum russowii* mire:

1  Wet peaty depressions among grassland and blanket bog on level or gently sloping ground on high plateaux and in high corries
2  Around edges of high-altitude lochans
3  Flushes among grasslands, springs, mires and snow-bed vegetation on sloping ground in high corries
4  Flushes among grasslands, springs and mires on high slopes

M7 is also shown in the picture for U7.

## Description
These small mires lie half-hidden in flushed hollows, surrounded by bogs or montane grasslands. The short, grey-green sward of *Carex echinata*, *C. curta*, *C. nigra*, and *Eriophorum angustifolium* has a dense, ochre-green underlay of *Sphagnum papillosum* and, in many stands, *S. russowii*. The associated flora is generally rather species-poor. There can be a few grasses such as *Nardus stricta* and *Agrostis canina*, and in most stands the pale-green, rounded leaves of *Viola palustris* are dotted over the carpet of Sphagna.

There are two sub-communities. The *Carex bigelowii-Sphagnum lindbergii* sub-community M7a has much *Carex bigelowii* and *Nardus*; *C. curta* can be absent. *Sphagnum subnitens* and *S. denticulatum* predominate in the moss layer. There is more *C. curta* in the *Carex aquatilis-Sphagnum fallax* sub-community M7b, and here the place of *C. bigelowii* is taken by *C. aquatilis*, and that of *S. lindbergii*, *S. subnitens* and *S. denticulatum* by *S. fallax*, *Calliergon stramineum* and *Polytrichum commune*.

## Differentiation from other communities

Montane species such as *Festuca vivipara*, *Saxifraga stellaris*, *Carex bigelowii*, *Sphagnum russowii* and *S. lindbergii* are quite common in *Carex-Sphagnum russowii* mire, but are generally absent from stands of the related *Carex echinata-Sphagnum* mire M6. *Carex-Sphagnum russowii* mire also has more *C. curta*, whereas some of the common species of *Carex echinata-Sphagnum* flushes, such as *Sphagnum palustre*, *Juncus effusus* and *Molinia caerulea*, are scarce. *Carex-Pinguicula* mire M10, which also occurs in the montane zone, is distinguished by its calcicolous flora; it also usually has a sparser sward than *Carex-Sphagnum russowii* mire. *Carex saxatilis* mire M12, another montane flush community, bears a superficial resemblance to *Carex-Sphagnum russowii* mire but is dominated by *C. saxatilis* and usually contains at least a few other calcicolous forbs or mosses.

## Ecology

*Carex-Sphagnum russowii* mires are the montane counterpart of the sedge-dominated forms of *Carex echinata-Sphagnum* mire. They occur above 600 m, occupying flushed hollows over wet peats on high slopes, in corries and on plateaux. They are commonly set in a matrix of montane blanket mire, but also occur around springs and snow-beds,

Most stands of *Carex-Sphagnum russowii* mire occur in the east-central and north-western High-lands where snow lies late on the high ground. The community also occurs in the Breadalbane range, extending south-west into the Cowal hills. There are similar mires in the Lake District, although they have fewer montane species. Some of the best and largest stands of the community are in the Cairn-gorm, Monadhliath and Caenlochan hills and on Lochnagar, but there are also good examples further west on Ben Alder, the Affric-Cannich hills, Beinn Dearg at the head of Loch Broom, and Ben More Assynt.

There is similar vegetation in Scandinavia. This community is an important link between montane vegetation in Great Britain and that in more conti-nental parts of Europe.

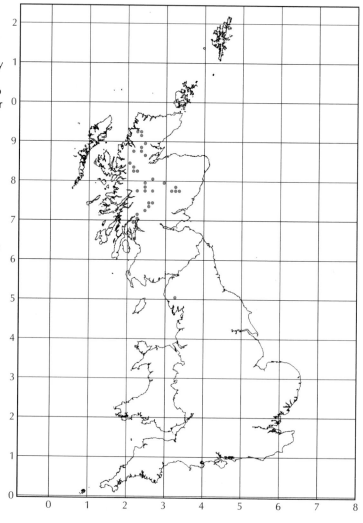

and around lochans and shallow pools. The community is apparently indifferent to geology, as the peat insulates it from the underlying rocks, but the irrigating water is generally acid. Most stands are covered by deep snow in the winter (McVean and Ratcliffe 1962), so the plants that grow there must tolerate a short growing-season and irrigation with icy water.

## Conservation interest

This community is of interest as an example of near-natural, montane flush vegetation. Flushes of this type provide one of the few British habitats for the rare *Carex rariflora*, *C. aquatilis*, *C. lachenalii*, *Sphagnum lindbergii* and *S. riparium*. They are also important breeding grounds for invertebrates, which in turn feed some of the birds that breed on high mountain plateaux in summer, including Ptarmigan *Lagopus mutus*, Dotterel *Charadrius morinellus*, Golden Plover *Pluvialis apricaria* and Dunlin *Calidris alpina*.

## Management

Most stands of *Carex-Sphagnum russowii* mire are grazed by deer or sheep. The community is not usually threatened by management activities, although disturbance is possible in skiing areas. At the southerly sites in England, some of the species are close to the limits of their geographical ranges and could be vulnerable to climate change.

# M8 *Carex rostrata-Sphagnum warnstorfii* mire

## Synonyms

*Carex rostrata-Sphagnum warnstorfianum* nodum (McVean and Ratcliffe 1962); Sub-montane *Carex rostrata-Sphagnum contortum* mire H3e (Birks and Ratcliffe 1980).

Typical habitats of M8 *Carex-Sphagnum warnstorfii* mire:

1   Wet peaty depressions among grassland and blanket bog on level or gently sloping ground on high plateaux and in high corries
2   Around edges of high-altitude lochans

## Description

This mire consists of a grey-green sward of *Carex rostrata*. Beneath the sedge there is a red-tinged carpet of *Sphagnum warnstorfii*, variegated with green-brown tufts of *S. teres*, golden-green, pointed shoots of *Calliergonella cuspidata*, golden, finely branched stems of *Hylocomium splendens*, and thick, yellow-green mats of *Aulacomnium palustre*. Small vascular plants, such as *Thalictrum alpinum*, *Carex pulicaris* and *Selaginella selaginoides*, grow through the moss carpet.

## Differentiation from other communities

This community is distinguished from the more lowland *Carex-Sphagnum squarrosum* mire M5 by the presence of montane species such as *Thalictrum alpinum* and *Persicaria vivipara*; other differences between the two communities are outlined on p. 145. *Carex-Sphagnum warnstorfii* mire is intermediate in its nutrient levels and flora between the oligotrophic *Carex rostrata-Sphagnum fallax* mire M4 and the mesotrophic *Carex-Calliergonella* mire M9. In comparison with *Carex-Calliergonella* mire, *Carex-Sphagnum warnstorfii* mire has much more *Sphagnum* and fewer calcicolous brown mosses such as *Drepanocladus revolvens* and *Scorpidium scorpioides*. *Sphagnum warnstorfii*, *Selaginella selaginoides*, *Carex pulicaris* and other species characteristic of mildly base-rich conditions distinguish *Carex-Sphagnum warnstorfii* mire from *Carex rostrata-Sphagnum fallax* mire.

## Ecology

*Carex-Sphagnum warnstorfii* mire is confined to wet hollows with stagnant, or percolating, moderately base-rich water at moderate to high altitudes. Most stands are above 400 m. Underlying the vegetation there is usually peat more than one metre deep and with a pH between 5.5 and 5.7 (McVean and Ratcliffe 1962). In many places this type of mire is sharply demarcated from the surrounding vegetation, which is usually some form of montane grassland.

## Conservation interest

*Carex-Sphagnum warnstorfii* mire is rare and localised in Great Britain and its total extent is small. It appears to be a near-natural community which constitutes an important floristic link between the vegetation of British mountains and that of more continental parts of Europe. Scarce species recorded in this community include the mosses *Sphagnum contortum*, *S. teres*, *S. platyphyllum* and *Tomenthypnum nitens*.

*Carex-Sphagnum warnstorfii* mire is rare in Great Britain, evidently because waterlogged hollows with base-rich water are scarce in wet upland regions where leaching usually leads to more acidic, oligotrophic soils. The community is most common on the Dalradian schist of the Breadalbane hills in central Scotland, where there are particularly good examples on Ben Lawers and Ben Vrackie. There are outliers in Sutherland, Lochaber, on base-rich Silurian rocks in the Southern Uplands, on limestone in the north Pennines, and on volcanic rocks in the Lake District and north Wales.

Similar vegetation occurs in Scandinavia where it is more common and more species-rich than it is in Scotland. Scandinavian stands usually have more montane species, particularly willows.

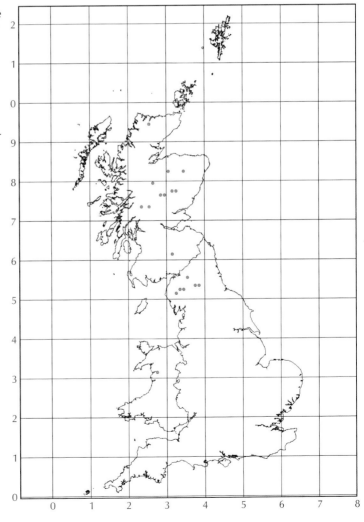

**Management**
The vegetation is generally grazed lightly by sheep or deer. *Salix aurita* has been seen in some stands (McVean and Ratcliffe 1962), and a reduction in the intensity of grazing might allow this and other shrubs to become established more generally. It is possible that acidic deposition may affect the flora of this community, particularly in the south of its range where there is more atmospheric pollution.

# M9 *Carex rostrata-Calliergonella cuspidata/Calliergon giganteum* mire

## Synonyms

*Carex rostrata*-brown moss provisional nodum (McVean and Ratcliffe 1962); *Carex rostrata-Scorpidium scorpioides* Association (Birks 1973); Sub-montane *Carex rostrata*-brown moss mire H3g (Birks and Ratcliffe 1980); *Carex rostrata-Calliergon cuspidatum/ giganteum* mire (Rodwell 1991b).

Typical habitats of M9 *Carex-Calliergonella/Calliergon giganteum* mire:

1 Wet peaty hollow
2 Along edges of stream
3 Around edge of pool
4 Hillside flushes

M9 is also shown in the picture for M26.

## Description

Like *Carex rostrata-Sphagnum fallax* M4, *Carex-Sphagnum squarrosum* M5 and *Carex-Sphagnum warnstorfii* M8 mires, this is a grey-green sward of *Carex rostrata* and *Eriophorum angustifolium* interleaved with *Menyanthes trifoliata*, *Potentilla palustris* and *Galium palustre*. These plants grow through a deep, wet mass of mosses including *Calliergonella cuspidata*, *Campylium stellatum*, *Scorpidium scorpioides* and other species, rather than the Sphagna that are characteristic of the other three communities.

There are two sub-communities. The *Campylium stellatum-Scorpidium scorpioides* sub-community M9a has a more open sward, in which *Carex limosa* and *C. echinata* are common, and mosses including *Campylium stellatum*, *Drepanocladus revolvens* and *Scorpidium scorpioides* grow in glossy wefts over the wet ground. The *Carex diandra-Calliergon giganteum* sub-community M9b has a taller, denser sward in which *C. diandra* and mesotrophic forbs such as *Caltha palustris*, *Mentha aquatica*, *Angelica sylvestris*, *Lychnis flos-cuculi* and *Filipendula ulmaria* are common.

## Differentiation from other communities

The superficially similar *Carex-Sphagnum warnstorfii* and *Carex-Sphagnum squarrosum* mires have Sphagna instead of the *Calliergonella cuspidata* and other mosses characteristic of *Carex-Calliergonella* mire. Some of the base-tolerant Sphagna, including *S. contortum* and *S. warnstorfii*, occur locally in *Carex-Calliergonella* mire but they are not common in this community.

In the uplands, *Carex-Calliergonella* mire is unlikely to be confused with related lowland swamps and fens. However, in more marginal hill ground, such as on the basalt of Mull and Skye, the *Carex-Calliergon* sub-community can be hard to separate from the *Lysimachia* sub-community of *Carex-Potentilla* tall-herb fen S27b. *Carex rostrata*, *Eriophorum angustifolium*, *Menyanthes trifoliata*, *Potentilla palustris* and *Galium palustre* are among the many species that are common to both forms of vegetation. *Carex-Potentilla* fen generally includes tall plants, such as *Lysimachia vulgaris*, *Phragmites australis*, *Lythrum salicaria* and *Iris pseudacorus*, that are rare in *Carex-Calliergonella* mire. *Carex-Calliergonella* mire generally has a more or less continuous layer of mosses below the vascular plants, whereas in *Carex-Potentilla* fen the vascular plants usually form dense mats standing in or floating on shallow water with a patchier

*Carex-Calliergonella* mire is widespread but scarce in the British uplands, where most stands belong to the *Campylium-Scorpidium* sub-community. In the uplands the community is perhaps most common on the calcareous Dalradian rocks of the Breadalbane region, on the limestone and basalt of Skye and Lismore, and on the basalt of Mull. It also occurs in the lowlands, for example in the fens of East Anglia and north-west Wales.

Similar vegetation occurs in topogenous mires in Norway and Sweden, although British examples generally have more oceanic species (McVean and Ratcliffe 1962).

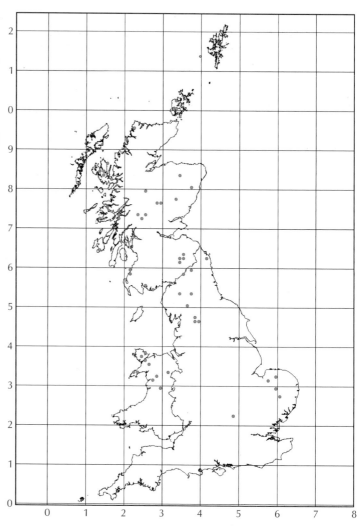

underlayer of mosses. However, many stands of vegetation cannot be assigned clearly to either community.

Some examples of *Carex-Calliergonella* mire resemble *Carex-Pinguicula* mire M10 except that they have a sedge sward consisting mainly of *C. rostrata*, rather than smaller species such as *C. viridula* ssp. *oedocarpa*, *C. panicea* and *C. dioica*.

## Ecology
This community is confined to places where base-rich water seeps through deep, wet peat. It occurs in hollows and seepage lines in blanket bogs, in calcareous fens, in topogenous mires, and around lochans, springs and raised mires. A characteristic setting is where base-rich water from calcareous springs runs over blanket peat (McVean and Ratcliffe 1962). Most upland stands are on limestone, calcareous schist or glacial drift. The water-level can fluctuate dramatically, but is rarely so low that the vegetation dries out altogether. Stands of *Carex-Calliergonella* mire are usually small and typically occur in mosaics with other mires where the distribution of the various communities is determined by local variations in base-status. The community has been recorded up to almost 800 m (McVean and Ratcliffe 1962). It also occurs in the lowlands of south-east England in fens and basin mires (Rodwell 1991b).

## Conservation interest
This is a scarce community, which can occur in mosaic with other local types of vegetation. Some stands harbour uncommon species, including *Thalictrum alpinum*, *Carex diandra*, *C. appropinquata*, *Utricularia intermedia*, *Sphagnum contortum*, *S. teres*, *S. subsecundum*, *Cinclidium stygium*, *Hamatocaulis vernicosus* and *Barbilophozia kunzeana*. As in *Carex-Sphagnum squarrosum* mire, the forbs and sedges are able to flower and set seed, and so maintain the genetic diversity of the populations.

## Management
Stands of *Carex-Calliergonella* mire are usually too wet to graze, although some lowland examples are mown (Rodwell 1991b), and are similarly protected from fire. The community is susceptible to drainage and could possibly be adversely affected by acidic deposition, although the base-rich water must afford some protection from acidification.

# M10 *Carex dioica-Pinguicula vulgaris* mire

## Synonyms

*Hypno-Caricetum alpinum, Carex panicea-Campylium stellatum* nodum and *Schoenus nigricans* provisional nodum (McVean and Ratcliffe 1962); *Carex panicea-Campylium stellatum* Association, *Eriophorum latifolium-Carex hostiana* Association and *Schoenus nigricans* Association (Birks 1973); *Carex nigra*-brown moss mire H3f, montane *Carex nigra*-brown moss mire H3i, *Schoenus nigricans* mire H4 and *Carex hostiana-Eriophorum latifolium* flush I1a (Birks and Ratcliffe 1980).

M10 *Carex-Pinguicula* mire occupying narrow, stony flushes among grassland and heath on hill slope. The flushes broaden out and coalesce on flatter ground at lower right of picture.

M10 is also shown in the pictures for M37 and M38.

## Description

These mires have a patchy sward of sedges and bryophytes, studded with the lime-green, sticky stars of *Pinguicula vulgaris*. The most common sedges are the bright-green *Carex viridula* ssp. *oedocarpa*, the grey-green *C. panicea* and the dark-green, fine-leaved *C. dioica*. Below the sparse canopy of sedges there are tufts and wefts of calcicolous mosses, including *Drepanocladus cossonii, D. revolvens, Scorpidium scorpioides, Campylium stellatum* and *Blindia acuta*. Many stands have little more than this, but the most species-rich examples have a rich array of small calcicoles such as *Thalictrum alpinum, Tofieldia pusilla, Selaginella selaginoides, Eleocharis quinqueflora* and *Saxifraga oppositifolia*. There can be some tufa formation. As with other base-rich mires there can be a strong smell of decomposing vegetation.

There are three sub-communities. The *Carex viridula* ssp. *oedocarpa-Juncus bulbosus* sub-community M10a is the most widespread and occurs on slightly more acid soils than the other sub-communities. In addition to *C. viridula* ssp. *oedocarpa* and *Juncus bulbosus*, it is defined by species such as *Erica tetralix, Eleocharis quinqueflora, Narthecium ossifragum* and *Drosera rotundifolia*; *Schoenus nigricans* can also be common here. The *Briza media-Primula farinosa* sub-community M10b is distinguished by more basiphilous species, including *Briza media*, *Primula farinosa*, *Linum catharticum* and *Carex flacca*. The *Hymenostylium recurvirostrum* sub-community M10c is an unusual assemblage of species including *Plantago maritima*, *Sagina nodosa* and *Minuartia verna*, as well as the montane *Carex capillaris* and *Juncus triglumis*, and the moss *Hymenostylium recurvirostrum*.

## Differentiation from other communities

Most calcicolous small-sedge mires in the uplands belong to either this community or *Carex-Saxifraga* mire M11. These two communities have many species in common and can be difficult to distinguish. The NVC tables suggest that the moss *Blindia acuta* is more characteristic of *Carex-Saxifraga* mire, but this species is actually also common in many stands of *Carex-Pinguicula* mire. The most reliable distinguishing species is *Saxifraga aizoides*, which is generally scarce in *Carex-Pinguicula* mire. There are usually more montane species in *Carex-Saxifraga* mire. Both mires can form very open and stony flushes; those on a more continuous layer of soil are mostly *Carex-Pinguicula* mire.

*Carex-Pinguicula* mires occur throughout the uplands of Scotland, northern England and Wales, wherever there are base-rich substrates and associated flushes. They are particularly common on the basic Dalradian rocks in the Breadalbane region of the Highlands. They are not recorded from south-west England. The *Carex-Juncus* sub-community is the most common form; the *Briza-Primula* sub-community occurs mainly in northern England and southern Scotland; the *Hymenostylium* sub-community has been recorded only from limestone hills in the Pennines.

There is vegetation similar to *Carex-Pinguicula* mire in mainland Europe, although the Scandinavian counterparts tend to be ungrazed and to contain montane *Salix* species. Essentially the same type of vegetation also occurs in the Faroe Islands (Hobbs and Averis 1991a).

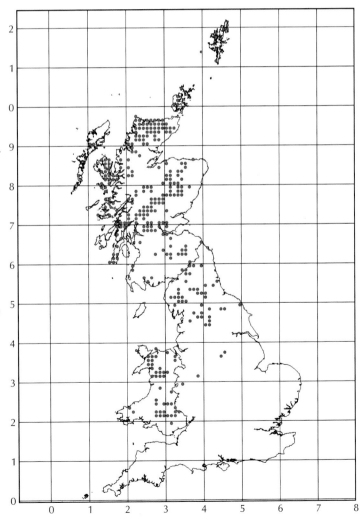

## Ecology

The community can occur wherever there is flushing with base-rich water, either below a springhead or where water emerges more diffusely from the ground. Most stands are constantly irrigated. They appear as elongated or oval patches, often in vertical strips running downslope from lines of springs. On flatter ground they can spread out in anastomosing channels covering a large area. The soils are saturated, silty muds with a surface of humus or shallow peat, and with a pH between 5.9 and 6.3 (McVean and Ratcliffe 1962). Stands can occur from sea-level (in the north-west) to over 900 m, especially in central Scotland where there is calcareous rock at high elevations. Most stands are set in a matrix of upland grassland or heath, but the community also occurs in upland woods.

## Conservation interest

These mires are of great value for nature conservation because although they are usually small they contribute greatly to the diversity of upland vegetation. They are also home to a number of scarce species, including *Bartsia alpina*, *Juncus alpinoarticulatus*, *Kobresia simpliciuscula*, *Minuartia stricta*, *Schoenus ferrugineus*, *Primula farinosa*, *P. scotica*, *Saxifraga hirculus*, *Meesia uliginosa*, *Catoscopium nigritum*, *Cinclidium stygium*, *Amblyodon dealbatus* and *Tritomaria polita*.

## Management

*Carex-Pinguicula* mires are damaged by drainage and by changes to the water-level and nutrient-status. Most stands are grazed by livestock, and in Scotland they can be used as wallows by deer. Some grazing and trampling helps to maintain the open structure and diversity of the vegetation (Pigott 1956), although some species may be lost if there is excessive grazing and trampling. In the absence of herbivores, most stands would probably become colonised by trees and shrubs, and would eventually develop into some form of damp woodland (e.g. *Alnus-Fraxinus-Lysimachia* woodland W7), but higher-altitude examples may be naturally open. Within the large deer-proof exclosures that have been erected by the National Trust for Scotland on Ben Lawers there are stands of *Carex-Pinguicula* mire and also *Carex-Saxifraga* mire close to a source of seed of montane willows. It will be interesting to see whether *Salix*-dominated mires similar to the Scandinavian examples develop here in the absence of grazing.

# M11 *Carex viridula* ssp. *oedocarpa-Saxifraga aizoides* mire

## Synonyms
*Cariceto-Saxifragetum aizoidis* (McVean and Ratcliffe 1962); *Carex-Saxifraga aizoides* nodum (Birks 1973); Sub-montane *Carex demissa-Saxifraga aizoides* flush I1b and Montane *Carex demissa-Saxifraga aizoides* flush I1c (Birks and Ratcliffe 1980); *Carex demissa-Saxifraga aizoides* mire (Rodwell 1991b).

M11 *Carex-Saxifraga* mire occupying narrow, stony flushes among grassland and heath on hill slope. The flushes broaden out and coalesce on flatter ground at lower right of picture.

M11 is also shown in the picture for M38.

## Description
These are stony flushes of base-rich hillsides at high altitudes. The grass-green *Carex viridula* ssp. *oedocarpa* and the grey-green *C. panicea* and *C. pulicaris* form a sparse sward mixed with the succulent-leaved *Saxifraga aizoides* which has conspicuous starry yellow flowers in summer. Between the vascular plants there are tufts and patches of calcicolous bryophytes growing on wet stones and bare, muddy soil; *Blindia acuta*, *Drepanocladus cossonii*, *D. revolvens*, *Campylium stellatum*, *Preissia quadrata*, *Calliergon trifarium* and *Fissidens adianthoides* are among the most characteristic species.

There are two sub-communities, dividing the more montane from the less montane stands. The *Thalictrum alpinum-Juncus triglumis* sub-community M11a is home to many montane species, of which the most common are *Juncus triglumis*, *Thalictrum alpinum* and *Persicaria vivipara*. In the *Palustriella commutata-Eleocharis quinqueflora* sub-community M11b there are fewer montane plants, and more *Eleocharis quinqueflora*, *Schoenus nigricans*, *Carex hostiana* and *Palustriella commutata*.

## Differentiation from other communities
*Carex-Saxifraga* mire is very similar to *Carex-Pinguicula* mire M10 but *Saxifraga aizoides* and other montane species are usually more common. However, the separation

of the two communities can be hard to justify on some hills where they would be better described as a single well-defined type of vegetation. *Carex-Saxifraga* mire also has close affinities with *Carex saxatilis* mire M12, but this community is typically dominated by *Carex saxatilis* and has less *S. aizoides* and *Carex pulicaris*. The more lowland *Palustriella-Eleocharis* sub-community can resemble *Palustriella-Carex* spring vegetation M37, and indeed the two can occur together, but it has more *Carex viridula* ssp. *oedocarpa*, *C. panicea*, *Blindia acuta*, *Drepanocladus revolvens*, *Campylium stellatum* and *Eleocharis quinqueflora*. *Carex* and *Juncus* species are more common in *Carex-Saxifraga* mire than in *Saxifraga-Alchemilla* vegetation U15.

## Ecology

*Carex-Saxifraga* mire is a spring-fed community of slopes or channels where the water usually flows rapidly over the surface. Some rocks provide more favourable substrates than others and the soft, easily weathered mica-schist of the Breadalbanes is particularly good, forming sheets of unstable, base-rich debris. Although the community can occur close to sea-level in the far north-west, most stands are at altitudes above 500 m and many of the constituent species are montane plants. Stands are usually small but can form mosaics with other communities over a large area.

*Carex-Saxifraga* mire occurs from north Wales to Orkney, but is most widespread and best developed in the Scottish Highlands. Some of the largest stands are on the calcareous Dalradian rocks of the Breadalbanes, but there are also fine examples on the limestone and basalt hills of the north-west Highlands and on the Borrowdale Volcanic rocks of the Lake District. Welsh stands have fewer montane species and lack *Saxifraga aizoides*.

There is similar vegetation elsewhere in the mountains of northern and central Europe.

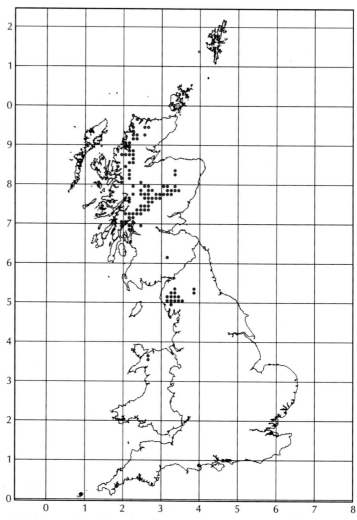

**Conservation interest**
Over much of its range, particularly at higher altitudes, *Carex-Saxifraga* mire appears to be a near-natural vegetation type. The community is a locus for many rare montane calcicoles, including *Carex atrofusca, C. vaginata, C. microglochin, C. norvegica, Equisetum variegatum, Juncus biglumis, J. triglumis, J. castaneus, J. alpinoarticulatus, Kobresia simpliciuscula, Tofieldia pusilla, Schoenus ferrugineus, Bartsia alpina, Meesia uliginosa, Tayloria lingulata, Calliergon trifarium, Scorpidium turgescens, Catoscopium nigritum, Amblyodon dealbatus, Oncophorus virens, O. wahlenbergii, Orthothecium rufescens, Leiocolea gillmanii, Scapania degenii, Tritomaria polita* and *Barbilophozia quadriloba*. It is also valuable for insects and birds.

**Management**
The mats of plants characteristic of *Carex-Saxifraga* flushes can be fragmented and some species may be lost if there is excessive grazing and trampling. However, light trampling may help to maintain the open sward and unstable soils that are favoured by some of the rarer species. The rare *Salix reticulata* has occasionally been recorded in the community (Rodwell 1991b); it would be interesting to know whether flushes would become colonised by willows if grazing was prevented or reduced. Vegetation similar to this type of mire can develop on artificially exposed, base-rich rock debris, for example at the edges of tracks and roads and on rocky slopes in abandoned quarries.

# M12 *Carex saxatilis* mire

## Synonyms
*Caricetum saxatilis* (McVean and Ratcliffe 1962); Montane *Carex saxatilis* mire H3j (Birks and Ratcliffe 1980).

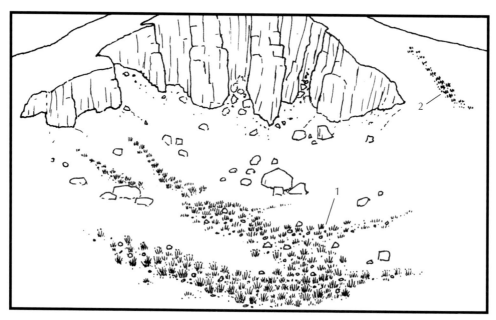

Typical habitats of M12 *Carex saxatilis* mire:

1   Flushes among grassland, snow-bed vegetation and mires in high corrie
2   Flushes among grassland, snow-bed vegetation and mires on other high slopes

M12 is also shown in the picture for U7.

## Description
This mire community consists of dark-green swards of *Carex saxatilis* in which the stiff, curved leaves of the sedge give a distinct texture to the vegetation. The flushes are set into small hollows or gullies on high hillsides or spread over irrigated slopes, and are studded in summer with the dark flowers of *C. saxatilis*. Other characteristic species include *Deschampsia cespitosa* and *C. viridula* ssp. *oedocarpa*. Hidden among the stems of the sedges is a rich array of small vascular plants, including scarce montane calcicoles such as *Persicaria vivipara*, *Saxifraga aizoides*, *S. oppositifolia*, *Thalictrum alpinum* and *Juncus triglumis*. At the base of the sward is a loose layer of *Campylium stellatum*, *Fissidens adianthoides*, *Ctenidium molluscum*, *Scorpidium scorpioides* and other calcicolous mosses.

## Differentiation from other communities
This community can be confused only with *Carex-Saxifraga* mire M11 and *Carex-Pinguicula* mire M10. Various differences between the three communities are apparent from the NVC tables, but they have so many species in common that the abundance of *Carex saxatilis* provides the most reliable means of distinguishing *Carex saxatilis* mire.

## Ecology

This is a montane community of high slopes above 700 m in corries where snow lies late and which are flushed with cold, base-rich water. It is most common downslope of north-facing cliffs, where there is some snow-lie in winter and where there is constant irrigation from dripping rocks above. The underlying soils are generally base-rich, humus-rich muds or peats with a pH between 4.6 and 5.7 (McVean and Ratcliffe 1962). Prolonged snow cover, strong flushing from melting snow, freeze-thaw and solifluction keep the soils unstable and maintain an open, stony habitat. This enables the rarer small sedges and forbs to find a niche and discourages colonisation by larger competitors.

## Conservation interest

This community is one of the rarest mire types in Great Britain, and is a valuable example of a montane, near-natural type of vegetation. *Carex saxatilis* is itself a scarce species, and the community provides a habitat for some of our rarest montane plants, including *Juncus biglumis*, *J. castaneus*, *J. alpinoarticulatus*, *Kobresia simpliciuscula*, *Carex microglochin*, *C. atrofusca* and *Scorpidium turgescens*. *C. microglochin* and *S. turgescens* are known only in this type of mire and some examples of *Carex-Saxifraga* mire on and near Ben Lawers. As with all montane mires, the insects which breed in the

*Carex saxatilis* mire is confined to the Scottish Highlands, between Ben More Assynt in the north, Ben Lui and Beinn Ime in the south-west, and Caenlochan in the east. It occurs only where there are base-rich rocks and so is most common on the Dalradian mica-schist of the Breadalbanes. In the north-west it occurs on base-rich outcrops of Moine or Lewisian rocks (McVean and Ratcliffe 1962).

There is similar vegetation in the Faroe Islands (Hobbs and Averis 1991a), Norway and Sweden.

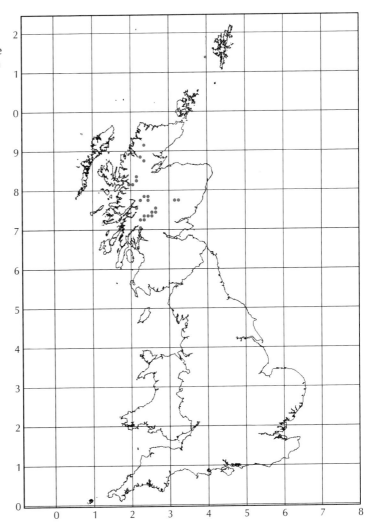

soft wet soils of *Carex saxatilis* mire are an important source of food for birds, such as Dotterel *Charadrius morinellus*, Dunlin *Calidris alpina* and Snow Bunting *Plectrophenax nivalis*.

## Management

*Carex saxatilis* mire is almost certainly a form of climax vegetation and is maintained by climate rather than management. Most stands experience only light grazing and some wallowing by deer in summer. These treatments may help to maintain unstable soils and an open sward in which the rarer species are able to flourish.

# M15 *Trichophorum cespitosum-Erica tetralix* wet heath

## Synonyms
*Trichophoreto-Eriophoretum caricetosum* (McVean and Ratcliffe 1962); *Molinia-Myrica* mire and *Trichophoreto-Callunetum* (McVean and Ratcliffe 1962; Birks 1973); *Trichophorum cespitosum-Carex panicea* Association (Birks 1973); *Myrica gale-Molinia caerulea* mire H1, *Erica tetralix-Carex panicea* mire H3a, *Narthecium ossifragum-Sphagnum* flush I3, *Scirpus cespitosus-Calluna vulgaris* mire G2 and *Molinia caerulea-Calluna vulgaris* mire G3 (Birks and Ratcliffe 1980); *Scirpus cespitosus-Erica tetralix* wet heath (Rodwell 1991b).

Typical view of M15 *Trichophorum-Erica* wet heath in the western Scottish Highlands. This drawing is of part of Beinn Eighe in Wester Ross.

Foreground vegetation and medium-stippled areas in middle distance (i.e. most of picture): M15 *Trichophorum-Erica* wet heath on moist peat on gentle to moderate lower and middle slopes.

Dark-stippled areas in middle distance: mainly H10 *Calluna-Erica* and H21 *Calluna-Vaccinium-Sphagnum* heaths on steeper well-drained slopes, with some U20 *Pteridium-Galium* vegetation.

Light-stippled areas among M15 on lower slopes: M17 *Trichophorum-Eriophorum* bog on level, deep, waterlogged peat.

Unstippled areas in upper part of picture: higher ground, mainly with H14 *Calluna-Racomitrium* heath, cliffs and scree.

M15 is also shown in the pictures for M21, H4, H10 and H19–H22.

## Description
Most of the ubiquitous wet heaths in the north and west of Scotland belong to this type of vegetation. These vast, ochre-brown tracts of moorland consist of mixtures of *Calluna vulgaris*, *Erica tetralix*, *Trichophorum cespitosum* and *Molinia caerulea*, entwined with

*Potentilla erecta*, and pricked through by the narrow, upright shoots of *Narthecium ossifragum* and the long, dark-green leaves of *Eriophorum angustifolium*.

There are four sub-communities, representing points fairly widely spaced within the continuum of variation of the community. *Trichophorum-Erica* wet heath is more variable than the NVC tables indicate. It seems to have been inadequately sampled, perhaps because it is so common and does not attract much attention from botanists in search of rare species.

The *Carex panicea* sub-community M15a is more of a soligenous mire than a wet heath, with a thin canopy of the characteristic species augmented by *Carex panicea* and other vascular plants such as *C. echinata*, *Juncus squarrosus* and *Drosera rotundifolia*. There is a patchy carpet of mosses over the wet peaty ground in which the most common species are usually *Sphagnum denticulatum* and *Campylopus atrovirens*. There can be small patches of other bryophytes, such as the mosses *Breutelia chrysocoma*, *Racomitrium lanuginosum*, *S. capillifolium* and *S. papillosum* and the liverwort *Pleurozia purpurea*. In the west Highlands and the Hebrides *Schoenus nigricans* can grow in thick, whitish tussocks in this type of mire, and on Skye and the Outer Hebrides some stands have bright-green sheets of the rare moss *Campylopus shawii*. Flushes with much *Campylopus atrovirens* and a sparse sward of *Narthecium*, *Trichophorum*, *Carex panicea* and *Nardus stricta* also belong to the *Carex* sub-community. They are more common at higher altitudes and occur in depressions in grasslands as well as in wet heaths and bogs. These correspond to the *Narthecium-Sphagnum* flush of Birks and Ratcliffe (1980).

The Typical sub-community M15b has a thick tufted sward of *Calluna*, *Erica tetralix*, *Molinia* and *Trichophorum*, and there can be a thick speckling of *Myrica gale* with its sweet spice-scented leaves. Under the canopy the ground may be clad with a pale fawn layer of *Molinia* litter. There are commonly small patches of *Sphagnum capillifolium*, *S. denticulatum*, *S. papillosum* and *Campylopus atrovirens*, and *Pleurozia purpurea* may also be common. *Schoenus nigricans* occurs locally in this type of vegetation in the far north-west. Some stands are dense, tall, species-poor swards dominated by *Calluna* and *Molinia*.

The *Cladonia* species sub-community M15c has a shorter and more open sward. *Myrica* is rare here, and *Erica cinerea* can be as common as *E. tetralix*. Despite the name of the sub-community, not all stands have a rich flora of lichens. More characteristic in western stands is the moss *Racomitrium lanuginosum*, which grows in a thin, silvery-green weft beneath the vascular plants. *Cladonia* lichens can be common, and in some places in the northern Highlands and Orkney they form a conspicuous creamy-white frosting over the ground. In a few localities in the north-west Highlands and Western Isles there is a form of the *Cladonia* sub-community with *Juniperus communis* ssp. *nana* (A B G Averis 1994; Averis and Averis 1998a, Averis *et al.* 2000).

The *Vaccinium myrtillus* sub-community M15d is a drier and grassier assemblage, characterised by *Nardus*, *Deschampsia flexuosa*, *Juncus squarrosus* and *Vaccinium myrtillus*. This is the driest type of *Trichophorum-Erica* wet heath, and has mosses such as *Dicranum scoparium*, *Pleurozium schreberi* and *Hypnum jutlandicum*, rather than Sphagna.

Locally in the western Highlands there are montane wet heaths resembling the *Cladonia* and *Vaccinium* sub-communities and including northern upland species such as *Antennaria dioica*, *Carex bigelowii*, *Diphasiastrum alpinum*, *Empetrum nigrum* ssp. *hermaphroditum*, *Vaccinium uliginosum*, the oceanic liverworts *Bazzania pearsonii* and *Anastrophyllum donnianum*, and the lichen *Cetraria islandica* (e.g. Averis and Averis 1998a; Averis and Averis 2003).

There are also stands of *Trichophorum-Erica* wet heath composed of a dense sward of *Trichophorum*, with few dwarf shrubs and almost no other species. These are mostly in places where there has been heavy grazing, especially by cattle or deer (e.g. Averis and Averis 1999a).

## Differentiation from other communities

Wet heaths, blanket mires and valley mires share a similar array of species, but the typical bog plants *Sphagnum papillosum* and *Eriophorum vaginatum* are generally rare in wet heaths. *S. papillosum* can be quite common in some stands of the *Carex* and Typical sub-communities of *Trichophorum-Erica* wet heath, but does not form the extensive sheets characteristic of *Trichophorum-Eriophorum* M17 and *Erica-Sphagnum* M18 blanket mires and *Narthecium-Sphagnum* valley mire M21.

The dry heaths *Calluna-Deschampsia* H9, *Calluna-Erica* H10 and *Calluna-Vaccinium* H12 have no *Erica tetralix* and less *Molinia* and *Trichophorum* than *Trichophorum-Erica* wet heath. The damp heaths *Calluna-Vaccinium-Sphagnum* H21 and *Vaccinium-Rubus* H22 have *Vaccinium myrtillus*, and could be confused with the *Vaccinium* sub-community of *Trichophorum-Erica* wet heath, but they have less *E. tetralix*, *Molinia* and *Trichophorum*, and more *Sphagnum capillifolium* and large pleurocarpous mosses such as *Rhytidiadelphus loreus* and *Hylocomium splendens*. All of these dry and damp heath communities have a thicker, darker canopy of heather than *Trichophorum-Erica* wet heath.

Within the wet heaths, this community can be confused with the floristically similar *Erica-Sphagnum compactum* heath M16, the counterpart of *Trichophorum-Erica* heath in the south and east of Great Britain. *Sphagnum compactum* and *S. tenellum* are generally more common in the *Erica-Sphagnum compactum* community; in addition, *Potentilla erecta* is generally scarce in this form of wet heath. However, the distinction can be difficult to make in practice. For example, the NVC floristic tables for the Typical sub-community of *Trichophorum-Erica* wet heath and the *Succisa-Carex* sub-community of *Erica-Sphagnum compactum* wet heath M16b are so similar that some stands of vegetation could be equally well classified as either of these two types. There are also important floristic and ecological divisions among the range of upland wet heath vegetation in Great Britain that are not aligned with the distinction between the two NVC communities. One such division is between the heathy soligenous mires or flushes (the *Carex* sub-community of *Trichophorum-Erica* heath and the *Rhynchospora-Drosera* sub-community of *Erica-Sphagnum compactum* heath M16c) and the less consistently waterlogged true wet heaths belonging to the other sub-communities. *Carex* species, *Drosera* species, *Pinguicula* species, *Rhynchospora alba* and *Eleocharis multicaulis* can all be common in the soligenous mires but are scarce in the true wet heaths.

In the western Highlands it is possible to confuse the *Cladonia* sub-community of *Trichophorum-Erica* wet heath with *Calluna-Cladonia* heath H13, *Calluna-Racomitrium* heath H14 and the *Racomitrium* sub-community of *Nardus-Galium* grassland U5e. *Calluna-Cladonia* and *Calluna-Racomitrium* heaths have prostrate or severely wind-pruned heather, more montane species, and rarely much *E. tetralix*, *Trichophorum* or *Sphagnum* species. The *Racomitrium* sub-community of *Nardus-Galium* grassland has more *Nardus* than *Trichophorum* and generally has some montane species.

## Ecology

*Trichophorum-Erica* wet heath is a community of shallow, wet or intermittently waterlogged, acid peat or peaty mineral soils on hillsides, over moraines, and within tracts of blanket mire. It also extends on to deep peat where the original bog vegetation has been damaged or modified by burning, grazing, drainage and peat cutting. The sub-communities occupy different terrain: the more soligenous *Carex* sub-community occurs in hollows, channels and soakways, the Typical sub-community on shallow slopes at low to moderate altitudes, the *Cladonia* sub-community on steeper slopes with thinner peat and at higher elevations, and the *Vaccinium* sub-community on drier substrates. Most stands with *Schoenus nigricans* are close to the west coast, where nutrient enrichment from sea-spray may allow *S. nigricans* to grow in otherwise acid heaths. The species-poor *Calluna-Molinia* form of the Typical sub-community typically occurs in places that are recovering from moderate to heavy grazing. Most stands of *Trichophorum-Erica* wet heath are at low to moderate altitudes, within the altitudinal range of woodland, but the *Cladonia* and *Vaccinium* sub-communities can occur at well over 600 m in the hills of the west Highlands.

## Conservation interest

*Trichophorum-Erica* wet heath is important for nature conservation because there is so much of it in the west Highlands and so little anywhere else in the world. Nowhere does it make such a contribution to the appearance and flora of the upland landscape as it does in Scotland. A few rare species have been recorded in this community, in particular the mosses *Campylopus setifolius*, *C. atrovirens* var. *falcatus* and, especially in stands of

*Trichophorum-Erica* wet heath is widespread in the north and west of Great Britain. It is most common in the western High-lands, where the Typical and *Cladonia* sub-communities cover very large areas. It is common but much less extensive in the eastern Highlands, the Southern Uplands, the Cheviot Hills, the Lake District, Wales and south-west England. It is not recorded in the Pennines, but does occur in the North York Moors.

Within Europe, *Trichophorum-Erica* wet heath has a strongly oceanic distribution and is rare outside Great Britain and Ire-land. Similar heaths occur in mainland western Europe from Sweden to Spain, but large stands are rare.

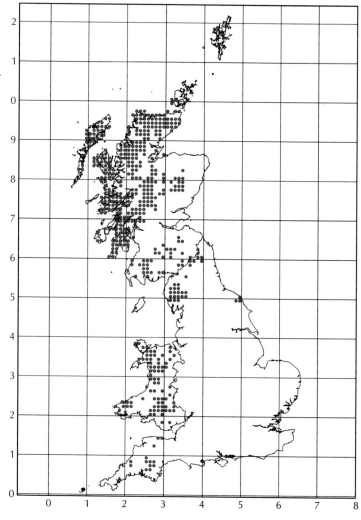

the *Carex* sub-community in the Outer Hebrides and on Skye, *C. shawii*. On some north-facing slopes there are patches of western liverworts, including *Herbertus aduncus* ssp. *hutchinsiae*, *Anastrophyllum donnianum*, *Bazzania tricrenata*, *B. pearsonii*, *Pleurozia purpurea*, *Plagiochila carringtonii*, *Scapania gracilis*, *S. ornithopodioides* and *Anastrepta orcadensis*. The form of the *Cladonia* sub-community with *Juniperus communis* ssp. *nana* is one of the main habitats of this shrub in Great Britain; heaths of this type on Beinn Eighe are also home to the liverwort *Herbertus borealis* at its only known British site. *Trichophorum-Erica* wet heath is not an outstanding habitat for upland birds, but these wide empty lands are hunted over by Buzzards *Buteo buteo*, Ravens *Corvus corax*, Short-eared Owls *Asio flammeus* and, in Scotland, Golden Eagles *Aquila chrysaetos*. In northern Scotland, the community is an important nesting habitat for Greenshank *Tringa nebularia*.

## Management

Very little *Trichophorum-Erica* wet heath is natural, although some higher-altitude stands of the *Cladonia* sub-community on rocky slopes may be nearly so, and wet heaths might always have occupied open, boggy glades even when the upland landscape was well wooded. There would once have been woodlands on most of the ground that is now covered with this type of vegetation. When woodlands are cleared and there are no trees to take up water, the soils can become waterlogged and wet heaths may develop on shallow slopes. Some examples of *Trichophorum-Erica* wet heath have been derived directly from blanket mire in response to burning and grazing.

Most stands of *Trichophorum-Erica* wet heath are grazed by deer and sheep and are sporadically managed by burning, usually in large patches. Without grazing and burning these heaths could potentially revert to woodland, although this might be a slow process on the impoverished acid soils, and without grazing the dwarf shrubs and *Molinia* might in some places grow so tall and dense that tree seedlings would not easily compete. Planted trees will grow in this community, especially if the ground is first prepared by drainage, and many commercial forestry plantations are on slopes where there was once *Trichophorum-Erica* wet heath. The reinstated deciduous woodlands on Rum were almost all established in this type of vegetation, and within ten years the leaf litter from the trees had enriched the ground enough for a recognisable woodland understorey to be able to flourish (Peter Wormell, pers. comm.). The community also occurs within the range of *Pinus-Hylocomium* woodland W18, and new pine woodlands have been established in *Trichophorum-Erica* wet heath in the Blackmount and Glen Orchy (Wormell 2000).

Grazing, especially by deer and cattle, seems to be necessary to maintain the structural and floristic diversity of the community by reducing competition from *Calluna* and *Molinia*. However, too much grazing can reduce the vegetation to a species-poor sward of *Trichophorum* with few dwarf shrubs (Averis and Averis 1999a), especially if the heaths are also burned. Such treatments, especially if combined with drainage, can eventually convert *Trichophorum-Erica* wet heath to grassland dominated by *Nardus*, *Juncus squarrosus* or *Molinia*. If the vegetation is not grazed at all, or if it is drained, *Calluna* and *Molinia* may come to dominate in a dense species-poor sward where little else has room to grow. Burning can help to maintain *Trichophorum-Erica* heaths, as long as it is not done too frequently. Severe burning on a short rotation can remove much of the peaty soil, producing a sparse sward of impoverished vegetation on dry, patchy peat interspersed with gravel and stones. If burning is carried out, the ideal method is to burn every 10–20 years with superficial fires, which burn away the old woody growth of the shrubs and the *Molinia* litter but which do not destroy the bryophytes or scorch the soil (Phillips *et al.* 1993, Moorland Working Group 1998, Scotland's Moorland Forum 2003).

However, some vegetation should be left unburnt to grow tall as nesting habitat for breeding birds. The diverse habitat that results from this kind of management is also valuable for invertebrates. Wet heaths should not be burned if the dwarf shrubs are wind-clipped, nor if they are on shallow rocky soils with a dense layer of mosses or lichens; consequently, most stands of the *Cladonia* sub-community should not be burned. Burning should also be avoided on the wetter stands of the *Carex* sub-community as they are important breeding grounds for insects and feeding sites for upland birds.

# M16 *Erica tetralix-Sphagnum compactum* wet heath

## Synonyms
*Campylopo-Ericetum tetralicis p.p.* (Birse and Robertson 1976).

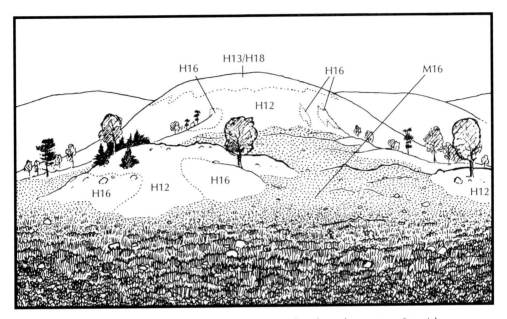

Typical view of M16 *Erica-Sphagnum compactum* wet heath in the eastern Scottish Highlands (foreground and stippled in middle distance): on moist peat on gentle to moderate slopes.

Vegetation types on drier ground nearby:

H12 *Calluna-Vaccinium* heath
H13 *Calluna-Cladonia* heath
H16 *Calluna-Arctostaphylos uva-ursi* heath
H18 *Vaccinium-Deschampsia* heath.

M16 is also shown in the pictures for M21 and H4.

## Description
This type of vegetation includes the wet heaths of the south and east of Great Britain. They have a short grey to brown-yellow sward of *Calluna vulgaris*, *Erica tetralix* and *Molinia caerulea*, with a patchy underlay of *Sphagnum compactum* and in some places *S. tenellum*.

There are four sub-communities. The Typical sub-community M16a has no particular distinguishing characteristics of its own. In the *Succisa pratensis-Carex panicea* sub-community M16b *S. compactum* is scarce and there is much *Potentilla erecta*, *Succisa pratensis* and *Sphagnum denticulatum*, and, in some places, *Polygala serpyllifolia* and *Carex panicea*. The *Rhynchospora alba-Drosera intermedia* sub-community M16c is essentially a soligenous mire of wetter peats, and has *Kurzia pauciflora*, *Drosera intermedia*, *D. rotundifolia* and *Rhynchospora alba*. The *Juncus squarrosus-Dicranum scoparium* sub-community M16d is the most common form in the north of Great Britain, and has *Juncus squarrosus*, *Dicranum scoparium*, *Hypnum jutlandicum*, *Racomitrium lanuginosum*, *Cladonia portentosa* and *C. uncialis* ssp. *biuncialis*.

## Differentiation from other communities

Like *Trichophorum-Erica* wet heath, *Erica-Sphagnum compactum* wet heath can be distinguished from the bog vegetation types *Trichophorum-Eriophorum* M17, *Erica-Sphagnum papillosum* M18, *Calluna-Eriophorum* M19, *Eriophorum vaginatum* M20 and *Narthecium-Sphagnum* M21 by the scarcity or absence of *Sphagnum papillosum* and *Eriophorum vaginatum*. In addition, *S. compactum* is common in *Erica-Sphagnum compactum* heath but rare in the bog communities. Separation of the two wet heath types is based mainly on *S. compactum* and *S. tenellum* being commoner in *Erica-Sphagnum compactum* heath, and *Potentilla erecta* being commoner in *Trichophorum-Erica* heath. The relationship between the two communities is discussed in more detail under *Trichophorum-Erica* wet heath (see p. 171). In south-west England, *Erica-Sphagnum compactum* wet heath may occur in association with the wetter forms of *Ulex gallii-Agrostis* heath H4, from which it differs in having more *S. compactum* and little or no *Erica cinerea*, *Ulex gallii* and *Agrostis curtisii*.

## Ecology

*Erica-Sphagnum compactum* wet heath typically occurs on shallow acid peat on sloping ground, although it can cover almost level ground in the eastern Highlands and in south-west England. The soils are moist and intermittently waterlogged. It is primarily a lowland community, but can occur above 500 m in the eastern Highlands where it can cover deep peats that one might expect to be clothed with *Calluna-Eriophorum* or *Erica-Sphagnum papillosum* blanket mires. It seems that the change from bog to wet heath has been caused by frequent or severe burning and heavy grazing and trampling. Most examples of *Erica-Sphagnum compactum* wet heath are semi-natural, but there are perhaps near-natural stands at high altitudes in the Cairngorms, where it occurs in hollows among montane *Calluna-Cladonia* heath H13.

## Conservation interest

In common with *Trichophorum-Erica* heath, *Erica-Sphagnum compactum* heath is of international importance because of its oceanic distribution and relative scarcity outside Great Britain and Ireland. Most upland stands are species-poor and have few notable species. *Rhynchospora fusca* and *Lycopodiella inundata* occur very rarely, and other species recorded include *Rhynchospora alba*, *Pinguicula lusitanica*, *Drosera intermedia*, *Dicranum spurium* and *Hypnum imponens*. Wide tracts of *Erica-Sphagnum compactum* heath can be important hunting grounds for upland raptors and other scavenging birds.

## Management

Almost all upland examples of *Erica-Sphagnum compactum* wet heath are grazed by sheep, and also by red deer in the Highlands and on Exmoor. The vegetation is typically managed by burning, often in small patches on the grouse moors of eastern Scotland. Management recommendations are much the same as for *Trichophorum-Erica* wet heath. Much *Erica-Sphagnum compactum* wet heath has been lost under conifer plantations, and large areas have been drained and over-grazed, especially in south-west England. Heavy grazing can convert the community to impoverished acid grassland, and this process is exacerbated by drainage. In the uplands the usual end result is *Nardus-Galium* grassland U5 or *Juncus-Festuca* grassland U6. Frequent or severe burning is also damaging, producing a species-poor sward of *Calluna*, *Molinia* and *Erica tetralix*. Frequent or severe fires soon eliminate the larger bryophytes, and their place is taken by opportunistic mosses, such as *Funaria hygrometrica*, *Pohlia nutans* and the non-native *Campylopus introflexus*.

This community replaces *Trichophorum-Erica* wet heath in the drier, more continental climate of the south and east. It is widespread but rather scarce in lowland Great Britain, and extends into upland areas in eastern Scotland, the Pennines, the North York Moors, south-west England, and parts of Wales. It is particularly common in the eastern Highlands and the North York Moors.

Vegetation resembling *Erica-Sphagnum compactum* wet heath is confined to the oceanic fringes of Europe between western Norway and Spain.

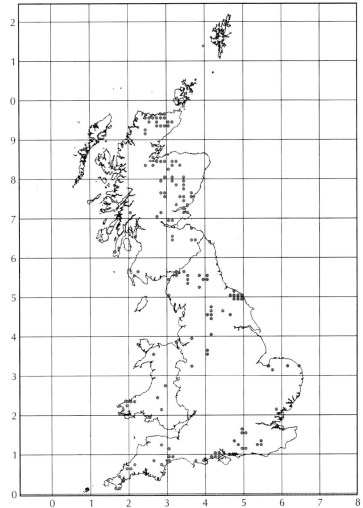

As with *Trichophorum-Erica* heath, almost all *Erica-Sphagnum compactum* heath is on ground that must once have been occupied by woodland, including pine woodland in the eastern Highlands. Indeed, *Erica-Sphagnum compactum* wet heath occurs in glades within some *Pinus-Hylocomium* woodland W18. Many stands would revert to woodland in the absence of burning and grazing, especially in the eastern Highlands where birch can colonise very rapidly.

# M17 *Trichophorum cespitosum-Eriophorum vaginatum* blanket mire

## Synonyms
*Trichophoreto-Eriophoretum* typicum (McVean and Ratcliffe 1962; Birks 1973); *Scirpus cespitosus-Myrica gale* mire G1 (Birks and Ratcliffe 1980); *Scirpus cespitosus-Eriophorum vaginatum* blanket mire (Rodwell 1991b).

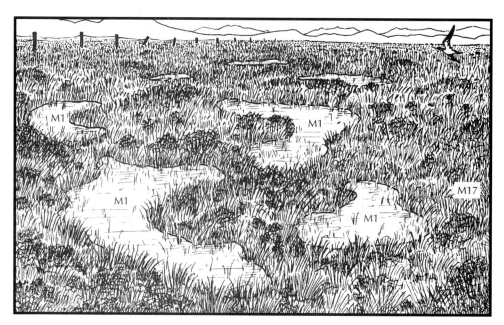

Typical view of M17 *Trichophorum-Eriophorum* blanket mire forming extensive area of bog on level, deep, waterlogged peat. The bog is dotted with small, wet depressions containing M1 *Sphagnum denticulatum* bog pool community.

M17 is also shown in the picture for M15.

## Description
The pale ochre-gold sheets of this mire are composed of *Eriophorum vaginatum*, *E. angustifolium*, *Trichophorum cespitosum* and *Molinia caerulea*, dotted with darker clumps of *Calluna vulgaris* and *Erica tetralix*. Beneath the vascular plants there are shallow spongy mats of *Sphagnum papillosum* and *S. capillifolium*. Small vascular plants prick up through the layer of mosses: the red-gold spikes of *Narthecium ossifragum*, the red, sticky rosettes of *Drosera* species, the trailing, rich-green shoots of *Potentilla erecta*, and in some places the stiff green leaves of *Dactylorhiza maculata* with its conspicuous spotted pale lilac flowers in early summer.

Small-scale variation in the species composition of this mire type is described by Lindsay (1995). The mire surface is corrugated into a system of pools and hummocks, each with characteristic assemblages of Sphagna and other plants. Some of the pools belong to the *Sphagnum denticulatum* bog pool community M1 or the *Sphagnum cuspidatum/fallax* bog pool community M2, but the vegetation of larger lochans and smaller hollows is not described in the NVC. The largest pools occur in the north and west of Scotland, where up to 50% of the mire surface may be open water. The pools can form spectacular ladder-systems on gentle slopes and reticulate patterns on shallow domes.

There are three sub-communities, corresponding to variation in the wetness of the peat. The *Drosera rotundifolia-Sphagnum* species sub-community M17a occurs in the most consistently wet conditions. *Drosera* species are especially common here, and *Sphagnum papillosum* and *S. capillifolium* cover most of the peat surface. *Myrica gale*, with its grey-green sweet-scented leaves, can grow thickly in this type of mire, and in Scottish stands there can be purple-red tufts of the oceanic liverwort *Pleurozia purpurea*. In the west Highlands and the Hebrides *Rhynchospora alba* can be common, and in a few places *Schoenus nigricans* grows on the mire surface. Where the peat is level and deep the mire surface may be broken by innumerable shining pools, forming the distinctive patterns so obvious in aerial views of the immense tracts of blanket mire in the northern and western Highlands. The *Cladonia* species sub-community M17b occurs on slightly drier peats, for example where the surface has been dried out by burning. Like the *Cladonia* sub-community of *Trichophorum-Erica* wet heath M15c, its name is deceptive, as in many places it is the moss *Racomitrium lanuginosum*, rather than *Cladonia* lichens, that defines this sub-community. *R. lanuginosum* grows in silvery-green patches and low hummocks which are often visible from afar, giving the mire a distinctive knobbed appearance. Sphagna are scarcer here, as are bog plants such as *Drosera* species. Lichens are common, and in the far north-west Highlands can grow thickly enough to make the vegetation look as if it is sprinkled with snow; the usual species are *Cladonia arbuscula*, *C. portentosa* and *C. uncialis* ssp. *biuncialis*. The *Juncus squarrosus-Rhytidiadelphus loreus* sub-community M17c is characteristic of drier peats and is mostly at higher altitudes. It has a mixed and tussocky sward, thick with the stout green rosettes of *Juncus squarrosus*, pale tufts of *Nardus stricta*, and pleurocarpous mosses such as *Rhytidiadelphus loreus*, *Pleurozium schreberi*, *Hylocomium splendens* and *Hypnum jutlandicum* as well as the characteristic Sphagna. There can also be a little *Vaccinium myrtillus* and *Deschampsia flexuosa*.

## Differentiation from other communities

This is an ombrogenous mire community. It can be distinguished from wet heaths, which are composed of a similar set of species, by *Eriophorum vaginatum* and peat-building Sphagna such as *Sphagnum papillosum*. Of the other ombrogenous mires, *Erica-Sphagnum papillosum* mire M18 most closely resembles *Trichophorum-Eriophorum* blanket mire, but on the deep, soft and strongly saturated peats that are the most characteristic habitat of *Erica-Sphagnum papillosum* mire there is generally less *Trichophorum*, *Potentilla erecta* and *Molinia*, and more *Sphagnum tenellum* and *Odontoschisma sphagni*. There is also usually more *Erica tetralix* in *Erica-Sphagnum papillosum* mire, giving a greyish tone to the vegetation, especially in stands with very little *Calluna*, and the carpet of Sphagna is typically more continuous. The wetter *Sphagnum-Andromeda* sub-community M18a is further distinguished from *Trichophorum-Eriophorum* mire by the presence of *Sphagnum magellanicum* and *Vaccinium oxycoccos* in many stands, and between mid-Wales and southern Scotland there can be some *Andromeda polifolia* too. The *Cladonia* sub-community of *Trichophorum-Eriophorum* mire can closely resemble the *Empetrum-Cladonia* sub-community of *Erica-Sphagnum papillosum* mire M18b, but has less *Empetrum nigrum*, and usually more *Molinia* and *Trichophorum*.

Despite what has just been said, the separation of *Trichophorum-Eriophorum* mire and *Erica-Sphagnum papillosum* mire can be difficult. The situation is confused by treating *V. oxycoccos* and *Andromeda* as indicators of the *Sphagnum-Andromeda* sub-community of *Erica-Sphagnum papillosum* mire. These are scarce species, and although they are commonest on the wet, soft peat surfaces of this form of mire they can also occur in quantity on less strongly waterlogged peat in the *Empetrum-Cladonia* sub-community of

*Erica-Sphagnum papillosum* mire, in *Calluna-Eriophorum* blanket bog M19, and in vegetation with much *Trichophorum* that otherwise resembles *Trichophorum-Eriophorum* mire. Perhaps these two species should be regarded as characteristic of geographical variants of various sub-communities of *Trichophorum-Eriophorum*, *Erica-Sphagnum papillosum* and *Calluna-Eriophorum* mires.

The *Erica* sub-community of *Calluna -Eriophorum* mire M19a shares some of the species of *Trichophorum-Eriophorum* mire: *Erica tetralix*, *Trichophorum*, *Hypnum jutlandicum*, and, in some stands, *Narthecium* and *Molinia*. Here, however, they are set in a dense dark, tussocky sward of *Calluna* and *Eriophorum vaginatum* lacking the sheets of *Sphagnum papillosum* that are typical of *Trichophorum-Eriophorum* mire. Drier stands of the *Cladonia* sub-community of *Trichophorum-Eriophorum* mire may have little *S. papillosum* and much *S. capillifolium*; they have more *Trichophorum* than *Calluna-Eriophorum* mire and less *Rhytidiadelphus loreus*, *Hylocomium splendens*, *Plagiothecium undulatum* and other pleurocarpous mosses.

The *Juncus-Rhytidiadelphus* sub-community of *Trichophorum-Eriophorum* mire can be confused with the *Sphagnum* sub-community of *Juncus-Festuca* grassland U6a, which shares the same habitat of shallow peats on plateaux, and which can have large amounts of *S. papillosum* and *S. capillifolium*. However, *Juncus-Festuca* grassland usually has much more *Juncus squarrosus* and *Festuca ovina*, and smaller quantities of *Calluna*, *E. tetralix* and bog plants such as *E. vaginatum*, *Drosera* species and *Narthecium*.

Where grazing and burning have suppressed the dwarf shrubs, *Trichophorum-Eriophorum* mire can resemble *Eriophorum angustifolium* mire M3, *Eriophorum vaginatum* mire M20 or *Molinia-Potentilla* mire M25. Dwarf shrubs and Sphagna are co-dominant or at least very common in *Trichophorum-Eriophorum* mire, whereas in the other vegetation types *Eriophorum* species or *Molinia* are clearly dominant, and dwarf shrubs are usually sparse or absent.

## Ecology

*Trichophorum-Eriophorum* mires are characteristic of the mild and wet climate of the western uplands. Rainfall exceeds evapotranspiration, the soils become waterlogged and anaerobic, and the dead remains of plants eventually form a thick layer of peat. This insulates the vegetation from the underlying rock and from ground water, and almost all nutrients are received from mist, rain and snow. Whole landscapes can become enveloped in peat, and *Trichophorum-Eriophorum* mire can be the prevailing type of vegetation over many square kilometres. Most stands are on level ground or gentle slopes, but in the Outer Hebrides and in north-west Sutherland the community clothes surprisingly steep slopes, perhaps because the climate is so cool and wet that deep, waterlogged peat can accumulate even on sloping ground. The peat is acid, with a pH of about 4 (McVean and Ratcliffe 1962; Rodwell 1991b). In the western Highlands, *Trichophorum-Eriophorum* mire is most common below about 450 m, but further east and south it occurs at slightly higher elevations. The upper altitudinal limit of the *Juncus-Rhytidiadelphus* sub-community is over 850 m (Rodwell 1991b). On Dartmoor in south-west England, the community occurs on the high plateau above 450 m, a habitat analogous to that of *Calluna-Eriophorum* mire in the Pennines.

## Conservation interest

Because it is so rare globally, *Trichophorum-Eriophorum* mire is one of the most important types of British upland vegetation. Lindsay *et al.* (1988) estimate that 13% of the blanket bog in the world occurs in Great Britain and Ireland; this includes *Erica-Sphagnum papillosum* and *Calluna-Eriophorum* mires, as well as *Trichophorum-*

*Trichophorum-Eriophorum* mire is widespread in upland areas of western Great Britain. It is most common and extensive in the western and northern Highlands and the Hebrides. It also occurs in the eastern Highlands, Galloway, the Lake District, Wales, south-west England and, very sparingly, in the north Pennines.

A related form of blanket mire is common in Ireland, but comparable vegetation is unknown elsewhere in Europe.

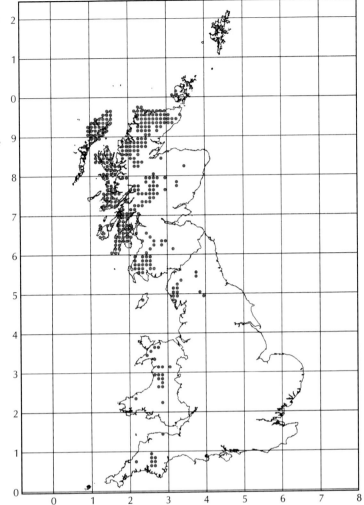

*Eriophorum* mire. This is all the more impressive given that Great Britain and Ireland comprise only 0.23% of the land area of the world. The wild, desolate, bird-haunted moorlands of the Flow Country have no equal anywhere else on earth.

These mires are not as poor in species as they often seem. The peat is covered by a living skin of vegetation, which can be home to a rich array of invertebrates. These in turn are the food of upland birds, and many *Trichophorum-Eriophorum* bogs in Scotland are the breeding grounds of internationally important populations of waders. Greenshank *Tringa nebularia*, Dunlin *Calidris alpina* and Golden Plover *Pluvialis apricaria* nest in this community, and in Shetland and the Outer Hebrides so do Whimbrels *Numenius phaeopus*. In the far north of Scotland, on the Outer Hebrides and Orkney and Shetland, Great Skuas *Catharacta skua* and Arctic Skuas *Stercorarius parasiticus* form loose nesting colonies in this form of vegetation, and Red-necked Phalaropes *Phalaropus lobatus* occasionally nest here in the Outer Hebrides, Sutherland and Caithness. Red-throated Divers *Gavia stellata* nest on the margins of lochans within blanket bog. Greylag Geese *Anser anser* also nest among boggy lochans, and other geese occasionally use these bogs as winter roosts. Golden Eagles *Aquila chrysaetos* and Short-eared Owls *Asio flammeus* hunt over the vast landscapes, and Buzzards *Buteo buteo* and Ravens *Corvus corax* scavenge for dead sheep and deer. A few rare plant species grow in

*Trichophorum-Eriophorum* mire, including the mosses *Sphagnum pulchrum*, *S. affine*, *S. austinii*, *Campylopus setifolius*, *C. atrovirens* var. *falcatus* and the Hebridean speciality *C. shawii*. This type of mire is also an important habitat for the liverworts *Calypogeia sphagnicola*, *Cephalozia connivens*, *C. macrostachya*, *Kurzia pauciflora* and *Mylia anomala*, which typically occur as scattered stems among *Sphagnum*.

## Management

*Trichophorum-Eriophorum* mires are grazed by deer, sheep, cattle and, in some places, ponies; this keeps the vegetation short and open. Where they are not grazed, the heather and *Myrica gale* grow much taller. Too much grazing can eliminate the dwarf shrubs, and the associated poaching can initiate peat erosion. According to Mackey *et al.* (1998), the area covered by blanket mire in Scotland declined by 21% between the 1940s and the 1980s. Afforestation caused 51% of this loss; the remainder was almost all caused by the conversion of blanket bog to rough grassland and heather moorland as a result of drainage and other management treatments. There were also minor losses to more improved forms of grassland.

   *Trichophorum-Eriophorum* mires are frequently burned, often in large patches, to provide a fresh flush of dwarf shrubs and grasses for grazing animals. It can take the vegetation 25 years to recover from a fire, and more frequent burning can impoverish the vegetation and dry out the surface of the peat. *Trichophorum* and *Molinia* are able to flourish under these conditions, being relatively resilient to burning, and the bog vegetation may change to wet heath, especially when the vegetation is also grazed. If burning is used as a management tool, fires should ideally sweep through the vascular plant layer leaving the bryophytes untouched, but it is difficult to manage fires in this way. For nature conservation purposes it is not necessary to burn *Trichophorum-Eriophorum* mires.

   In the western Highlands and Islands many of these mires have been cut over for peat to be used as domestic fuel. This was, and in many places still is, done on a local scale, with the peats cut and stacked by hand. This is sustainable; the mire plants are able to recolonise the newly exposed peat surfaces, especially when the turves of vegetation are laid back on the cut-over surface. Commercial cutting by machine for fuel or for horticulture is far more damaging. *Trichophorum-Eriophorum* mires are susceptible to damage by atmospheric pollution, but most stands are in regions where little airborne pollution is deposited.

# M18 *Erica tetralix-Sphagnum papillosum* raised and blanket mire

Typical view of M18 *Erica-Sphagnum papillosum* raised and blanket mire forming extensive area of bog on level, deep, waterlogged peat. The bog is dotted with small, wet depressions containing the M2 *Sphagnum cuspidatum/fallax* bog pool community.

## Description

This type of mire consists of a sparse low sward of *Erica tetralix*, *Calluna vulgaris*, *Eriophorum angustifolium*, *E. vaginatum* and in some places *Trichophorum cespitosum* standing over a varied carpet of bryophytes and lichens. As in the *Trichophorum-Eriophorum* mire M17, the mire surface can be crinkled into a series of hummocks and hollows, each with its typical Sphagna and other plants (see p. 178).

There are two sub-communities. The *Sphagnum magellanicum-Andromeda polifolia* sub-community M18a occupies saturated peats. It has an almost continuous carpet of Sphagna: the ochre *S. papillosum*, the rich red *S. capillifolium*, the pale-green *S. tenellum* and *S. cuspidatum*, and in many stands the fat-leaved, wine-red *S. magellanicum*. The Sphagna usually grow with a few common pleurocarpous mosses and with a sprinkling of *Aulacomnium palustre* and tufts of *Polytrichum strictum*. The layer of mosses is pierced by the sharp curved leaves of *Narthecium ossifragum* and speckled with the sticky red rosettes of *Drosera* species. In many stands the thin dark stems and shiny oval leaves of *Vaccinium oxycoccos* creep in a wiry tangle over the mosses; the surprisingly large pink flowers are a cheering sight in early summer. In Wales, northern England and southern Scotland, the *Sphagnum-Andromeda* sub-community is home to the scarce *Andromeda polifolia* with its upright shoots bearing narrow grey-green leaves. In the drier *Empetrum nigrum* ssp. *nigrum-Cladonia* sub-community M18b there can be much *Empetrum nigrum* ssp. *nigrum*, although this species is also common in some stands of the *Sphagnum-Andromeda* sub-community. Sphagna are scarcer in the *Empetrum-Cladonia* sub-community, and there can be a thin frosting of *Cladonia* lichens, as well as more continuous carpets of large pleurocarpous mosses such as *Rhytidiadelphus loreus* and *Pleurozium schreberi*. In some stands *Racomitrium lanuginosum* grows in silver-green patches and hummocks on the peat. Bogs belonging to this sub-community can be hagged.

## Differentiation from other communities

This community is most likely to be confused with *Trichophorum-Eriophorum* blanket mire M17, as the two communities share a similar set of species (see p. 179). Some stands of *Erica-Sphagnum papillosum* mire have a dense dark sward of *Calluna* and *Eriophorum vaginatum* in which *Erica tetralix* is less common than usual, and look more like *Calluna-Eriophorum* mire M19 from a distance. However, at close quarters, the *Erica-Sphagnum papillosum* mire has *Sphagnum papillosum* and in many places *S. magellanicum* mixed with other Sphagna under the heather, rather than just *S. capillifolium*, as is typical in *Calluna-Eriophorum* mire. *Drosera rotundifolia* and *Narthecium* are also much more characteristic of *Erica-Sphagnum papillosum* mire.

## Ecology

*Erica-Sphagnum papillosum* mire occurs on deep, wet, ombrogenous peat. It is the characteristic plant community of raised mires in the lowlands, but also occurs in the

The community is widespread but local from Wales north to Orkney in both upland and lowland Great Britain. It is most common in Sutherland, Caithness, southern Scotland, northern England and Wales.

There is similar vegetation in the lowlands of western Europe and in Ireland.

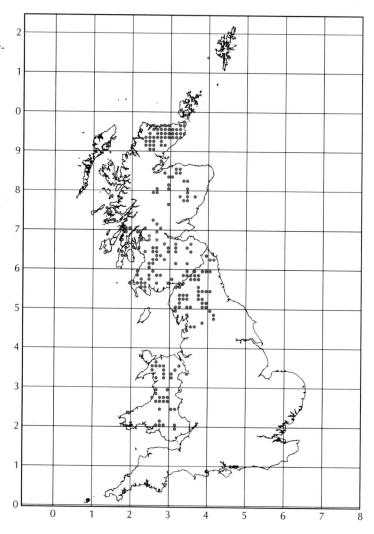

uplands in basins and on concave slopes as well as on level, poorly drained plateaux. Here it usually forms mosaics with *Calluna-Eriophorum* blanket mire and *Eriophorum vaginatum* mire M20, marking out localised areas of deep saturated peat on cols and in shallow hollows. Blanket mire forms of the community occur from low altitudes to over 500 m. The thick layer of peat insulates the vegetation from the underlying rock, so the mire can occur over any substrate from limestone to granite. The peat has a pH of around 4.0 (Rodwell 1991b) and is largely composed of *Sphagnum* remains.

## Conservation interest

Lowland raised bog is a rare and declining habitat, both in Great Britain and in the rest of Europe. Remnant examples are considered to be of great value for nature conservation, and are well-represented within SSSIs and candidate SACs. Upland stands of *Erica-Sphagnum papillosum* mire are also of considerable interest, as they are a component of the internationally important blanket bog vegetation of Great Britain.

*Erica-Sphagnum papillosum* mire is the main British habitat for *Andromeda polifolia*, *Sphagnum austinii* and *S. pulchrum*, and for a group of bog hepatics including *Calypogeia sphagnicola*, *Cephalozia macrostachya*, *C. connivens*, *C. loitlesbergeri*, *C. lunulifolia*, *Kurzia pauciflora* and *Mylia anomala*. These peatlands are rich wildlife habitats and have a characteristic fauna of invertebrates, reptiles, amphibians, mammals and birds. The peat is usually deeper under these mires than under other types, and so the deposits are extremely valuable historical records of past flora, vegetation and climate.

## Management

Lowland raised bogs composed of *Erica-Sphagnum papillosum* mire have suffered more than any other type of bog from drainage to produce agricultural land, and from exploitation for peat. Upland examples of the community, remote and surrounded by other types of mire, have rarely been drained or exploited for peat, but many have been damaged by heavy grazing or burning or both, and the original vegetation has been replaced by wet or dry heath. Fires may occur accidentally or burning may be used as a management tool. Some stands have been lost under commercial conifer plantations. There is severe and continuing erosion in some upland bogs of this type. This can be exacerbated by human interference, but whether human activity or natural processes have initiated erosion is not clear.

# M19 *Calluna vulgaris-Eriophorum vaginatum* blanket mire

## Synonyms

*Calluneto-Eriophoretum* (McVean and Ratcliffe 1962; Birks 1973); *Empetreto-Eriophoretum* (McVean and Ratcliffe 1962); *Calluna vulgaris-Eriophorum vaginatum* mire G4 (Birks and Ratcliffe 1980); *Vaccinio-Ericetum tetralicis* and *Rhytidiadelphus loreus-Sphagnum fuscum* community (Birse and Robertson 1976).

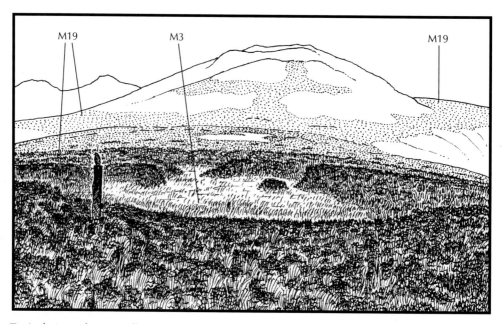

Typical view of M19 *Calluna-Eriophorum* blanket mire (foreground and stippled) on peat-covered watershed at medium altitude, extending onto gentle slopes higher up in the distance. M3 *Eriophorum angustifolium* bog-pool community occurs in wet depressions among hagged M19 in centre.

M19 is also shown in the pictures for H12, H18, H19–H22 and U4–U6.

## Description

These are mires with a dense, shaggy, purple-brown and dark-green, tussocky sward of *Calluna vulgaris* and *Eriophorum vaginatum*, speckled with the long, shining, deep-green leaves of *E. angustifolium*, straggling shoots of *Vaccinium myrtillus*, and low clumps of *Empetrum nigrum* ssp. *nigrum*. Over the ground there is a deep rich-red-gold quilt of *Sphagnum capillifolium*, *S. subnitens* and large mosses such as *Hylocomium splendens*, *Pleurozium schreberi*, *Hypnum jutlandicum*, *Rhytidiadelphus loreus* and *Plagiothecium undulatum*. In many places the vegetation is broken by hags, with great spreads of bare peat, especially in larger stands. The hummocks, hollows and pools characteristic of the wetter *Trichophorum-Eriophorum* M17 and *Erica-Sphagnum papillosum* M18 blanket mires are rare in *Calluna-Eriophorum* vegetation but do occur locally, for example in the Inner Hebrides and on Orkney.

Three sub-communities are described in the NVC, but they do not represent the most obvious patterns of floristic and ecological variation among British *Calluna-Eriophorum* mires. In broad terms, the sub-communities form a series from oceanic, southern or western vegetation to northern, boreal and montane vegetation. The *Erica tetralix* sub-community M19a, which descends to the lowest altitudes and is the most common form in the far west, has *Erica tetralix*, *Trichophorum cespitosum* and *Molinia caerulea*, in some places with a sprinkling of *Narthecium ossifragum* or *Drosera rotundifolia*; *Rubus chamaemorus* is generally absent. The moss *Hypnum jutlandicum* is listed as a characteristic species of this sub-community in the NVC table, but it is so common in *Calluna-Eriophorum* mire generally that it cannot be used as an indicator of any particular sub-type. It also seems likely that most of the records of *H. cupressiforme* in the NVC table actually refer to *H. jutlandicum*. The more northern and boreal *Empetrum nigrum* ssp. *nigrum* sub-community M19b is defined in the NVC as having more *Empetrum nigrum* ssp. *nigrum*, with fewer of the species which define the *Erica* sub-community. However, *E. nigrum* ssp. *nigrum* is also common in the *Erica* sub-community, blurring the distinction between the two types. *R. chamaemorus* can be common in the *Empetrum* sub-community. Although it is a deciduous herb, the dead leaves persist under the canopy all winter in brown drifts which are still conspicuous in spring when the new leaves start to unfurl. The *Vaccinium vitis-idaea-Hylocomium splendens* sub-community M19c takes in an assortment of more northern or montane mires, generally with *Vaccinium vitis-idaea*, *V. myrtillus*, *Empetrum nigrum* ssp. *hermaphroditum* and *Cladonia arbuscula*. *Sphagnum fuscum* can be common here, and there can also be a few montane species such as *Carex bigelowii*, *Vaccinium uliginosum*, *Polytrichum alpinum* and *Cetraria islandica*. At the highest altitudes *E. nigrum* ssp. *hermaphroditum* takes the place of *Calluna*, growing in a distinctive low, bright-green sward. The rare shrub *Betula nana* occurs locally in this sub-community at moderate to high altitudes. In some stands in the northern and eastern Highlands, *Cladonia* lichens grow thickly enough to make the ground under the vascular plants look as if it is dusted with snow (McVean and Ratcliffe 1962).

Locally in Wales there are small patches of bog with a canopy of *Calluna* and *E. vaginatum* over a ground layer that resembles *Carex echinata-Sphagnum* mire M6 in comprising *Sphagnum fallax* and *Polytrichum commune*. This vegetation is not described in the NVC. Although it has a superficial resemblance to *Calluna-Eriophorum* mire it may actually be derived from *Erica-Sphagnum papillosum* bog.

### Differentiation from other communities

*Calluna-Eriophorum* mire as a whole is a well-defined and distinct type of vegetation. Stands of *Erica-Sphagnum papillosum* mire with a thick canopy of *Calluna* and rather little *Erica tetralix* can resemble it from a distance, but are very different at close quarters (see p. 184). The *Erica* sub-community of *Calluna-Eriophorum* mire shares several species with *Trichophorum-Eriophorum* mire, but the two communities usually look very different. *Calluna-Eriophorum* mire has a dark-coloured, tussocky sward of *Calluna* and *Eriophorum vaginatum*, whereas *Trichophorum-Eriophorum* mire consists of pale, open spreads of graminoids and Sphagna. *E. tetralix*, *Trichophorum*, *Molinia* and *Sphagnum papillosum* are all more common in *Trichophorum-Eriophorum* mire than they are in *Calluna-Eriophorum* mire. The *Calluna-Cladonia* sub-community of *Eriophorum vaginatum* mire M20b has much in common with *Calluna-Eriophorum* mire, but there is usually less *Sphagnum*. *Calluna-Eriophorum* mire has more *E. vaginatum* and Sphagna than *Trichophorum-Erica* M15 and *Erica-Sphagnum compactum* M16 wet heaths. Forms of *Calluna-Eriophorum* mire with carpets of *Sphagnum capillifolium* and pleurocarpous mosses under the *Calluna* have much in common with *Calluna-Vaccinium-Sphagnum* H21 and *Vaccinium-Rubus* H22 damp heaths. However, the blanket bog occurs on deeper peat and has more *E. vaginatum*. This species occurs as sparse tufts, if at all, in the damp heath communities, and it is never co-dominant with *Calluna*. *E. vaginatum* also serves to distinguish the form of the *Vaccinium-Hylocomium* sub-community that is dominated by *Empetrum nigrum* ssp. *hermaphroditum* from *Sphagnum*-rich forms of *Vaccinium-Racomitrium* heath H20.

## Ecology

*Calluna-Eriophorum* mire covers watersheds and gentle slopes where a deep layer of peat has been able to accumulate. It occurs on drier peats than either *Trichophorum-Eriophorum* mire or *Erica-Sphagnum papillosum* mire. Although the mire surface can be ragged with hags and wet peaty channels containing *Eriophorum angustifolium*, there are rarely the pools and hollows characteristic of wetter mires, nor is there often water lying over the peat surface. The peat itself is generally firm, moist and fibrous rather than wet and slimy.

*Calluna-Eriophorum* mire is a more northern, boreal and montane type of vegetation than *Trichophorum-Eriophorum* mire. Although it occurs locally below 100 m in north-west Scotland, most stands are at higher altitudes. In the west of Great Britain, it generally replaces *Trichophorum-Eriophorum* mire above about 350 m. The more montane forms of the *Vaccinium-Hylocomium splendens* sub-community extend the altitudinal range of the community to over 900 m on the high plateaux of the Cairngorms, Lochnagar and Caenlochan.

## Conservation interest

*Calluna-Eriophorum* mires form part of Great Britain's blanket bog vegetation and are internationally important. They have elements in common with the wet tundra vegetation of high latitudes in Norway, Sweden, Finland and Russia (in contrast with the *Trichophorum-Eriophorum* and *Erica-Sphagnum papillosum* mires which have fewer such affinities).

This form of blanket mire is the home of the rare moss *Dicranum elongatum*, and is the most important habitat in Great Britain for the uncommon *Betula nana*, *Rubus chamaemorus* and *Listera cordata*. *Cornus suecica* and *Arctostaphylos alpinus* can also occur in this community. Merlin *Falco columbarius*, Hen Harrier *Circus cyaneus*, Dunlin *Calidris alpina*, Curlew *Numenius arquata* and Golden Plover *Pluvialis apricaria* nest in *Calluna-Eriophorum* mire, and it is one of the more important habitats of Red Grouse *Lagopus lagopus*. There are usually large populations of Meadow Pipits *Anthus pratensis*; these are a common host for Cuckoos *Cuculus canorus*, so these too often frequent the bogs. Various species of gull nest in *Calluna-Eriophorum* mire, mainly in Scotland; some Scottish sites have been used for so long that the places have been named after the birds.

## Management

Almost all stands of *Calluna-Eriophorum* mire are grazed by sheep or deer or both. In eastern Scotland, the Pennines and Wales the community constitutes parts of grouse moors and the vegetation is burnt at intervals. This form of blanket bog vegetation has been intensively studied around Moor House in the northern Pennines for many years, and much is known about how it responds to various forms of management (Hobbs and Gimingham 1987). In general, repeated burning on a short rotation combined with grazing tends to eliminate dwarf shrubs, Sphagna and large pleurocarpous mosses, and to change the vegetation to *Eriophorum vaginatum* mire. This is a rather drier type of vegetation, dominated by cottongrass and in some places with only a few mosses and lichens clothing the bare, crusty peat between the vascular plants. This process is exacerbated by the deposition of atmospheric sulphur and nitrogen pollutants, which can kill the Sphagna. The conversion to *Eriophorum vaginatum* mire is reversible if the bogs are fenced against grazing animals and not burned, as long as there is not too much

*Calluna-Eriophorum* mire occurs throughout the British uplands from Wales northwards. Extensive tracts of the community occur in the central and eastern Highlands and in the northern Pennines and the Cheviot. There are also large stands on the stepped basalt hills of Skye and Mull. In the rugged hills of the west Highlands and further south in the Lake District and Wales, *Calluna-Eriophorum* mire tends to occur in smaller stands. In south-west England its place is taken by *Trichophorum-Eriophorum* mire. In the southern Pennines, which have a long history of burning, grazing and air pollution, much *Calluna-Eriophorum* vegetation has been replaced by *Eriophorum vaginatum* mire M20.

There is vegetation similar to *Calluna-Eriophorum* mire in western and central Norway but the stands are small (Lindsay 1995).

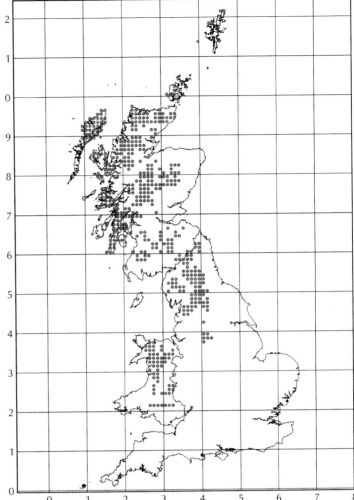

unvegetated peat. The reversion back to *Calluna-Eriophorum* mire can occur in as few as 15 years or so, but obviously the more impoverished stands are likely to take longer, especially where there are still high levels of atmospheric pollution (Rodwell 1991b). In the western Highlands, frequent burning usually converts *Calluna-Eriophorum* mire to wet heath with much *Trichophorum* rather than to *Eriophorum*-dominated vegetation.

If *Calluna-Eriophorum* mires are over-grazed the *Sphagnum* carpet is broken up by poaching and bare peat is exposed. Although erosion may be a natural process in the severe climate at higher altitudes, it is possible that some of the spectacular erosion seen in stands of this community has been initiated when burning and grazing have opened up the sward.

*Calluna-Eriophorum* vegetation is one of the drier types of bog and it is quite easy to drain it by digging ditches. Drainage, especially accompanied by regular burning, is likely to lead to the development of wet or dry heath. In the southern Pennines and the North York Moors there are virtual monocultures of *Calluna* on deep peat that have been produced in this way. At lower altitudes many stands of *Calluna-Eriophorum* mire have been lost to commercial forestry plantations.

Blanket bogs seem to be able to persist without human intervention. *Calluna* is able to rejuvenate itself by adventitious rooting into the bryophyte carpet (Forrest 1971; Hobbs

1984; Hobbs and Gimingham 1987) and burning is not necessary. Trees do not grow readily on the waterlogged peat. The most sensible form of management is light grazing that will not suppress the dwarf shrubs or damage the carpet of bryophytes.

# M20 *Eriophorum vaginatum* blanket and raised mire

## Synonyms
*Eriophorum*-dominated mire G4f (Birks and Ratcliffe 1980).

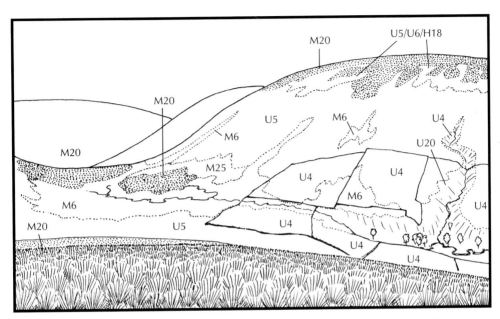

Typical habitats of M20 *Eriophorum vaginatum* mire (stippled): on deep peat on plateau (upper right); in wide depression/col (middle left); and on gentle slopes (foreground).

Associated vegetation types:

M6 *Carex echinata-Sphagnum* mire
M25 *Molinia-Potentilla* mire
H18 *Vaccinium-Deschampsia* heath
U4 *Festuca-Agrostis-Galium* grassland
U5 *Nardus-Galium* grassland
U6 *Juncus squarrosus-Festuca ovina* grassland
U20 *Pteridium-Galium* community.

## Description
*Eriophorum vaginatum* mire appears as a rather dull looking grey-green expanses of tus-socky *E. vaginatum*, dotted in summer with its white feathery fruiting heads like flakes of snow. There are few other characteristic species.

There are two sub-communities. The Species-poor sub-community M20a is made up of tussocks of *E. vaginatum* intermingled with shoots of *E. angustifolium* and in some places a little *Deschampsia flexuosa*. The peat surface is typically glazed with a thin crust of algae and can be furred over with the silvery shoots of the introduced moss *Campylopus introflexus*. The *Calluna vulgaris-Cladonia* species sub-community M20b is a little richer in species. The pale tussocky swards of *E. vaginatum* are enlivened by *Vaccinium myrtillus*, *Empetrum nigrum* ssp. *nigrum*, and in some places grazed, dis-torted shoots of *Calluna*. Locally, *Calluna* may thicken up to share dominance with *E. vaginatum*. The tussocks of *E. vaginatum* are interleaved with grasses, especially *Agrostis canina*, *Nardus stricta* and *Deschampsia flexuosa*, and there can be conspicuous patches of the lichen *Cladonia arbuscula*.

It is quite common to find bogs dominated by *E. vaginatum* that do not correspond well to either sub-community. In some bogs the tussocks of *E. vaginatum* are overgrown with great masses of *Hypnum jutlandicum* and *Pleurozium schreberi*. Another type has much *Sphagnum papillosum*, *S. capillifolium* and *Polytrichum strictum*, and in some places scattered shoots of *Vaccinium oxycoccos* or *Andromeda polifolia*; this appears to be derived from *Erica-Sphagnum papillosum* mire M18. A third type has a moss layer composed mainly of *Sphagnum fallax* and *Polytrichum commune*, as in *Carex echinata-Sphagnum* mire M6.

## Differentiation from other communities

The dominance of *Eriophorum vaginatum* means that this vegetation type can only be an ombrogenous mire. This distinguishes it from wet and dry heath communities. The species-poor sub-community can hardly be mistaken for any other type of vegetation. The *Calluna-Cladonia* sub-community may be confused with *Calluna-Eriophorum* mire M19, but often has less *Calluna* and almost no Sphagna. The other types of *Eriophorum vaginatum*-dominated bog described above are distinguished from other bog communities by the dominance of *E. vaginatum* and the scarcity of dwarf shrubs.

## Ecology

*Eriophorum vaginatum* mires cover watersheds and gentle slopes, level plateaux and peat-filled hollows: terrain where *Calluna-Eriophorum* mire would be the more natural type of mire on drier peats and *Erica-Sphagnum papillosum* mire in more saturated situations. It occurs from about 300 m up to over 900 m in Scotland. *Eriophorum vaginatum* mire is probably derived from *Calluna-Eriophorum* mire because of a prolonged regime of grazing and burning coupled with atmospheric pollution. Gradations between the two communities can be found in many areas. The community is less common on raised mires, where it is probably derived from *Erica-Sphagnum papillosum* mire.

Many stands in the southern Pennines are broken up by vast tracts of bare, hagged, peat, memorably described as a landscape looking as if it was covered in the droppings of dinosaurs (Hillaby 1970). Hagging is widespread in blanket peat in Great Britain and may be a natural phenomenon but exacerbated by many years of sheep grazing, accidental fires, atmospheric pollution and burning for management purposes. All of these are factors that operate especially intensively at the southern end of the range of blanket mire in England, where *Eriophorum vaginatum* mire is so extensive.

## Conservation interest

*Eriophorum vaginatum* mires are generally less valuable for nature conservation than the stands of less modified and impoverished blanket bog from which they have been derived. However, Red Grouse *Lagopus lagopus*, Curlew *Numenius arquata*, Golden Plover *Pluvialis apricaria* and Meadow Pipit *Anthus pratensis* can nest here, and Ravens *Corvus corax*, Buzzards *Buteo buteo*, Short-eared Owls *Asio flammeus* and Hen Harrier *Circus cyaneus* will hunt for small birds and mammals.

## Management

This vegetation is produced and maintained by injudicious management: injudicious because the end result of intensive grazing and repeated burning is an unproductive and unpalatable sward that is of little use to grazing animals. The tussocks are also difficult for sheep to negotiate. Some stands are managed as grouse moors, especially where the vegetation belongs to the more heathy *Calluna-Cladonia* sub-community and where there are more heathery moorlands close at hand. *Calluna* and *Vaccinium myrtillus* can be eliminated and replaced by *Eriophorum vaginatum* when *Calluna-Eriophorum* mires

*Eriophorum vaginatum* mire is most common in the Pennines, especially in the south on the hills between the industrial towns of Yorkshire and Lancashire. It is also common in Wales and the Lake District. In Scotland it occurs more sparingly in the Southern Uplands and in the southern and eastern Highlands.

There is similar vegetation in eastern Ireland and in the most polluted regions of south-west Norway.

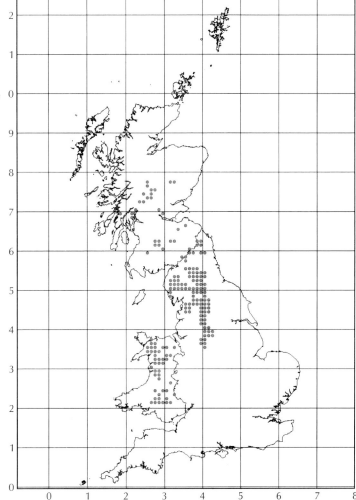

are grazed by as few as one ewe to every two hectares (Rawes and Hobbs 1979). The unpalatable *Empetrum nigrum* can also thrive in these grazed mires, although it is less common on the wettest peats. The aim of management for nature conservation should be to reverse this process of degeneration. Studies at Moor House in the northern Pennines (summarised by Rodwell 1991b) suggest that vegetation dominated by *E. vaginatum* would revert to *Calluna-Eriophorum* mire in the absence of burning and grazing in about 15 years or so, especially now that atmospheric pollution is less than it was in the early 20th century.

# M21 *Narthecium ossifragum-Sphagnum papillosum* valley mire

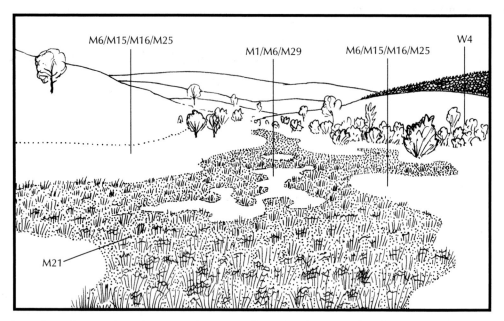

M21 Typical view of *Narthecium-Sphagnum* valley mire (foreground and stippled) on deep, waterlogged peat in shallow peaty depression.

Vegetation types found in the wettest area in the middle of M21:

M1 *Sphagnum denticulatum* bog pools
M6 *Carex echinata-Sphagnum* mire
M29 *Hypericum-Potamogeton* soakway

Vegetation types on moist ground around the edge of M21:

M6 *Carex echinata-Sphagnum* mire
M15 *Trichophorum-Erica* wet heath
M16 *Erica-Sphagnum compactum* wet heath
M25 *Molinia-Potentilla* mire.

W4 *Betula-Molinia* woodland (right of the picture) among M6/M15/M16/M25 mire/wet heath mosaics.

## Description

These are richly coloured, attractive stands of vegetation in shallow valleys adjoining sluggish streams. They consist of warm golden-yellow and green-yellow sheets of *Sphagnum papillosum*, *S. denticulatum* and *S. fallax*, interspersed with thick patches of *Narthecium ossifragum* with its orange-tipped leaves and golden flowers in summer, and overtopped by a thin green sward of *Molinia caerulea* and *Eriophorum angustifolium*. There are scattered bushes of *Erica tetralix* and *Calluna vulgaris*. The Sphagna are dotted with the glandular red rosettes of *Drosera rotundifolia* and in some places are entwined with the slender stems of *Anagallis tenella*, with its light-green pairs of neat round leaves and pale pink flowers.

There are two sub-communities. The *Rhynchospora alba-Sphagnum denticulatum* sub-community M21a has much *S. denticulatum*, and in the summer there are usually spreads of *Rhynchospora alba*, its stems topped with sharp creamy-white flowers. *Myrica gale* is common here, as are the liverworts *Odontoschisma sphagni* and *Kurzia pauciflora*. The *Vaccinium oxycoccos-Sphagnum fallax* sub-community M21b has, in addition to *Vaccinium oxycoccos* and *S. fallax*, much *Potentilla erecta*, *Carex echinata*, *C. panicea* and *Aulacomnium palustre*.

## Differentiation from other communities

*Narthecium-Sphagnum* valley mires resemble both *Trichophorum-Eriophorum* blanket mire M17 and *Erica-Sphagnum papillosum* mire M18. All three communities have extensive lawns of Sphagna under a low, open sward of vascular plants. The peat is wetter in *Narthecium-Sphagnum* vegetation than in either of the other two communities, and the dwarf shrubs are sparse and poorly grown as a result. In contrast to raised and blanket mires, *Narthecium-Sphagnum* vegetation typically has neither *Eriophorum vaginatum* nor *Trichophorum cespitosum*. Neither does it have the surface patterning that can occur in *Trichophorum-Eriophorum* and *Erica-Sphagnum papillosum* mires, although there are pools in some of the larger stands. The community has no upland or montane species such as *Vaccinium myrtillus* and *Rubus chamaemorus*, and also lacks the pleurocarpous mosses characteristic of *Calluna-Eriophorum* mire M19.

*Narthecium-Sphagnum* valley mire has a patchy distribution in Great Britain. It forms part of the vegetation of valley mires in southern England: in lowland areas, such as the New Forest, and also in upland areas, such as Dartmoor and Exmoor (Rodwell 1991b). There are scattered records in Wales and northern England, and two outlying records from the Scottish Highlands.

There is similar vegetation in Ireland, western France, and parts of Spain and Portugal. A related type of mire with more northern species occurs locally in western Norway.

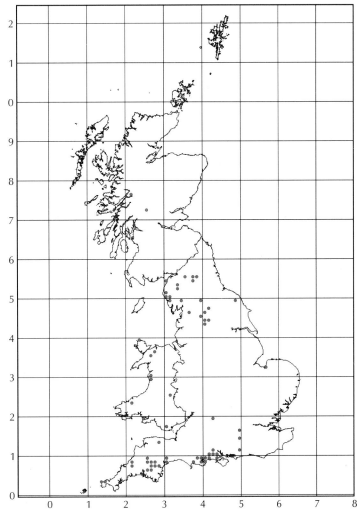

## Ecology
This community fills the bottoms of waterlogged valleys and hollows and rarely occurs above 200 m. It occurs on peat over impervious acid rocks, such as granite. The peat is usually less than 150 cm deep, and is acid with a pH between 3.5 and 4.5 (Rodwell 1991b). It is always saturated, so much so that unlike other upland bogs it is difficult or even impossible to walk across. The wetness is maintained by a high water-table and by streams flowing through the vegetation. Unlike blanket and raised mires, valley mires are sustained by water moving laterally through the peat as well as that supplied by rain.

## Conservation interest
Valley mires are very rare in Great Britain, and many examples have been severely modified by human influences. Remnant stands of *Narthecium-Sphagnum* mire are consequently of great value for nature conservation. In the south-western uplands, the community is one of the most natural types of vegetation, and is also of interest as a primarily lowland vegetation type occurring at the edge of its ecological and geographical range. Rare species recorded in *Narthecium-Sphagnum* valley mire include *Hammarbya paludosa* and, especially in Cornish stands, *Osmunda regalis*.

## Management
In its natural state the peat beneath *Narthecium-Sphagnum* valley mire is so wet and unstable that the vegetation is hardly grazed. The community is seriously damaged by drainage and burning. These treatments dry out the peat, make conditions less suitable for the peat-building *Sphagnum* species, and allow grazing animals to encroach. In combination, drainage and burning typically lead to the replacement of the original vegetation by various forms of heath and grassland, and may encourage invasion by trees and shrubs. Some stands are apparently drained because they are dangerous to livestock; fencing would be an equally effective and less damaging solution.

# M23 *Juncus effusus/acutiflorus-Galium palustre* rush-pasture

## Synonyms
*Juncus acutiflorus-Acrocladium cuspidatum* nodum *p.p.* (McVean and Ratcliffe 1962); *Juncus acutiflorus-Filipendula ulmaria* Association (Birks 1973); *Potentillo-Juncetum acutiflora* (Birse and Robertson 1976); *Juncus acutiflorus* – herb- and moss-rich mire H2c (Birks and Ratcliffe 1980).

Typical habitats of M23 *Juncus-Galium* rush-pasture:

1   Hillside flushes
2   Wet level ground in valley floor
3   In ditch
4   Along edges of stream
5   Around edges of pool

M23 is also shown in the pictures for M26, CG10, MG5 and U4.

## Description
This community comprises tall, deep-green swards of *Juncus effusus* or *J. acutiflorus* or both, entwined with grasses such as *Holcus lanatus, Molinia caerulea, Agrostis canina* and *Anthoxanthum odoratum*, the scrambling *Galium palustre*, and in most places a mixture of other mesotrophic herbs. On the wet ground under the rushes there is a thin weft of bryophytes such as *Calliergonella cuspidata, Rhizomnium punctatum, Brachythecium rutabulum* and *Chiloscyphus polyanthos*.

There are two sub-communities, one of which is usually more species-rich than the other. The *Juncus acutiflorus* sub-community M23a is the richer form. *J. acutiflorus* predominates in the sward, although there can be a fair amount of *J. effusus* as well. Just as noticeable as the rushes are the lush spreads of tall mesotrophic forbs that grow among them. Where grazing is not too heavy they are bright with flowers in summer. Among the most common species are *Filipendula ulmaria, Ranunculus acris, R. flammula, Galium palustre, Lotus pedunculatus, Geum rivale, Angelica sylvestris, Mentha aquatica, Myosotis secunda, Senecio aquaticus, Lychnis flos-cuculi, Succisa pratensis* and *Valeriana* officinalis. The *Juncus effusus* sub-community M23b usually has fewer forbs. Many stands are rather grassy or weedy, with *Cirsium palustre*, *Rumex acetosa* and lowland grassland species such as *Ranunculus repens* and *Poa trivialis*.

## Differentiation from other communities

Most stands of *Juncus-Galium* rush-pasture are easy to distinguish from the superficially similar rush-dominated sub-communities of *Carex echinata-Sphagnum* mire M6c and M6d, because they generally lack acidophilous Sphagna and *Polytrichum commune*. The most herb-rich stands of the *Juncus acutiflorus* sub-community might be confused with *Filipendula-Angelica* mire M27 or *Iris-Filipendula* mire M28. However, these mire types are dominated by *Filipendula ulmaria* and *Iris pseudacorus* respectively; although these plants grow in *Juncus acutiflorus*-dominated forms of *Juncus-Galium* rush-pasture, they are part of a diverse mixture of species and are not dominant. *J. acutiflorus*, *J. effusus* and many of the small forbs typical of *Juncus-Galium* rush-pasture occur in some forms of *Molinia-Potentilla* mire M25 and *Molinia-Crepis* mire M26, but in these communities *Molinia* is commoner than *Juncus* species and is usually dominant, whereas in *Juncus-Galium* vegetation it is less common than the rushes. The *Juncus effusus* sub-community of *Juncus-Galium* rush-pasture can be similar to the Typical sub-community of *Holcus-Juncus* rush-pasture MG10a, but fen species such as *Galium palustre*, *Lotus pedunculatus*, *Cirsium palustre* and *Ranunculus flammula*, and the moss *Calliergonella cuspidata*, are more common. However, some species-poor vegetation dominated by *Juncus effusus* cannot be assigned clearly to either community.

## Ecology

The dark-green, spiky-leaved swards of these mires are easy to pick out on gently sloping hillsides, along the margins of streams, and in marshy valleys. In the western Highlands and the Inner Hebrides, *Juncus-Galium* rush-pasture is common on level marshy ground close to the shore. Both types, but especially the *Juncus effusus* sub-community, are also common in neglected damp pastures and in ditches around fields and settlements in the upland margins. The community is sub-montane and occurs from high-tide level in the western Highlands to just over 400 m. In northern Scotland it is rare above 200 m.

Juncus-Galium rush-pasture occurs on peaty mineral soils and stagnogleys, often with a strong smell of decomposing vegetation. The soils are acid to neutral with a pH between 4 and 6. They are kept wet throughout the year by flushing and seepage, and there can be some standing water in winter. The *Juncus acutiflorus* sub-community tends to occur on wetter substrates than the *Juncus effusus* sub-community.

## Conservation interest

No particularly rare plants have been recorded in upland examples of *Juncus-Galium* rush-pasture, but some stands are home to uncommon oceanic species such as *Carum verticillatum*, *Scutellaria minor* and *Wahlenbergia hederacea*. Even without rare species, mires of this type contribute to the diversity of flora and vegetation structure around the upland fringes, and they add colour and texture to the landscape. The herb-rich stands of the *Juncus acutiflorus* sub-community are valuable centres of genetic diversity because the vascular plants are usually able to flower and set seed. Both sub-communities are a fine habitat for invertebrates and birds; Curlew *Numenius arquata*, Lapwing *Vanellus vanellus*, Snipe *Gallinago gallinago* and Redshank *Tringa totanus* often nest in rushy pastures where there is a mosaic of *Juncus-Galium* vegetation and unimproved or slightly improved grassland.

*Juncus-Galium* rush-pasture occurs throughout the west and north of Great Britain from Cornwall to Orkney. In the uplands it is especially common in south-west Scotland and the Inner Hebrides, and is scarce in the northern Highlands, the southern Pennines and the North York Moors. The richer *Juncus acutiflorus* sub-community is especially common over the basic Dalradian rocks of the western Breadalbanes, the basalt of Mull and Skye, and the limestone of Lismore (e.g. Averis and Averis 1995a, 1995b, 1996, 1999a, 1999b).

There is similar vegetation in Ireland, France, Spain and Portugal but not in Scandinavia nor apparently further east in Europe.

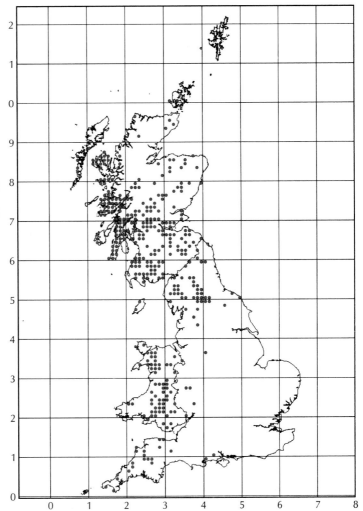

## Management

Almost all of these mires are probably derived from woodland or scrub. They occur well within the altitudinal range of woodland but the invasion of woody species is held in check by grazing. If stands were fenced against livestock they might well be colonised by trees, perhaps initially by *Salix cinerea* and *S. aurita*. The *Juncus acutiflorus* sub-community has a flora similar to that of herb-rich *Alnus-Fraxinus-Lysimachia* woodland W7, and many plants of this form of wet woodland are able to persist in the mire vegetation without a canopy of trees. The *Juncus effusus* sub-community seems to have less in common with the ground vegetation of semi-natural woodlands.

Most stands are grazed by sheep, cattle or deer; the *Juncus effusus* sub-community is usually more heavily grazed than the *Juncus acutiflorus* sub-community because it tends to occur on drier soils and to be more intimately associated with pastures. Many stands of the *Juncus acutiflorus* sub-community are grazed by cattle in autumn, when the old flowering stems of the herbs form a sort of standing hay. It is fairly easy to drain mires of this type and reclaim them for agriculture, and many stands must have been lost in this way. They can readily be converted to *Festuca-Agrostis-Galium* grassland U4 and, if this is then limed and fertilised, to *Lolium perenne-Cynosurus cristatus* grassland MG6.

# M25 *Molinia caerulea-Potentilla erecta* mire

## Synonyms

*Molinia-Myrica* nodum and *Molinia caerulea* grassland (McVean and Ratcliffe 1962); *Molinia caerulea-Myrica gale* association (Birks 1973); *Molinia caerulea* grasslands/bog C4 and *Myrica gale-Molinia caerulea* mire H1 (Birks and Ratcliffe 1980).

Typical habitats of M25 *Molinia-Potentilla* mire (foreground and stippled):

1   Moist peaty plateaux, basins and gentle slopes
2   Hillside flushes, mainly among grassland, wet heath and rush-dominated mire
3   Bog, especially that which has been repeatedly burned in the past
4   Wet peaty ground around stream in boggy valley floor

M25 is also shown in the pictures for M20, CG10 and U4.

## Description

These are the wet grasslands that can make walking in the hills of Wales, Galloway and the western Highlands and Inner Hebrides so wearisome. The tall dense tussocks of *Molinia caerulea*, with long leaves blown into waves by the wind and rain, conceal a treacherous network of peaty channels and in some places small winding streams. The grasslands look attractive in summer, with the long, lax *Molinia* leaves in shades of silvery green, but in winter the leaves collapse into dun-coloured, dank arrays of tussocks. There is usually a little *Potentilla erecta*, well-hidden beneath the *Molinia*. The habitat can be very diverse on a fine scale, with different species growing on the upper parts of *Molinia* tussocks, the sides of older tussocks and on the ground in between the tussocks.

Three sub-communities are described in the NVC: one heathy, one grassy and one herb-rich. The *Erica tetralix* sub-community M25a is the heathy form and also the most common of the three. Here *Erica tetralix* and a little *Calluna vulgaris* dot the sea of *Molinia*. There are usually a few other bog species too, such as *Eriophorum angustifolium*, *Myrica gale*, *Trichophorum cespitosum* and Sphagna. The *Anthoxanthum odoratum* sub-community M25b has a more mixed sward of grasses with *Anthoxanthum odoratum, Agrostis canina, Nardus stricta, Festuca ovina* and *Holcus lanatus* interleaved in the sward of *Molinia*. Other typical species include *Luzula multiflora, Succisa pratensis* and *Viola palustris*. The *Angelica sylvestris* sub-community M25c is moderately herb-rich and tends to cover wetter soils than the *Anthoxanthum* sub-community. Among the tussocks of *Molinia* there is a sprinkling of poor-fen forbs such as *Angelica sylvestris, Geum rivale, Valeriana officinalis, Filipendula ulmaria, Mentha aquatica, Succisa pratensis, Parnassia palustris* and *Caltha palustris*. Where grazing is not too heavy the swards can be quite a colourful sight in summer. Some flushes and damp cliff ledges in the Hebrides hold vegetation co-dominated by *Molinia* and *Schoenus nigricans*; this can be classed as a distinctive, extremely oceanic form of the *Angelica* sub-community.

There are also vast tracts of very impoverished *Molinia* grassland, especially in the Southern Uplands and Wales. They have so few species other than *Molinia* that the vegetation cannot be assigned to any of the three sub-communities.

### Differentiation from other communities

These *Molinia* grasslands are quite distinctive. Their flora and structure separate them from most other types of upland vegetation. *Molinia* can be very common in *Trichophorum-Erica* wet heath M15, *Erica-Sphagnum compactum* wet heath M16 and *Trichophorum-Eriophorum* mire M17, but in these communities it is usually mixed with *Erica tetralix, Calluna, Trichophorum cespitosum* and Sphagna, rather than dominating the sward as it does in *Molinia-Potentilla* mire. *Molinia*-rich stands of *Trichophorum-Eriophorum* mire and other forms of blanket bog have much *Sphagnum capillifolium, S. papillosum* and *Eriophorum vaginatum*, and usually have a thicker canopy of dwarf shrubs than *Molinia-Potentilla* mire.

*Molinia-Crepis* mire M26 might be confused with the herb-rich *Angelica* sub-community, but is generally a richer assemblage of species, with more *Crepis paludosa, Valeriana dioica, Sanguisorba officinalis* and *Briza media*.

The *Molinia-Potentilla* community includes two broad types of *Molinia* vegetation: the acidic and boggy *Erica* sub-community on the one hand and the herb-rich *Angelica* sub-community on the other. In many ways the *Angelica* sub-community is closer to *Molinia-Crepis* mire than to the *Erica* sub-community. The third British *Molinia*-dominated community described in the NVC is the lowland *Molinia caerulea-Cirsium dissectum* fen-meadow M24, which is distinguished primarily by *Cirsium dissectum*. However, its three sub-communities also span this same broad division between boggy and herb-rich types.

### Ecology

*Molinia-Potentilla* mire is a grassland of shallow wet peats on concave slopes, peaty mineral soils and wet gleyed muds. It can cover huge areas of ill-drained hillsides, fill the level floors of glens and valleys, or occur in narrow linear stands along the sides of streams. It is predominantly a community of the upland fringes but it can occur up to almost 600 m in the mild oceanic climate of Wales, the west Highlands and the Hebrides.

The soils are usually acid, with a pH ranging from 4.0 to 5.5, although the herb-rich *Angelica* sub-community shows signs of moderate nutrient enrichment, and is locally common on basalt and other basic rocks in the western Highlands and on Mull and Skye. The soils are well-aerated and are kept wet by moving water, although stands can be inundated in winter.

### Conservation interest

Although *Molinia* can be overwhelmingly dominant and the vegetation much impoverished, this community is not devoid of interest for nature conservation. Some stands,

## M25 *Molinia caerulea-Potentilla erecta* mire

*Molinia-Potentilla* mire occurs throughout the western uplands from Cornwall northwards, and is particularly extensive in Wales, Galloway and parts of the western Highlands. It has not been recorded in Orkney or Shetland, and is scarce in the Outer Hebrides, the North York Moors and parts of the Pennines. It is widely distributed in lowland Great Britain, especially in the west.

There are similar *Molinia*-dominated grasslands in western Ireland but Great Britain is the world headquarters for this type of vegetation.

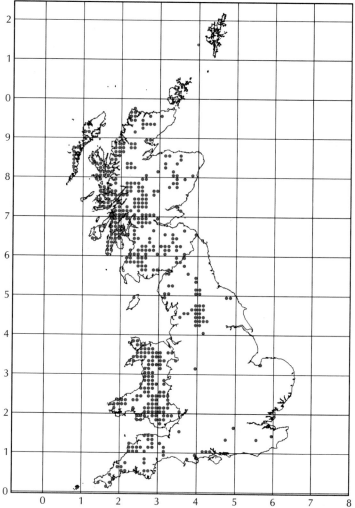

especially at lower altitudes, are moderately species-rich and contain scarce plants. For example, *Agrostis curtisii* occurs in south Wales and south-west England, *Carum verticillatum* in Wales and western Scotland, and the rare orchid *Spiranthes romanzoffiana* grows in *Molinia-Potentilla* mire near Loch Shiel in the western Highlands (Lusby and Wright 1996). These tussocky grasslands are also among the most important upland habitats for Field Voles *Microtus agrestis*, which in turn sustain populations of Short-eared Owls *Asio flammeus*, Kestrels *Falco tinnunculus* and Hen Harriers *Circus cyaneus*.

### Management

*Molinia-Potentilla* mire clothes ground where there would once have been woodland or where there is the potential for woodland to develop. Many square kilometres of *Molinia* grassland have been lost under forestry plantations. When the land is fenced and planted there can be a flush of ericaceous dwarf shrubs before the canopy of trees closes, suggesting that many stands of *Molinia-Potentilla* mire would, if ungrazed, revert first to wet heath and then to woodland or scrub. In many situations, this process is likely to culminate in *Betula-Molinia* woodland W4, the field and ground layers of which can resemble *Molinia-Potentilla* mire.

Frequent burning and grazing can convert wet heath and blanket bog to *Molinia-Potentilla* mire, especially when these treatments are combined with artificial drainage. When moorland vegetation is burnt in late winter, the living buds of the dormant *Molinia* are hidden in leaf sheaths within the fibrous depths of the tussocks, and are insulated by a thick damp blanket of dead leaves. Dwarf shrubs take a while to re-grow after fires, whereas the *Molinia* can spring up unchecked. If burning is repeated too often and if the vegetation is also grazed, *Molinia* can easily come to dominate at the expense of the dwarf shrubs.

*Molinia-Potentilla* grasslands are grazed by sheep, red deer, cattle and, more locally, ponies. Some stands on the upland margins have been drained, fertilised, and re-seeded to convert them to more productive swards, such as *Lolium perenne-Cynosurus cristatus* grassland MG6. Many stands are burned as often as every four years to clear the accumulation of litter and to encourage a flush of new shoots. This effectively prevents dwarf shrubs from becoming established. For the purposes of nature conservation it would be better not to burn these grasslands but to graze them at an intensity that would prevent the development of scrub but would allow dwarf shrubs and tall herbs to thrive.

# M26 *Molinia caerulea-Crepis paludosa* mire

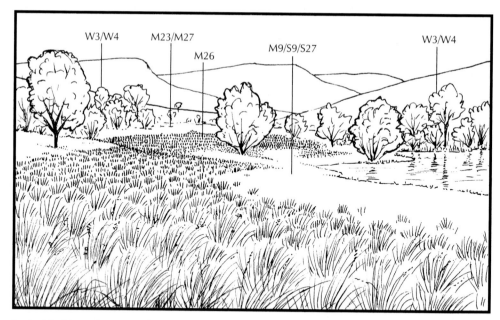

Typical habitat of M26 *Molinia-Crepis* mire (foreground and stippled), on level to gently sloping peaty ground.

Associated vegetation types:

W3 *Salix-Carex* woodland
W4 *Betula-Molinia* woodland
M9 *Carex-Calliergonella* mire
M23 *Juncus-Galium* rush-pasture
M27 *Filipendula-Angelica* tall-herb fen
S9 *Carex rostrata* swamp
S27 *Carex-Potentilla* tall-herb fen

## Description

This is species-rich, tall, tussocky vegetation. The dense, uneven sward of *Molinia caerulea* is patterned with clumps of *Carex nigra*, or, in some stands, *Juncus acutiflorus* and *J. conglomeratus*. It is scattered with forbs such as *Caltha palustris*, *Valeriana dioica*, *Succisa pratensis* and *Filipendula ulmaria*, among which there are usually some upland species including *Trollius europaeus*, *Crepis paludosa* and in some places *Primula farinosa*. There is little opportunity for a lush growth of bryophytes under the shade of the taller vascular species, but there is usually some *Calliergonella cuspidata*, its pointed shoots meshed together over the wet surface of the soil.

There are two sub-communities. The *Sanguisorba officinalis* sub-community M26a is defined by *Sanguisorba officinalis* and *Angelica sylvestris*, and also has *Galium palustre*, *Serratula tinctoria*, *Ctenidium molluscum*, *Campylium stellatum* and *Plagiochila asplenioides*. The more grassy *Festuca rubra* sub-community M26b has *Festuca rubra*, *Briza media*, *Holcus lanatus*, *Deschampsia cespitosa*, *Anthoxanthum odoratum*, *Lathyrus pratensis*, *Geum rivale* and *Juncus acutiflorus*.

## Differentiation from other communities

*Molinia-Crepis* mire is most likely to be confused with the *Angelica* sub-community of *Molinia-Potentilla* mire M25c and the predominantly lowland *Molinia caerulea-Cirsium dissectum* fen-meadow M24. All of these forms of *Molinia* grassland can have a rich flora of herbs. *Molinia-Crepis* mire can be distinguished from *Molinia-Cirsium* fen-meadow because it has *Crepis paludosa*, *Trollius europaeus* or *Sanguisorba officinalis*, and lacks *Cirsium dissectum*. In comparison with *Molinia-Potentilla* mire, *Molinia-Crepis* mire has more *Valeriana dioica*, *S. officinalis*, *C. paludosa*, *Briza media*, *Carex pulicaris* and *C. hostiana*.

## Ecology

*Molinia-Crepis* mire is confined to peats or peaty soils that are enriched with base-rich water, usually over calcareous rocks such as limestone. It occurs on flushed slopes or around open water, typically as small stands set in a matrix of other wet grasslands and mires. The peats are wet but not waterlogged except in winter. The pH of the substrate is generally between 5.5 and 7.0. This is a sub-montane type of vegetation and most stands are below 500 m.

This scarce community has been recorded only at a few sites in north Wales, the northern Pennines, the Lake District, the Southern Uplands, the High-lands and the eastern Scottish lowlands.

There is related vegetation in parts of Scandinavia and central Europe.

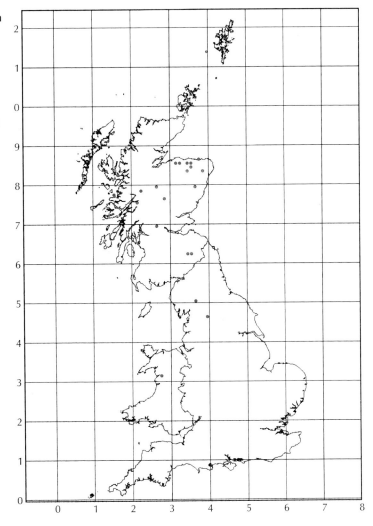

## Conservation interest

*Molinia-Crepis* mire is scarce in Great Britain. Many stands occur as components of undisturbed mosaics of uncommon vegetation types. As a wetland community of the upland margins it has suffered in the past from drainage, heavy grazing and agricultural improvement, and many stands have presumably been lost because of this (Rodwell 1991b). The community is one of the habitats of the scarce *Primula farinosa*. It is a herb-rich mire, and the herbs are generally able to flower, set seed and so maintain the genetic diversity of their populations.

## Management

Light grazing helps to keep the sward open and allows many species to flourish. When grazing is heavy, rushes tend to prevail and some of the more sensitive species die out, eventually leading to the development of *Juncus-Galium* rush-pasture. Wetter stands of *Molinia-Crepis* mire may be near-natural, but those on drier ground would probably revert to some kind of damp woodland if grazing ceased. Some stands of the *Festuca* sub-community in England are subjected to hay-meadow management: they are grazed in spring, and then ungrazed for several weeks before hay is cut in summer, after which the subsequent young plant growth (aftermath) is grazed. This management varies according to the weather and in wetter years the vegetation may not be cut at all.

# M27 *Filipendula ulmaria-Angelica sylvestris* tall-herb fen

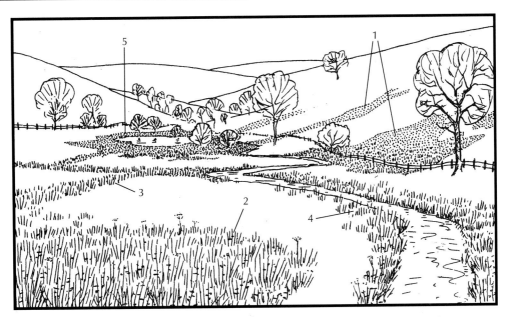

Typical habitats of M27 *Filipendula-Angelica* tall-herb fen (foreground and stippled):

1  Flushes on low-altitude slopes
2  Wet level ground in valley floor
3  In ditch
4  Along edges of stream
5  Around edges of pool

M27 fen is also shown in the picture for M26.

## Description

This is a tall, lush, herb-rich mire dominated by *Filipendula ulmaria*. In summer the dark-green sward is scattered with dense creamy patches of sweet-scented flowers – a colourful sight against the sombre greens and browns of other upland grasslands, heaths and mires.

The *Valeriana officinalis-Rumex acetosa* sub-community M27a has more tall herbs than the others; characteristic species include *Angelica sylvestris*, *Valeriana officinalis*, *Caltha palustris* and *Lychnis flos-cuculi*, and there can be some northern or upland plants such as *Crepis paludosa*, *Alchemilla glabra*, *Trollius europaeus* and *Cirsium heterophyllum*. The *Urtica dioica-Vicia cracca* sub-community M27b is a more lowland type, with weedy plants of disturbed eutrophic soils including *Urtica dioica*, *Galium aparine* and *Cirsium arvense*. It could be regarded as transitional between tall-herb fen and coarse grassland, *Phragmites* fen and certain weedy vegetation types. The *Juncus effusus-Holcus lanatus* sub-community M27c resembles the *Juncus effusus* sub-community of *Juncus-Galium* rush-pasture M23b, and is characterised by *Juncus effusus*, *Holcus lanatus*, *Mentha aquatica* and *Lotus uliginosus*.

## Differentiation from other communities

*Filipendula-Angelica* fen can be confused only with the other tall-herb mires and grasslands in which *Filipendula ulmaria* plays a part: *Juncus-Galium* rush-pasture M23, *Iris-*

207

*Filipendula* mire M28 and *Filipendula-Arrhenatherum* tall-herb grassland MG2. In comparison with the first two communities, *Filipendula-Angelica* fen is dominated by *F. ulmaria*, rather than rushes or *Iris pseudacorus*. *Filipendula-Arrhenatherum* grassland has less *F. ulmaria* than *Filipendula-Angelica* fen, and has more grasses and woodland forbs such as *Mercurialis perennis* and *Heracleum sphondylium*.

## Ecology

This is a mire of damp mesotrophic soils at low to moderate altitudes, extending up to about 400 m. It generally occurs in mosaics with other tall-herb mires and swamps, in glens and wet hollows, alongside slow-moving streams, at the edges of lochs, and on flushed slopes close to sea-level. These are all places where the water-table fluctuates widely over the year.

## Conservation interest

Although *Filipendula-Angelica* fen is not rare in Great Britain, stands are valuable for their rich flora of tall herbs. There are no particularly rare species in most stands, but the herbs are generally able to flower and set seed, so maintaining their genetic diversity.

*Filipendula-Angelica* fen is widespread throughout Great Britain, especially in the lowlands. In the uplands it occurs in south-west England, Wales, the Lake District and Scotland. There are regional differences in the distributions of the sub-communities: the *Valeriana-Rumex* sub-community is the most common form in the north, the *Juncus-Holcus* sub-community is most common in the south and west, and the *Urtica-Vicia* sub-community is primarily a lowland type.

There are mires of this type in Ireland, and vegetation essentially similar to the herb-rich *Valeriana-Rumex* sub-community occurs in Germany, Holland, Belgium and France.

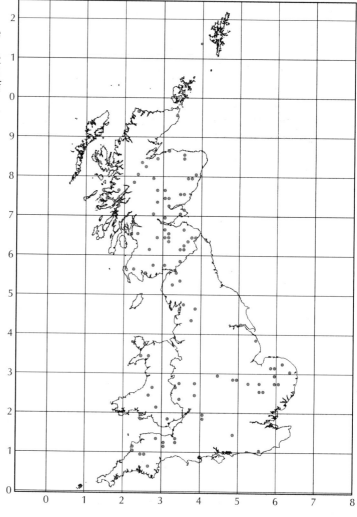

## Management

Most examples of *Filipendula-Angelica* fen are lightly grazed or ungrazed. The tall herbs are palatable, and the ground is not wet enough at all times of the year to prevent access by grazing animals. Under heavy grazing *F. ulmaria* is easily lost and other species take over, leading to the development of wet pasture dominated by rushes or *Molinia caerulea*. Many stands survive either because they are fenced or because they are isolated by water or by wetter types of mire and swamp. Even in the complete absence of grazing, it seems that succession to woodland may be prevented by the dense canopy of *F. ulmaria*. The characteristic tall herbs die out when the community is drained.

# M28 *Iris pseudacorus-Filipendula ulmaria* mire

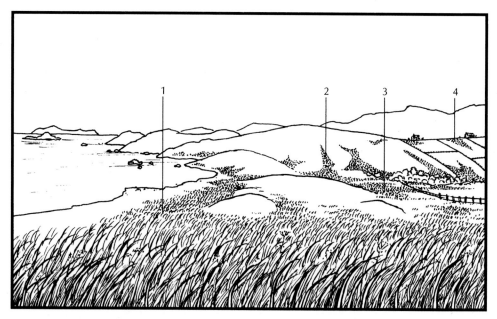

M28 Typical habitats of *Iris-Filipendula* mire (foreground and stippled) :

1   Along shoreline, just above upper limit of saltmarsh
2   In flushes among grassland, bracken and heath on gentle slopes at low altitude
3   Among rush-dominated mire and woodland in low-lying damp depressions and valley
    floors
4   Among grassland and rush-dominated mire in enclosed farmland at low altitude

## Description

This mire community comprises tall green swards of *Iris pseudacorus*, scattered with large untidy yellow flowers in summer and interleaved with other tall herbs such as *Filipendula ulmaria* and *Oenanthe crocata*. Other associated species include *Ranunculus acris*, *Cirsium palustre* and *Rumex acetosa*, and the grasses *Deschampsia cespitosa*, *Poa trivialis*, *Agrostis stolonifera* and *Holcus lanatus*. Under the vascular plants there is a thin layer of *Calliergonella cuspidata*, *Eurhynchium praelongum* and *Brachythecium rutabulum*. Stands near the coast are often entangled with seaweed, nylon rope, plastic bottles, driftwood and other debris washed up by high tides.

There are three sub-communities. The *Juncus* species sub-community M28a is moderately species-rich, and includes *Juncus effusus*, *J. acutiflorus*, *Ranunculus acris* and *Caltha palustris*. The *Urtica dioica-Galium aparine* sub-community M28b has more weedy species, such as *Urtica dioica*, *Galium aparine* and *Cirsium arvense*. The *Atriplex prostrata-Samolus valerandi* sub-community M28c is distinguished by maritime species including *Atriplex prostrata*, *Samolus valerandi* and, in some places, *Triglochin maritimum* and *Glaux maritima*. Grazed stands of *Iris-Filipendula* mire can be very grassy with few forbs, and can be impossible to assign to a sub-community.

## Differentiation from other communities

This is the only type of vegetation dominated by *Iris pseudacorus*. *I. pseudacorus* can grow in *Juncus-Galium* rush-pasture M23, *Holcus-Juncus* grassland MG10 and

*Filipendula-Angelica* tall-herb fen M27, but these communities are dominated by *Juncus* species, *Holcus lanatus* or *Filipendula ulmaria*.

## Ecology

*Iris-Filipendula* mire is floristically and ecologically related to *Juncus-Galium* rush-pasture and *Filipendula-Angelica* tall-herb fen. The dominant species vary between the different communities, but they all share the same associated flora of tall mesotrophic herbs. The three communities can form complex and interpenetrating mosaics, but the tall sword-like leaves of *Iris pseudacorus* stand out in well-defined patches clothing damp, more or less neutral soils at very low altitudes. The distribution and relative proportions of the different rush-pasture and fen types are determined by local variation in soil conditions, grazing and probably also cultivation; some stands of *Iris-Filipendula* mire are within or very close to old lazy-beds. *I. pseudacorus* can flourish only where the climate is mild and oceanic, and most stands of *Iris-Filipendula* mire are close to the sea, extending only a few kilometres inland and rarely occurring above 150 m.

*Iris-Filipendula* mire is common in coastal areas of the western Highlands and the Hebrides. It occurs much less commonly along the coasts of Shetland, Orkney, northern Scotland and Ayrshire. There are also outlying records in south-west England (not shown on the distribution map) and south-west Wales.

Fragmentary stands of similar vegetation occur locally in south-west Scandinavia.

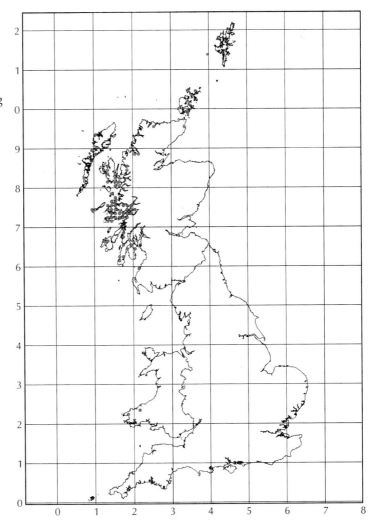

## Conservation interest

This attractive community is a rare, near-natural form of vegetation which forms an important element in the varied mosaics of grasslands and mires along the west Highland and Hebridean coasts. In the Hebrides, *Iris-Filipendula* mire is an important habitat for breeding Corncrakes *Crex crex*.

## Management

Most stands of *Iris-Filipendula* mire are subject to light grazing at most. If the vegetation is grazed hard, the *Iris* does not grow so tall and may be reduced to scattered, shorn-off stumps. The characteristic tall herbs are grazed out and the community is eventually converted to an open, grassy and impoverished sward. Drainage can also convert *Iris-Filipendula* mire into grassland. As with the *Filipendula-Angelica* fen, invasion by trees is possible but appears to take place very slowly, if at all.

Typical habitats of M29 *Hypericum-Potamogeton* soakway (foreground and stippled):

1   Soakways in the wettest parts of wet depressions, among various types of mire
2   Springs, seepages and soakways among grassland, mire and wet heath on gentle slopes

M29 is also shown in the picture for M21.

## Description

The hairy, silvery-grey shoots of *Hypericum elodes* and the shiny green leaves of *Potamogeton polygonifolius* grow here in loose drifts in shallow running water. *H. elodes* has clusters of bright yellow flowers which enliven the sward in summer. Surrounding the vascular plants is an open weft of mosses consisting mainly of *Sphagnum denticulatum*, *S. fallax* and *Calliergonella cuspidata*. Rising above this low mat of mosses are shoots of *Molinia caerulea*, *Carex viridula* ssp. *oedocarpa*, *C. panicea*, *Eriophorum angustifolium* and *Ranunculus flammula*. At their feet, often hard to see against the mosses, there may be delicate trailing shoots of *Anagallis tenella* and the dainty pale-green leaves of *Wahlenbergia hederacea*.

Some soakways lack *H. elodes* but otherwise have a flora similar to the *Hypericum-Potamogeton* community; most of these stands are beyond the geographical range of *H. elodes* but could perhaps be considered as forms of the community.

## Differentiation from other communities

This distinctive community, with its characteristic dominant species, is hard to confuse with other upland vegetation types. Species-poor forms without *Hypericum elodes* may be confused with *Sphagnum denticulatum* bog pools M1, which have a similar array of bryophytes and can have much *Potamogeton polygonifolius*, *Menyanthes trifoliata* and *Juncus bulbosus*. *Hypericum-Potamogeton* soakways have more *Carex viridula*, *C. panicea*, *Ranunculus flammula* and *Calliergonella cuspidata*, whereas bog pools have a more continuous layer of *Sphagnum* and more *Narthecium ossifragum* and *Drosera* species.

213

## Ecology

*Hypericum-Potamogeton* soakways occur where water emerges from the ground and flows more or less permanently over the surface in runnels and seepages. Some stands are associated with springs or stream-edges and form isolated patches within a matrix of drier grasslands and heaths. Others occur within mosaics of different mire types or mark out tracks of stronger seepage within valley bogs. The community also occurs in peaty pools with fluctuating water levels within blanket bogs. A few stands are in situations that can become rather dry in summer. The soils are peats or peaty gleys, and the water is acid, with a pH between 4.0 and 5.5 (Rodwell 1991b). The community occurs at low to moderate altitudes.

## Conservation interest

This community can occur in diverse mosaics of mires, grasslands and heaths around the lower fringes of the uplands. In northern areas, such as Mull, *Hypericum-Potamogeton* soakways are an important habitat for southern plants, including *Hypericum elodes, Anagallis tenella* and *Eleogiton fluitans*. Some stands are home to the rare *Pilularia globulifera*.

*Hypericum-Potamogeton* soakways are distributed throughout the oceanic west of Great Britain, from south-west England as far north as the Inner Hebrides and mid-west Highlands. They are most common in Wales and south-west England. The community is also recorded in Dorset and Hampshire.

Similar vegetation occurs in Ireland and elsewhere in western Europe.

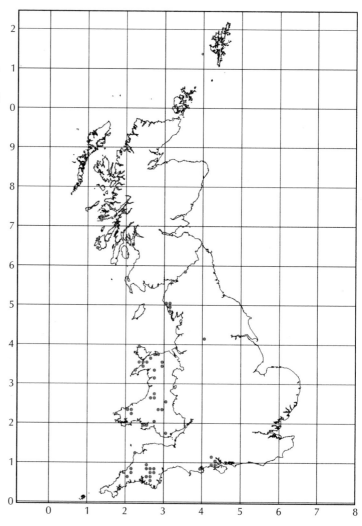

**Management**
Most stands of *Hypericum-Potamogeton* vegetation are open to grazing animals and are often trampled. In the absence of grazing, the community may include woody species such as *Salix cinerea* (Rodwell 1991b), but it is not known whether stands of this mire would ever develop into woodland or scrub. Along with most mire and flush communities that occur at lower altitudes, *Hypericum-Potamogeton* soakways are damaged by drainage, afforestation and agricultural improvement. They may also be adversely affected by acidic deposition, especially in north Wales and the Southern Uplands of Scotland.

# M31 *Anthelia julacea-Sphagnum denticulatum* spring

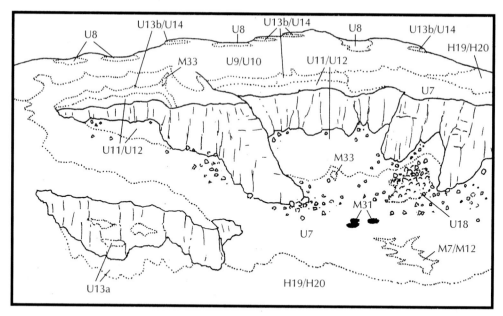

Typical habitat of M31 *Anthelia-Sphagnum* spring (dark fill) with montane grassland and mires on gentle slopes at high altitude.

Other vegetation types shown in this picture:

U7 *Nardus-Carex* grass-heath
U8 *Carex-Polytrichum* heath
U9 *Juncus-Racomitrium* rush-heath
U10 *Carex-Racomitrium* moss-heath
U11 *Polytrichum-Kiaeria* snow-bed
U12 *Salix-Racomitrium* snow-bed
U13 *Deschampsia-Galium* grassland
U14 *Alchemilla-Sibbaldia* dwarf-herb community
U18 *Cryptogramma-Athyrium* snow-bed
M7 *Carex-Sphagnum russowii* mire
M12 *Carex saxatilis* mire
M33 *Pohlia wahlenbergii* spring
H19 *Vaccinium-Cladonia* heath
H20 *Vaccinium-Racomitrium* heath

## Synonyms

*Anthelia-Deschampsia caespitosa* provisional nodum (McVean and Ratcliffe 1962); *Anthelia julacea* banks (Birks 1973); *Anthelia julacea* springs I4b (Birks and Ratcliffe 1980); *Anthelia julacea-Sphagnum auriculatum* spring (Rodwell 1991b).

## Description

These bryophyte-dominated springs form patches that stand out distinctly from the surrounding vegetation. The dark silvery-green tight mats and cushions of *Anthelia julacea* are interleaved with the twisted red-gold shoots of *Sphagnum denticulatum*, the diminutive dark-red stems of *Marsupella emarginata*, and the dark red-purple tufts of *Scapania undulata*. Rooted in this dense carpet of bryophytes are small vascular plants: *Viola palustris, Nardus stricta, Narthecium ossifragum, Saxifraga stellaris* and *Deschampsia cespitosa*.

## Differentiation from other communities
This community has more *Anthelia julacea* than any other bryophyte spring. The dense mats of *A. julacea* can resemble the liverwort crusts of *Salix-Racomitrium* snow-beds U12, where the related *A. juratzkana* is common, but the snow-beds are generally drier, and have a more mixed mat of liverworts studded with *Salix herbacea* and montane mosses.

## Ecology
*Anthelia-Sphagnum* springs are virtually restricted to the montane zone above 600 m and are commonly associated with *Nardus-Carex* snow-beds U7. They can occur on slopes of any aspect. The habitat varies from level ground around large snow-beds in the corries of the north-west Highlands and on the Cairngorm plateaux, to steep banks and rocky slopes below cliffs.

The waters that feed the springs often run from melting snow and are very cold. The water flows more slowly than in the *Philonotis-Saxifraga* M32 or *Pohlia wahlenbergii*

The community is common in the Scottish Highlands, and also occurs on some of the higher hills in south-west Scotland, the Lake District and north Wales.

There is similar vegetation in Norway and in the Faroes (Hobbs and Averis 1991a).

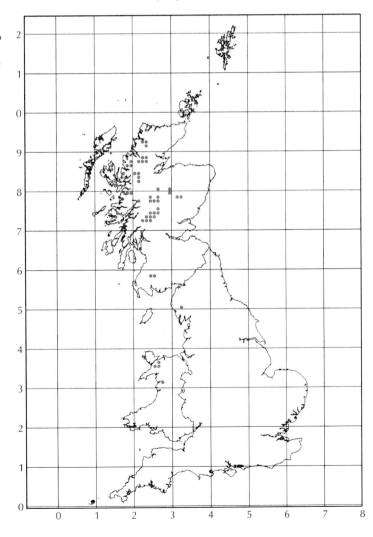

M33 springs (McVean and Ratcliffe 1962). Soils are fragmentary accumulations of debris trapped under the vegetation. The community occurs on a wide range of rock types, and the irrigating water is acid.

### Conservation interest

This is one of several near-natural, montane spring and flush communities that are scarce in the British uplands. In the montane zone *Anthelia-Sphagnum* springs are an important source of water and invertebrate prey for breeding birds. They can include rare plant species such as *Deschampsia cespitosa* ssp. *alpina*, the moss *Pohlia ludwigii*, and the liverworts *Anthelia juratzkana*, *Lophozia wenzelii* and *Marsupella sphacelata*.

### Management

This is a near-natural type of vegetation and is rarely affected by land management, although some stands may have been damaged by skiing developments. The outlying southern and western stands might be especially vulnerable to the effects of a warmer climate.

# M32 *Philonotis fontana-Saxifraga stellaris* spring

## Synonyms
*Philonoto-Saxifragetum stellaris* (McVean and Ratcliffe 1962; Birks 1973); *Philonotis fontana-Saxifraga stellaris* springs I4a (Birks and Ratcliffe 1980).

Typical habitats of M32 *Philonotis-Saxifraga* spring (stippled):

1 Among heaths and grasslands on hill slopes
2 Among blanket bog
3 Among snow-bed vegetation on high slopes

## Description
These small springs spangle the hillsides with deep rich colours. Bryophytes predominate, growing in deep spongy mats and patches. Any one of a number of species can prevail here. The most common are the apple-green *Philonotis fontana*, the almost impossibly bright yellow-green *Dicranella palustris*, the red-gold *Sphagnum denticulatum* and the red-purple *Scapania undulata*, but other possible dominants include the reddish *Calliergon sarmentosum*, the mid-green *Warnstorfia fluitans*, and the green, red or purple *Nardia compressa* and *Scapania uliginosa*. The bryophyte carpets are studded with small vascular species, especially sprawling plants of *Montia fontana* and *Chrysosplenium oppositifolium*, rosettes of *Saxifraga stellaris* with its loose clusters of white flowers, *Epilobium palustre* and in some places the scarcer *E. anagallidifolium* or *E. alsinifolium*. Isolated stands rarely cover more than a few square metres, and are sharply demarcated from the drier ground around them, but on high slopes and in corries they can coalesce to form a brightly coloured network along the sides of springs and shallow rills.

There are two sub-communities. *Sphagnum denticulatum* is common in the *Sphagnum denticulatum* sub-community M32a, which is rather species-poor. In contrast, the *Montia fontana-Chrysosplenium oppositifolium* sub-community M32b takes in more varied, species-rich stands. Around the old copper mines in Snowdonia there are

*Philonotis-Saxifraga* springs with a distinctive flora of species that can tolerate heavy metals, including *Armeria maritima*, *Cerastium arcticum* and *Minuartia verna*.

### Differentiation from other communities
This vegetation can be confused with a few other types of spring or rill. The more montane *Pohlia wahlenbergii* spring M33 is dominated by the bright-green moss *Pohlia wahlenbergii* var. *glacialis*. The community's southern counterpart is the *Ranunculus-Montia* spring M35; this has *Ranunculus omiophyllus* and no montane species. *Palustriella commutata* or *Cratoneuron filicinum* dominate *Palustriella-Festuca* M37 and *Palustriella-Carex* M38 springs, but occur sparsely, if at all, in *Philonotis-Saxifraga* springs.

### Ecology
This community is confined to the immediate vicinity of springs and to the margins of small streams. It can be associated with almost any type of upland vegetation. In many places it forms mosaics with other springs, such as the *Anthelia-Sphagnum* M31 and *Pohlia wahlenbergii* types, especially on flat ground around late snow-beds. Downstream, these spring communities are generally replaced by *Carex*-dominated mires.

*Philonotis-Saxifraga* springs are associated with a cool, wet climate. The distribution of the community shows a noticeable northern bias, although it is widespread in the uplands from mid-Wales northwards. Further south, springs tend to be dominated by more thermophilous species, and the *Philonotis-Saxifraga* community is replaced by the *Ranunculus-Montia* spring. In general, the montane element in the flora is most strongly developed at higher altitudes and northerly latitudes.

Almost identical vegetation has been recorded in Ireland, Scandinavia, Greenland, Iceland, the Faroes and central Europe.

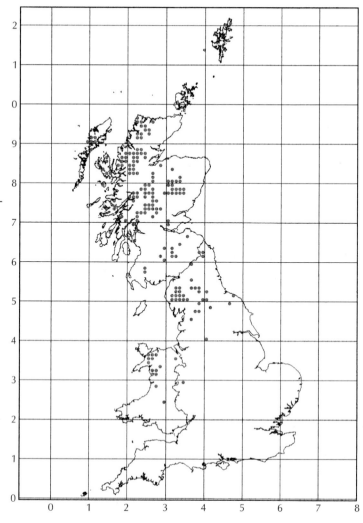

There are *Philonotis-Saxifraga* springs at all altitudes but they are scarce below 400 m (McVean and Ratcliffe 1962). They occur on a wide range of rock types, although on limestone *Philonotis fontana* is generally replaced by *P. calcarea*. The pH of the irrigating water can vary from less than 5.0 to more than 7.5 (A M Averis unpublished): this probably accounts for the great diversity of the flora. Soils are usually fragmentary – silt and humus trapped among stones – but the community can occur on flushed peats or gleys.

### Conservation interest
Many rare plants grow in this community, including *Saxifraga rivularis*, *Alopecurus borealis*, *Cerastium cerastoides*, *Epilobium alsinifolium*, *E. anagallidifolium*, *Myosotis stolonifera*, *Phleum alpinum* and *Sedum villosum*, the mosses *Bryum schleicheri* var. *latifolium*, *B. weigelii*, *Cinclidium stygium*, *Oncophorus virens*, *O. wahlenbergii*, *Philonotis seriata*, *Pohlia ludwigii* and *Splachnum vasculosum*, and the liverworts *Harpanthus flotovianus*, *Scapania paludosa*, *S. uliginosa* and *Tritomaria polita*. *Philonotis-Saxifraga* springs are important breeding grounds for insects, and can be among the few permanently wet habitats with soft soils in large expanses of moorland. Consequently they are valuable feeding areas for birds, including Ptarmigan *Lagopus mutus*, Red Grouse *Lagopus lagopus*, Dunlin *Calidris alpina* and Dotterel *Charadrius morinellus*.

### Management
*Philonotis-Saxifraga* springs are usually too wet to burn but they are grazed by sheep and deer, and wallowed in by deer. Springs that are fenced off from grazing have taller vegetation, and can become overgrown with vascular species such as *Agrostis canina* and *Rumex acetosa*; in such situations the carpet of bryophytes is neither so continuous nor as diverse as in grazed examples of the community. It is possible that grazing helps to prevent the establishment of woody species, especially at lower altitudes.

# M33 *Pohlia wahlenbergii* var. *glacialis* spring

## Synonyms

*Pohlietum glacialis* (McVean and Ratcliffe 1962); *Pohlia wahlenbergii* var. *glacialis* springs I4c (Birks and Ratcliffe 1980).

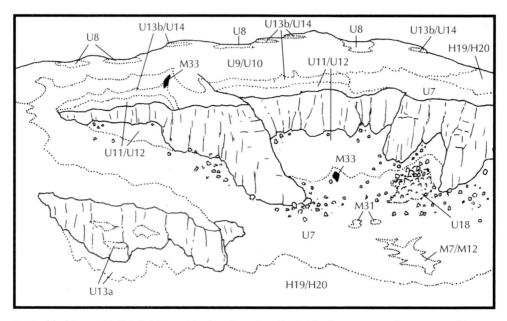

Typical habitats of M33 *Pohlia wahlenbergii* spring (dark filled), among various types of late snow-bed vegetation in corries and on other slopes at high altitude.

Other vegetation types shown in this picture:

U7 *Nardus-Carex* grass-heath
U8 *Carex-Polytrichum* heath
U9 *Juncus-Racomitrium* rush-heath
U10 *Carex-Racomitrium* moss-heath
U11 *Polytrichum-Kiaeria* snow-bed
U12 *Salix-Racomitrium* snow-bed
U13 *Deschampsia-Galium* grassland
U14 *Alchemilla-Sibbaldia* dwarf-herb community
U18 *Cryptogramma-Athyrium* snow-bed
M7 *Carex-Sphagnum russowii* mire
M12 *Carex saxatilis* mire
M31 *Anthelia-Sphagnum* spring
H19 *Vaccinium-Cladonia* heath
H20 *Vaccinium-Racomitrium* heath

## Description

These are attractive springs in which the pale-green plants of *Pohlia wahlenbergii* var. *glacialis*, covered with shining drops of water, stand out clearly from the surrounding vegetation in patches of almost luminous green. The colour of the stands is so distinctive that they are noticeable from a great distance. The springs are composed of the dense, glaucous shoots of *P. wahlenbergii* var. *glacialis*, in some places with a little *Philonotis fontana* or *Scapania undulata*. The deep-green *Pohlia ludwigii* may also be common, especially on slightly drier ground around the edges of springs. The bryophyte mats are speckled with a few small montane vascular plants such as *Saxifraga stellaris*, *Deschampsia cespitosa* ssp. *alpina* and *Cerastium cerastoides*. There is not much variation in the flora or the habitat of this community throughout its British range.

## Differentiation from other communities

This community is easily distinguished from other types of springhead vegetation because it is dominated by *Pohlia wahlenbergii* var. *glacialis*.

## Ecology

In Great Britain, *Pohlia wahlenbergii* springs are associated with late snow-beds at high altitudes in the Scottish Highlands; most stands are at altitudes above 900 m. They are generally on the steep upper slopes of corries or in gullies, but can also occur on flatter ground in mosaics with other springs, such as *Philonotis-Saxifraga* springs M32. Where the community is associated with very late snow-beds, it can form the only continuous vegetation, being surrounded by silty unstable banks with a sparse carpet of *Pohlia ludwigii*. The substrate is usually base-poor and the soil is a thin, poorly-structured accumulation of silt and humus. The water which feeds these springs is very cold – always below 4 °C (McVean and Ratcliffe 1962).

The main centre of distribution is the central and eastern Highlands – the areas with the most late-lying snow-beds. The community is also represented, though less commonly, throughout the high mountain ranges as far north as Ben More Assynt in Sutherland and as far west as Ben Lui and Ben Nevis. It appears to be absent from the Hebrides. *Pohlia glacialis* var. *wahlenbergii* grows very locally on wet gravelly soils at high altitudes in the hills of the Lake District and north Wales, but the community has not been recorded there.

There is spring vegetation similar to this community in Norway and the Faroes, where it is usually but not exclusively associated with late-lying snow. In the milder climate of Great Britain, snow-beds are evidently the only places where the microclimate is cold enough for the community to develop. *Pohlia*-dominated springs form one of the important links between British vegetation and that of the colder northern mountain regions in other parts of Europe.

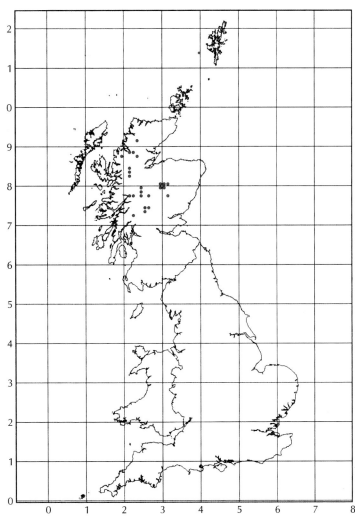

## Conservation interest

*Pohlia wahlenbergii* springs are rare in Great Britain, and typically occur in close associ-ation with snow-beds and other montane plant communities. These intricate mosaics of high-altitude, climax vegetation are of considerable value for nature conservation. Sev-eral rare species can grow in this community, including *Alopecurus borealis*, *Cerastium cerastoides*, *Deschampsia cespitosa* ssp. *alpina*, *Epilobium alsinifolium*, *Phleum alpinum*, *Saxifraga rivularis* and *Pohlia ludwigii*.

## Management

This is evidently a near-natural plant community, with a distribution owing more to the natural environment than to management. A changing climate might affect it, and some springs could perhaps be damaged by skiing developments. Some *Pohlia wahlenbergii* springs are wallowed in by deer during the summer.

**Plate 1.** The northern Highlands, Scotland

**A.** The Torridonian sandstone peaks of west Sutherland set on the ancient base of Lewisian gneiss. The foreground shows the characteristic 'knob and lochan' landform below Ben More Coigach. Distant hills include Stac Pollaidh and Suilven. (*Photo*: Pat MacDonald/SNH).

**B.** The tor-crowned ridge of Ben Loyal beyond the patterned *Trichophorum-Eriophorum* and *Erica-Sphagnum* blanket bogs of the Flow Country. (*Photo*: Steve Moore/SNH).

**Plate 2.** The western Highlands and Islands, Scotland

**A.** The characteristic steep, ice-carved slopes and narrow ridges of the West Highlands. An Teallach, Wester Ross. (*Photo*: Stephen Whitehorne/SNH).

**B.** *Trichophorum-Erica tetralix* wet heaths and *Molinia-Potentilla* mire on rocky slopes, and *Trichophorum-Eriophorum* blanket mire in peat-filled hollows. Askival and Trallival, Isle of Rum, Inner Hebrides. (Photo: Lorne Gill/SNH).

**Plate 3.** Skye and Arran, Scotland

**A.** A classic view of cliffs and landslips as part of the basalt scenery on Skye. The Storr from Loch Fada. (*Photo*: John Birks).

**B.** Granite hills eroded down to narrow ridges and corries. *Trichophorum-Erica tetralix* wet heath on the lower slopes gives way to *Calluna-Erica cinerea* heath on the steeper ground. Caisteal Abhail from Goat Fell, Arran. (*Photo*: Lorne Gill/SNH).

## Plate 4. Eastern Scotland

**A.** Woodland of native Scots pine *Pinus sylvestris* at Rothiemurchus, Cairngorms. Smooth rounded hills with extensive plateaux are cut through by the pass of the Lairig Ghru. (*Photo*: Lorne Gill/SNH).

**B.** *Calluna-Eriophorum* blanket mire and *Calluna-Vaccinium* heath showing the typical quiltwork pattern of muirburn to establish nesting and feeding habitat for red grouse *Lagopus lagopus scoticus*. Ravengill Dod, Southern Uplands. (*Photo*: Pat MacDonald/SNH).

**Plate 5.** Southern Scotland and Northumberland

**A.** *Calluna-Vaccinium* heath on the steep slopes has been burnt in patches for red grouse *Lagopus lagopus scoticus*. The sheep-grazed *Festuca-Agrostis-Galium* grassland and a stone-built sheep-fank dominate the foreground. Moorfoot Hills. (*Photo*: Glyn Satterley/SNH).

**B.** Smooth, rolling, grassy and heathy hills rising above a wide and level glen with patches of bracken *Pteridium aquilinum*. Many of the accessible and gentle slopes have been afforested. The Cheviot Hills. (*Photo*: John Birks).

**Plate 6.** The Lake District and the Pennines

**A.** The steep, glaciated slopes of the Lake District are clothed largely with *Nardus-Galium* grassland, and have been grazed heavily by sheep from the valley farms. Bracken *Pteridium aquilinum* is extensive. Upper Kentmere Valley. (*Photo*: John Birks).

**B.** Limestone pavement at Ingleborough, North Yorkshire. Trees and other woodland plants grow in the cool, shaded cracks (grikes), and there are small patches of *Fraxinus-Sorbus-Mercurialis* woodland. (*Photo*: John Birks).

**Plate 7.** Mid-Wales

**A.** The grassy, sheep-grazed rocky slopes of Pumlumon, Mid-Wales, are typical of the Welsh uplands. This area has prevalent *Festuca-Agrostis-Galium* grassland and *Nardus-Galium* grassland with species-poor patches of *Juncus effusus*. (*Photo*: Jeremy Moore/CCW).

**B.** Eroding *Erica-Sphagnum* and *Trichophorum-Eriophorum* blanket-mire on the Central Wales Plateau, with *Eriophorum angustifolium* recolonising the peat. Cambrian Mountains. (*Photo*: Jeremy Moore/CCW).

**Plate 8.** South Wales and south-west England

**A.** *Nardus stricta* grassland, bracken *Pteridium aquilinum* and blanket bog on smooth hills punctuated by rocky tors. Bedd Arthur, Mynydd Preseli, Pembrokeshire. (*Photo*: Jeremy Moore/CCW).

**B.** *Calluna-Ulex gallii* heath is a typical and colourful form of vegetation in the moorlands of south-west England. A granite tor is in the distance. Penwith Moors, Cornwall. (*Photo*: Alison Averis).

**Plate 9.** Montane (alpine) vegetation

**A.** Prostrate, montane *Calluna-Cladonia* heath pressed tightly against the stony ground of solifluction terraces, on a broad wind-exposed slope in the Cairngorms. (*Photo*: Lorne Gill/ SNH).

**B.** Montane *Salix* scrub with a silvery-grey canopy of the rare woolly willow *Salix lanata* growing over an understorey of tall herbs and grasses. Coire Sharroch, Caenlochan, Angus. (*Photo*: Lorne Gill/SNH).

**Plate 10.** Contrasting forms of wetland vegetation

**A.** *Trichophorum-Eriophorum* blanket mire on wet, acid peat. Heather *Calluna vulgaris* and cotton-grass *Eriophorum* grow through a red and green carpet of *Sphagnum* dotted with white lichens. Creag Meagaidh, Inverness-shire. (*Photo*: Lorne Gill/SNH).

**B.** A lime-rich *Palustriella-Festuca* spring with swollen golden masses of the moss *Palustriella commutata* speckled with small herbs and grasses. Inchnadamph, Sutherland. (*Photo*: Des Thompson).

# Plate 11. Upland woodlands

**A.** Scots pine woodland *Pinus sylvestris* with an understorey of *Calluna vulgaris* and *Vaccinium* species, characteristic of the cool climate of the eastern and central Highlands. Strath Mashie, Laggan, Central Highlands. (*Photo*: Ben Averis).

**B.** Bryophyte-rich birch *Betula* woodland on rocky slopes, typical of the oceanic climate of the west Highlands. Glen Coe. (*Photo*: Ben Averis).

**Plate 12.** The Cairngorms – affinities with arctic/alpine tundra

Montane three-leaved rush *Juncus trifidus* heaths, *Nardus*-dominated grasslands and mossy snow-beds, springs and wet flushes on the high plateau at the head of Loch Avon. This type of environment has affinities with arctic/alpine tundra in Scandinavia. Shelter Stone Crags, Cairngorms. (*Photo*: Lorne Gill/SNH).

**Plate 13.** Upland plants

**A.** A colourful carpet of *Sphagnum* mosses in blanket bog, consisting mainly of *S. magellanicum* (red) and *S. papillosum* (ochre). Cannich, Inverness-shire (*Photo*: N. Benvie).

**B.** *Racomitrium lanuginosum* montane heath with fir clubmoss *Huperzia selago*. North Corrie, Ben Lui, Scotland. (*Photo*: Angus MacDonald).

**Plate 14.** Contrasts in heathland vegetation in the uplands

**A.** *Ulex gallii* with *Erica cinerea* in *Calluna-Ulex gallii* heath in the mild climate of southern Great Britain. North Wales. (*Photo*: John Birks).

**B.** Heath of *Calluna vulgaris* and *Vaccinium myrtillus* being colonised by birch *Betula* in the uplands of the Central Highlands. Creag Meagaidh, Inverness-shire. (*Photo*: Lorne Gill/SNH).

**Plate 15.** Heather management in the uplands

**A.** Small-patch burning of heather (predominantly *Calluna*) moorland to sustain high densities of red grouse *Lagopus lagopus scoticus*. Some of the sloping terrain is dominated by bracken (the *Pteridium aquilinum–Galium saxatile* community). Note how sheep congregate on the burnt patches with short, young heather. Lammermuir Hills, Scotland. (*Photo:* Laurie Campbell).

**B.** Sheep-grazed *Nardus–Galium* and *Festuca-Agrostis-Galium* grasslands to the right of the fence contrast with heathery *Calluna-Vaccinium* heathery grouse-moor to the left. Native deciduous woodland on the valley slope and conifers planted in a block on the moorland beyond. Near Tomintoul, Eastern Highlands. (*Photo:* Lorne Gill/SNH).

**Plate 16.** Forestry conflicted with nature conservation interests in the uplands until the mid-1990s

Extensive afforestation in the form of unnatural, straight-sided blocks on blanket bog and wet heath, along watercourses and around much of the loch catchment. Extensive damage to lodgepole pine *Pinus contorta* was caused by pine beauty moth *Panolis flammea*. Loch an Tairbeart, Lewis. (*Photo*: Pat MacDonald/SNH).

# M34 *Carex viridula* ssp. *oedocarpa-Koenigia islandica* flush

## Synonyms
*Koenigia islandica-Carex demissa* nodum (Birks 1973); *Carex demissa-Koenigia islandica* flush (Rodwell 1991b).

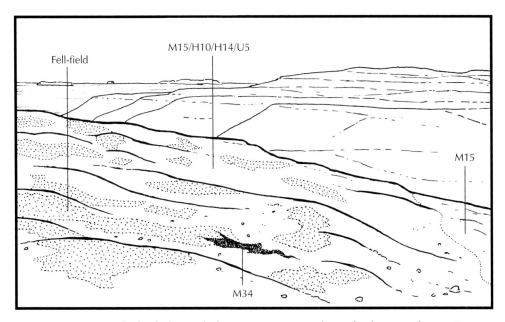

M34 *Carex-Koenigia* flush (dark stippled): on stony, exposed, gentle slopes and summit plateaux.

Associated vegetation types shown in this picture:

M15 *Trichophorum-Erica* wet heath
H10 *Calluna-Erica* heath
H14 *Calluna-Racomitrium* heath
U5 *Nardus-Galium* grassland

The light stippled areas are patches of stony fell-field.

## Description
This is an open flush community containing the annual *Koenigia islandica*. There is a thin, sparse sward of *Carex viridula* ssp. *oedocarpa* and *C. panicea*, usually with scattered plants of *Deschampsia cespitosa*, *Saxifraga stellaris*, *Juncus triglumis* and, in some places, other montane calcicoles such as *J. biglumis*, *Persicaria vivipara* and *Luzula spicata*. Among these vascular plants are tufts and wefts of bryophytes, most commonly *Blindia acuta*, *Hylocomium splendens* and *Marsupella emarginata*. The tiny plants of *Koenigia* are inconspicuous except in autumn, when the leaves and stems turn a rich bright red.

## Differentiation from other communities
In the summer months the presence of *Koenigia* distinguishes this community, although it grows just as commonly on gravelly fell-fields. During the winter, when *Koenigia* cannot be seen, the vegetation could be assigned to the *Carex-Saxifraga* mire M11.

## Ecology

*Carex-Koenigia* flushes occur on basalt gravel that is irrigated by cool acid or neutral water. Most stands are above 500 m in the montane zone. They occur among grasslands, sedge mires or summit heaths, and on rocky detritus below cliffs. In some places the community is associated with *Philonotis-Saxifraga* springs M32. *Koenigia* is sensitive to competition and tolerates open, unstable soils that are suitable for few other species.

## Conservation interest

In addition to *Koenigia* itself, this community provides a suitable habitat for a range of scarce plants, including *Juncus triglumis, J. biglumis, Deschampsia cespitosa* ssp. *alpina, Sagina saginoides, Sedum villosum, Racomitrium ellipticum, Oncophorus virens, O. wahlenbergii* and *Paraleptodontium recurvifolium*. It is an important type of vegetation, partly because it is so rare and partly because it forms a link between the vegetation of Great Britain and that of arctic Europe.

## Management

The *Carex-Koenigia* community is essentially a climax vegetation type and is little affected by management. Grazing and trampling by sheep probably have little effect on

This community has been found only on basalt in the Trotternish region of Skye and the Ardmeanach peninsula of Mull (Birks 1973; Averis and Averis 1997b, 1999b). *Koenigia* seems to require basic rocks and grows particularly well on basalt.

Similar vegetation occurs in Norway, northern Sweden, Iceland, the Faroes and Spitzbergen.

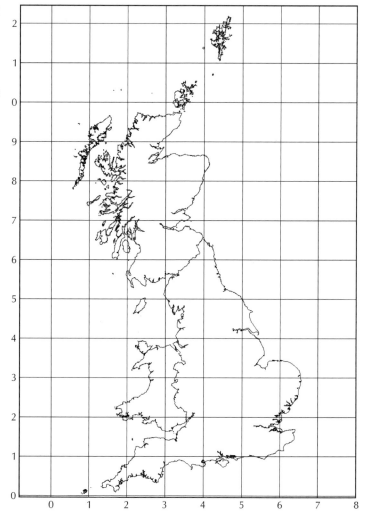

*Koenigia* as it exists as seed for much of the year. In the Faroes *Koenigia* grows on the gravel of car parks and tracks as well as on fell-fields that are subjected to upheaval by solifluction during the winter (Hobbs and Averis 1991a).

# M35 *Ranunculus omiophyllus-Montia fontana* rill

Typical habitats of M35 *Ranunculus-Montia* rill:

1   Soakways in the wettest parts of wet depressions, among various types of mire
2   Springs, seepages and soakways among grassland and mire on gentle slopes

## Description
This community comprises open or close-crowded patches of vegetation emerging from shallow running water or on irrigated wet mud. *Ranunculus omiophyllus* or *Montia fontana* are generally the most common species, and in summer the rounded crenate leaves and white fragile flowers of *R. omiophyllus* mark out the stands amid the surrounding heaths and grasslands. *Potamogeton polygonifolius* is common, growing with its shiny oval leaves held almost vertically where the sward of other plants is very dense, and there can be a few patches of *Sphagnum denticulatum*.

## Differentiation from other communities
The *Ranunculus-Montia* rill community resembles the *Philonotis-Saxifraga* spring M32, but lacks montane plants such as *Saxifraga stellaris, Epilobium alsinifolium, E. anagallidifolium* and *Scapania uliginosa*. In their place it has *Ranunculus omiophyllus* and other lowland species such as *Callitriche stagnalis* and *Poa annua*. *Hypericum-Potamogeton* soakways M29 can look rather similar, but have *Hypericum elodes* and little or no *R. omiophyllus* or *Montia fontana*.

## Ecology
*Ranunculus-Montia* rills occur in running water in springs, shallow streams and channels with peaty or muddy substrates. They usually occur as small, well-defined stands among mires, heaths and grasslands. The pH of the soils ranges from 4.5 to 6.5 (Rodwell 1991b), and the underlying rocks are usually acid igneous or sedimentary types, such as granite, slate and shale. This is a community of low to moderate altitudes and has not been recorded above 450 m (Rodwell 1991b).

This community replaces the *Philonotis-Saxifraga* spring in warmer oceanic environments, and is most common in the uplands of south-west England, Wales and the Pennines. It had not been described from Great Britain prior to the publication of the NVC, and for this reason is probably under-recorded.

Similar vegetation occurs in Ireland and in the mountains of Spain and Portugal.

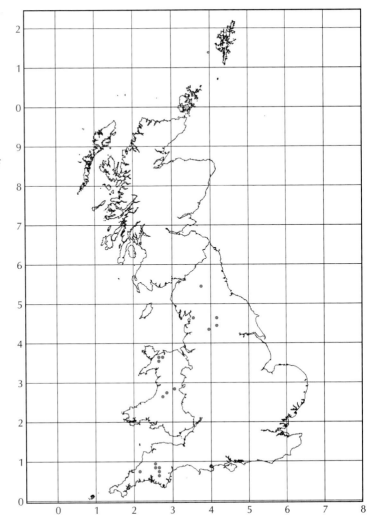

## Conservation interest

The *Ranunculus-Montia* community comprises mixtures of common plants, and is not known as a habitat for rarities. However, in many stands the flora and structure are evidently determined more by the wetness and instability of the habitat than by land management; such stands are valuable examples of apparently near-natural vegetation in regions where most of the vegetation is clearly anthropogenic. The community is also of interest as one of the few examples of an upland vegetation type with a predominantly south-western distribution in Great Britain.

## Management

*Ranunculus-Montia* rills are threatened by drainage and agricultural improvement of the low moorlands in which they occur. However, much of this vegetation in south-west England is now protected within statutory conservation sites. Most stands are grazed, and the open structure of the vegetation is undoubtedly maintained by trampling as well as by constant irrigation. However, poaching by cattle and sheep has disturbed the margins of some stands of *Ranunculus-Montia* vegetation, and excessive grazing is a potential threat.

# M37 *Palustriella commutata-Festuca rubra* spring

## Synonyms

*Cratoneuron commutatum-Saxifraga aizoides* nodum (McVean and Ratcliffe 1962; Birks 1973); *Cratoneuron commutatum* springs I4e *p.p.* (Birks and Ratcliffe 1980); *Cratoneuron commutatum-Festuca rubra* spring (Rodwell 1991b).

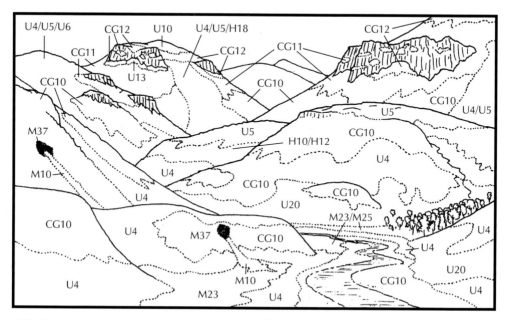

M37 *Palustriella-Festuca* spring vegetation (dark filled): occurring among calcicolous grassland on hill slopes.

Associated vegetation types shown in this picture:

U4 *Festuca-Agrostis-Galium* grassland
U5 *Nardus-Galium* grassland
U6 *Juncus-Festuca* grassland
U10 *Carex-Racomitrium* moss-heath
U13 *Deschampsia-Galium* grassland
U20 *Pteridium-Galium* community
CG10 *Festuca-Agrostis-Thymus* grassland
CG11 *Festuca-Agrostis-Alchemilla* grassland
CG12 *Festuca-Alchemilla-Silene* community
H10 *Calluna-Erica* heath
H12 *Calluna-Vaccinium* heath
H18 *Vaccinium–Deschampsia* heath
M10 *Carex-Pinguicula* mire
M23 *Juncus-Galium* rush-pasture
M25 *Molinia-Potentilla* mire

## Description

The swelling golden-brown cushions or mounds of the moss *Palustriella commutata* grow here in springheads where water issues from the ground, or hang in dripping curtains over wet calcareous rock faces. In some places *Cratoneuron filicinum* replaces *P. commutata*, and across the Breadalbanes the rare *P. decipiens* can grow here too. The bryophyte mats are usually species-poor, with *P. commutata* and *C. filicinum* accompanied by little other than *Bryum pseudotriquetrum*, *Philonotis fontana*, *Aneura pinguis* and *Scapania undulata*. There is a light sprinkling of vascular plants, including sparse clusters of *Festuca rubra* and *Agrostis canina*, and in some places rich-green clumps of *Carex viridula* ssp. *oedocarpa* and glaucous tufts of *C. panicea* and *C. flacca*. In many stands the bryophyte mats are studded with yellow-green stars of *Pinguicula vulgaris*, and there may also be the fat grey shoots and clustered golden flowers of *Saxifraga aizoides* or the neat rosettes and white delicate flowers of *S. hypnoides*. Other common species include *Filipendula ulmaria*, *Trollius europaeus* and *Cardamine pratensis*. Some *Palustriella-Festuca* springs contain deposits of calcium carbonate (known as tufa) around the plants, giving a whitish appearance and a crunchy, almost gravelly feel underfoot.

### Differentiation from other communities
*Palustriella-Festuca* springs are usually very distinctive. They can resemble the more exclusively montane *Palustriella-Carex* spring M38, but contain fewer montane species and have a more open and less diverse layer of vascular plants. In some cases it can be difficult to draw a dividing line between this community and *Carex-Pinguicula* M10 and *Carex-Saxifraga* M11 mires. In general, the short-sedge flushes have more sedges, such as *Carex viridula* ssp. *oedocarpa*, *C. pulicaris* and *C. panicea*, and more bryophyte and

*Palustriella-Festuca* springs occur throughout the British uplands from south Wales northwards, but are restricted to areas with highly calcareous rocks. They are most common on the Carboniferous limestone of northern England (including the Orton and Craven Fells) and on the Dalradian rocks of the Breadalbanes. There are also some fine stands on the Moine rocks of the northern Highlands, the Durness limestone on Skye and in the north-west Highlands, the calcareous igneous and sedimentary rocks of the Lake District and north Wales, and the Silurian shales of the Southern Uplands.

There are similar springs in the Norwegian and Swedish mountains.

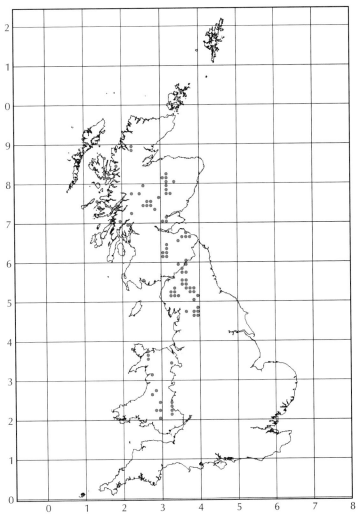

calcicolous forb species. Stands with a continuous carpet of bryophytes dominated by either *Palustriella commutata* or *Cratoneuron filicinum* over at least a square metre or so of ground, and where sedges and other small plants are sparse or absent, are *Palustriella-Festuca* springs rather than sedge-mires.

## Ecology
The community typically occurs as small stands in springs, on wet rocks and on flushed banks where the irrigating water is base-rich. It is confined to calcareous rocks. The underlying soil is usually a humus-rich silty mud. The altitudinal range is extremely wide, and the community occurs in essentially the same form in low-altitude woodlands as well as in montane grasslands. However, it is rarer at high altitudes where soils are more leached.

## Conservation interest
This community is of interest as an example of a near-natural vegetation type. Fewer rare species grow here than in the more montane *Palustriella-Carex* spring, but there can be a few uncommon plants, such as *Persicaria vivipara*, *Epilobium alsinifolium* and *Crepis paludosa*. In the Breadalbanes and north-west Highlands, if the vegetation was not grazed, montane willows might grow around the edges of *Palustriella-Festuca* springs, and add a great deal to their value for nature conservation.

## Management
*Palustriella-Festuca* vegetation appears to be near-natural, and is probably not dependent on management for its maintenance. The bryophyte mats and soft calcareous soils are easily damaged by trampling, which may break up the cushions of *Palustriella commutata*, leading to the development of *Carex-Pinguicula* mire or other base-rich flushes. Most stands seem to be at least lightly grazed, so that taller forbs such as *Cardamine pratensis* and *Filipendula ulmaria* are held in check.

# M38 *Palustriella commutata-Carex nigra* spring

## Synonyms
*Carex panicea-Campylium stellatum* nodum *p.p.* (McVean and Ratcliffe 1962); *Cratoneuron commutatum* springs I4e *p.p.* (Birks and Ratcliffe 1980); *Cratoneuron commutatum-Carex nigra* spring (Rodwell 1991b).

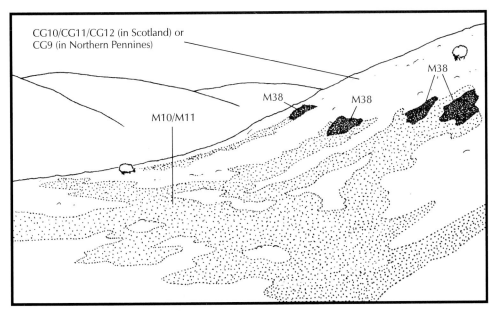

M38 *Palustriella-Carex* spring vegetation (dark stippled): occurring upslope of sedge flushes among calcicolous grassland on high hill inclines.

Associated vegetation types shown in this picture:

CG9 *Sesleria-Galium* grassland
CG10 *Festuca-Agrostis-Thymus* grassland
CG11 *Festuca-Agrostis-Alchemilla* grassland
CG12 *Festuca-Alchemilla-Silene* community
M10 *Carex-Pinguicula* mire
M11 *Carex-Saxifraga* mire

## Description
*Palustriella-Carex* springs consist of golden-brown mats of *Palustriella commutata*. In comparison to *Palustriella-Festuca* springs M37, they have a thicker sward of vascular plants over a more diverse carpet of bryophytes. *Bryum pseudotriquetrum*, *Philonotis fontana*, *P. calcarea* and *Aneura pinguis* are more conspicuous among the *P. commutata*, and there is a rich spread of small sedges and other vascular plants, including *Carex nigra*, *C. panicea*, *C. viridula* ssp. *oedocarpa*, *Selaginella selaginoides*, *Persicaria vivipara*, *Juncus triglumis* and *Sagina nodosa*. In some stands, *Cratoneuron filicinum* takes the place of *P. commutata*.

## Differentiation from other communities
These springs have more montane species and a more continuous sward of vascular plants than the *Palustriella-Festuca* community. They can be hard to separate from

233

examples of *Carex-Pinguicula* M10 or *Carex-Saxifraga* M11 flushes with much *Palustriella commutata*, but they have a deeper and more continuous bryophyte layer dominated by *P. commutata* or *Cratoneuron filicinum*, and the vascular sward is usually sparser.

## Ecology

This is a montane community, occurring mainly at altitudes over 600 m. It is invariably associated with calcareous rocks where the emerging water is highly base-rich. The soils are saturated stagnogleys, commonly with a surface layer of peat and in places crunchy with precipitated calcium carbonate (Rodwell 1991b). Individual stands are generally small, corresponding to the area of fast-flowing water, but there can be many patches scattered across large areas of upland: for example, networks of springs along the side of a valley. The springs can grade into *Carex-Pinguicula* mire or *Festuca-Agrostis-Thymus* grassland CG10 with increasing distance from the springhead. Other stands are sharply delimited from the surrounding vegetation.

*Palustriella-Carex* mire is a scarce community and has been recorded mainly around the Carboniferous limestone of Upper Teesdale and on the Dalradian rocks of the Breadalbane region of the central Highlands. There are outlying stands in the Pentland Hills, the Southern Uplands and in the Orton area of Cumbria.

Similar vegetation occurs in northern Scandinavia and in montane areas of central Europe.

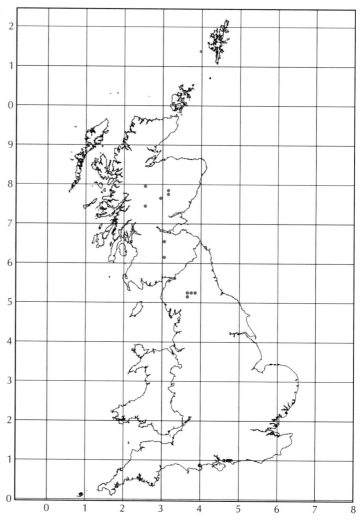

**Conservation interest**
This type of spring is one of the rarest mire types in the British uplands. It is valuable for its rich flora; notable species include *Saxifraga hirculus* and the moss *Oncophorus virens*.

**Management**
The soils beneath *Palustriella-Carex* springs are constantly wet and strongly flushed, but some grazing may be necessary to maintain the open and species-rich sward characteristic of the community. Colonisation by trees or shrubs is unlikely in the harsh climate at these high altitudes, but tall herbs may overwhelm smaller plants in the absence of grazing.

# H4 *Ulex gallii-Agrostis curtisii* heath

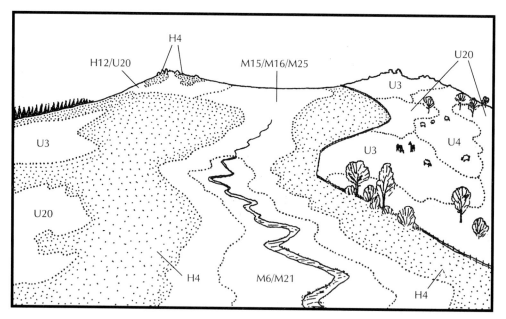

H4 *Ulex gallii-Agrostis curtisii* heath (stippled): on dry acid soils on gentle to moderate slopes (H4 around left summit), and fairly dry to damp acid soils on gentle slopes (H4 elsewhere).

Associated vegetation types in this drawing are:

H12 *Calluna-Vaccinium* heath
M6 *Carex-Sphagnum* mire
M15 *Trichophorum-Erica* wet heath
M16 *Erica-Sphagnum compactum* wet heath
M21 *Narthecium-Sphagnum* valley mire
M25 *Molinia-Potentilla* mire
U3 *Agrostis curtisii* grassland
U4 *Festuca-Agrostis-Galium* grassland
U20 *Pteridium-Galium* community

## Description

This is a dense mixed heathland, which becomes a splendid blaze of purple and gold when the dwarf shrubs are in flower in late summer. *Calluna vulgaris*, *Ulex gallii*, *Molinia caerulea*, *Erica tetralix*, *E. cinerea*, and in some stands *Vaccinium myrtillus*, grow in a thick sward together with the delicate flowers and narrow grey-green leaves of the grass *Agrostis curtisii*. In some examples the dwarf shrubs are bound together by the red sinuous shoots of *Cuscuta epithymum*.

There are four sub-communities. The *Agrostis curtisii-Erica cinerea* sub-community H4a stands out from the others because it has hardly any *E. tetralix*. The *Festuca ovina* sub-community H4b is grassy, with species such as *Festuca ovina*, *Danthonia decumbens* and *Agrostis canina* either growing in a fine-scale mosaic with the dwarf shrubs or forming more intimate mixtures; *E. cinerea* and *E. tetralix* occur only in small quantity. *Molinia* is common in some stands of the *Agrostis-Erica* and *Festuca* sub-communities but very sparse in others; it is more consistently common in the other two sub-communities. The *Erica tetralix* sub-community H4c is a short, rather sparse and open damp heath, in which the vegetation is commonly broken by patches of bare peat and humus; it is given a greyish tinge by *E. tetralix* and *Carex panicea*, which in some places make up the bulk of the sward. The *Trichophorum cespitosum* sub-community H4d is also a damp heath, typically with *Trichophorum cespitosum*, *Molinia* and *E. tetralix*.

## Differentiation from other communities

This community looks very similar to the strictly lowland *Ulex minor-Agrostis curtisii* heath H3, but the characteristic gorse is *Ulex gallii* rather than *U. minor*. Another similar-looking vegetation type is *Calluna-Ulex gallii* heath H8, which differs from *Ulex gallii-Agrostis* heath in having little or no *Agrostis curtisii*, *Molinia* and *Erica tetralix*. The damper forms of *Ulex gallii-Agrostis* heath (mainly the *Erica* and *Trichophorum* sub-communities) can resemble *Trichophorum-Erica* wet heath M15 and *Erica-Sphagnum compactum* wet heath M16, but have much *A. curtisii* and *U. gallii*, both of which are rare in the wet heath communities. In Ireland and Wales there are forms of damp heath consisting mainly of *Calluna*, *U. gallii*, *Molinia*, *E. tetralix* and *E. cinerea* (Hobbs and Averis 1991b; Derek Ratcliffe pers. comm.; Averis 2001a); these closely resemble *Ulex gallii-Agrostis* heath but lack *A. curtisii*.

The NVC treatment of *Ulex gallii-Agrostis* heath unites various forms of heathland into a single NVC community on the basis of the combined presence of the south-western species *A. curtisii* and *U. gallii*. The community is described as a damp rather than a dry heath, but it includes some vegetation with sufficiently little *Molinia*, *E. tetralix* or *Trichophorum* to be better described as dry heath. Much of the vegetation assigned to the *Agrostis-Erica* and *Festuca* sub-communities might be better understood as south-western '*Agrostis curtisii-Ulex gallii*' forms of *Calluna-Ulex gallii* H8, *Calluna-Erica* H10 or *Calluna-Vaccinium* H12 dry heaths. Conversely, the *Erica* and *Trichophorum* sub-communities could be regarded as south-western '*Agrostis curtisii-Ulex gallii*' forms of *Trichophorum-Erica* or *Erica-Sphagnum compactum* wet heaths.

## Ecology

*Ulex gallii-Agrostis* heath is described as a community of moist gleyed soils or moist podsols over acid rocks, although certain details in the floristic table contained in the NVC suggest that some of the vegetation is actually on drier soils (see above). It occurs from sea-level to over 500 m in the uplands of south-west England. Most of the stands at lower altitudes are on level moorland. The higher examples are usually on steeper slopes because in the wet upland climate the gentler slopes and flat plateaux are blanketed with peat.

## Conservation interest

*Ulex gallii-Agrostis* heath is the principal heathland community of very mild oceanic climates in Great Britain, and, in common with other types of dwarf-shrub heath, is of great value for nature conservation. It includes uncommon plants such as *Agrostis curtisii*, and, more locally, *Erica ciliaris* and *E. vagans*, as well as being a visually attractive component of the upland landscape. Like *Narthecium-Sphagnum* valley mire, it is of interest as a form of vegetation which occurs mainly at low altitudes, and which is at the northern limit of its distribution in the uplands of south-west Great Britain.

## Management

This community occurs below the upper altitudinal limit of woodland. Some coastal stands are exposed to westerly winds, notably on the Lizard Peninsula and the Penwith Moors, and may never have been completely wooded, but most stands would probably revert to secondary woodland if they ceased to be managed. The ideal management is careful, controlled burning and light grazing. Burning is often done casually, without

## H4 *Ulex gallii-Agrostis curtisii* heath

This type of heathland is confined to south-west England and south Wales, where it is recorded in both upland and lowland areas. In the uplands it is well represented on Dartmoor, Exmoor and the Penwith moorlands, and sparingly on moorland in the valleys of south Wales. The grassy swards with a scattering of shrubs on Bodmin Moor are a form of this community impoverished by grazing. Damp heaths with a similar flora to *Ulex gallii-Agrostis* heath, but lying beyond the range of *Agrostis curtisii*, have been recorded in mid- and north Wales (see above).

Heathland similar to this community occurs in the oceanic regions of western France, Portugal and Spain.

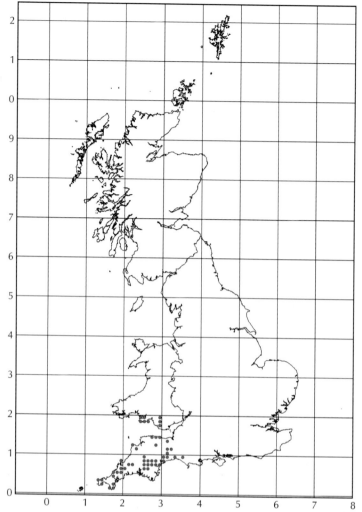

considering how the frequency of burning and the size of the burnt patches is likely to affect the vegetation, and in many places the number of grazing animals is too high, especially on common land (Rodwell 1991b). Excessive burning (including accidental fires) and grazing tend to impoverish the heath vegetation, eventually converting it to grassland (e.g. *Agrostis* U3, *Festuca-Agrostis-Galium* U4 or *Molinia-Potentilla* M25 swards), as on Bodmin Moor and in south Wales. *Ulex gallii-Agrostis* heaths are also threatened by agricultural improvement and forestry (Rodwell 1991b).

# H8 *Calluna vulgaris-Ulex gallii* heath

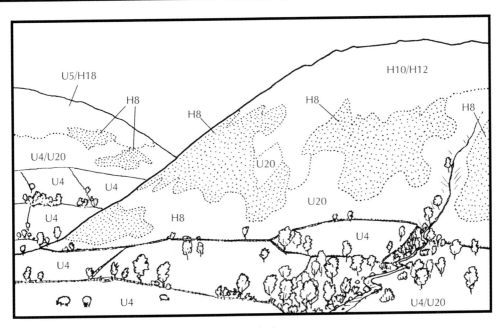

Typical habitats of H8 *Calluna-Ulex* heath (stippled):

Middle and right of picture: extensive H8 with U20 *Pteridium-Galium* community and H10 *Calluna-Erica* and H12 *Calluna-Vaccinium* heaths on well-drained, lightly grazed slope. H8 mostly downslope from H10 and H12.

More distant hill at left of picture: smaller patches of H8 among U4 *Festuca-Agrostis-Galium* grassland, U5 *Nardus-Galium* grassland, H18 *Vaccinium-Deschampsia* heath and U20 *Pteridium-Galium* community on heavily grazed slope.

## Description

These are mixed heaths of *Calluna vulgaris*, *Ulex gallii* and *Erica cinerea*, all of which flower in splendid tones of purple and gold in late summer.

There are five sub-communities, of which only three are likely to occur in the uplands. The Species-poor sub-community H8a has a dense canopy of *Calluna*, *U. gallii* and *E. cinerea* which few other plants are able to penetrate, although in some stands there can be much *Agrostis capillaris* and mosses such as *Hypnum cupressiforme*, *H. jutlandicum*, *Dicranum scoparium* and *Campylopus flexuosus*. In the grassy *Danthonia decumbens* sub-community H8b the dwarf shrubs are set in a matrix of acid grassland species, such as *Danthonia decumbens*, *Anthoxanthum odoratum*, *Festuca rubra*, *Agrostis canina*, *A. capillaris* and *Potentilla erecta*. In some stands the shrubs are grazed into tight round islands within a sea of grasses, and are hard to see from a distance when they are not in flower. The *Sanguisorba minor* sub-community H8c is a lowland heath of base-rich soils with species including *Sanguisorba minor*, *Helianthemum nummularium*, *Stachys officinalis* and *Galium verum*. The *Scilla verna* sub-community H8d is a coastal heath with *Scilla verna*, *Thymus polytrichus* and *Plantago maritima*. The *Vaccinium myrtillus* sub-community H8e occurs on acid soils in the uplands, and has *Deschampsia flexuosa*, *Nardus stricta*, *Rhytidiadelphus loreus* and *Pleurozium schreberi* under a thick prickly canopy of dwarf shrubs containing *V. myrtillus* in addition to *Calluna*, *U. gallii* and *E. cinerea*.

There are also species-poor heaths with much *U. gallii* but no *Calluna* or *E. cinerea*. The *U. gallii* grows either in dense swards or in a matrix of grasses. These heaths are not represented in the NVC, but clearly have a close affinity with *Calluna-Ulex gallii* heath. In Ireland and Wales there are damp heaths of *Calluna*, *U. gallii*, *Molinia*, *Erica tetralix* and *E. cinerea* (Hobbs and Averis 1991b; Derek Ratcliffe pers. comm.; Averis 2001a), which are also not described in the NVC. This vegetation might be regarded as a wet heath counterpart of *Calluna-Ulex gallii* heath or as being closely allied to *Ulex gallii-Agrostis* heath H4.

## Differentiation from other communities

*Calluna-Ulex gallii* heath is one of the few heathland types in which *Ulex gallii* is either dominant or co-dominant with other dwarf shrubs. Another such heath is *Ulex gallii-Agrostis* heath. This is distinguished by *Agrostis curtisii*, which is rare in *Calluna-Ulex gallii* heath and occurs only in stands in south-west England and south Wales. *U. gallii* also occurs in the strictly lowland *Erica vagans-Schoenus nigricans* H5 and *Erica vagans-Ulex europaeus* H6 heaths, but these contain the rare *Erica vagans* and are restricted to Cornwall. The undescribed damp heath recorded from Ireland and Wales differs from *Calluna-Ulex gallii* heath in that both *Molinia* and *E. tetralix* are common in the sward.

## Ecology

*Calluna-Ulex gallii* heath occurs on free-draining soils which range from acid to mildly basic. The soils are derived from a wide variety of parent rocks, including sandstone and igneous and metamorphic rocks. Where the soils are derived from calcareous drift the heath can be moderately species-rich, and the *Sanguisorba* sub-community is characteristic. Where the soils are leached acid sand or peat the community is usually rather species-poor.

*Calluna-Ulex gallii* heath occurs in warm oceanic parts of the lowlands, including coastal cliffs and slopes, but also penetrates into the uplands. Most stands are below 350 m, but it occurs above 400 m on sheltered slopes. The community typically forms patches where steep slopes with mineral soils break through blanketing peat, and along the steep sides of gullies and stream valleys. In some places there are broad bands of it around the lower hill slopes; there are good examples of this in the Carneddau in north Wales, and also in the Mourne Mountains in eastern Ireland.

## Conservation interest

*Calluna-Ulex gallii* heaths are an important component of the mosaics of semi-natural vegetation in the upland fringes of southern Great Britain. As with other forms of dwarf-shrub heath, this community is of international importance, but in contrast to most other heath communities in the British uplands it is largely restricted to England and Wales, and scarcely extends into Scotland. Stonechats *Saxicola torquata* breed in the gorse, and Red Grouse *Lagopus lagopus* may nest among the dwarf shrubs, although this form of heathland is rare on moorlands managed for sport. The flowers, which are almost overpoweringly sweet-scented on warm days in late summer, feed a great array of insects. Coastal stands provide one of the most important British habitats for *Scilla verna*.

## Management

*Calluna-Ulex gallii* heath is maintained by grazing and burning and is probably originally derived from woodland or scrub. In the absence of management, most stands would eventually become colonised by trees. Only perhaps in maritime sites is it a natural community, where salt spray and exposure prevent the growth of trees. Stands are

This form of dry heath is most common in Wales, but also occurs in south-west England, East Anglia, Lincolnshire, the south Pennines, the Isle of Man, Cumbria and the Scottish side of the Solway Firth.

There are floristically similar heaths in Ireland, for example in Connemara (Horsfield *et al.* 1991) and in Northern Ireland (Hobbs and Averis 1991b). Outside Great Britain and Ireland heaths of this type are known only from the western seaboards of France, Spain and Portugal.

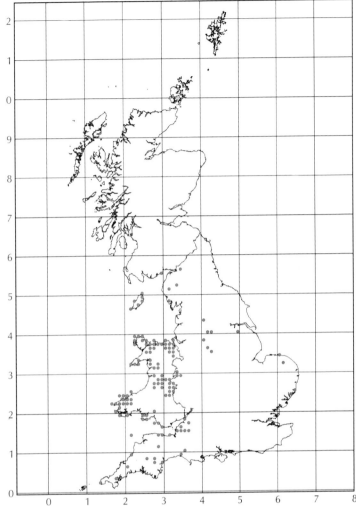

typically grazed by sheep, cattle, ponies or rabbits. For nature conservation purposes, grazing should be sufficiently heavy to maintain the herbaceous element of the flora. Stands on the upland fringes are often burnt to produce a flush of more nutritious growth for grazing livestock but many isolated lowland or coastal stands are probably not deliberately burnt. Small-patch burning is recommended to maintain small-scale mosaics with a range of structure and flora. If carried out judiciously, grazing and burning can open up the canopy of dwarf shrubs, allowing other species to flourish and preventing the development of dense, species-poor stands in which herbs and mosses are shaded out. However, some patches of dense gorse are useful to provide shelter for livestock in bad weather. If the intensity of grazing is too high, grasses and small forbs proliferate at the expense of the dwarf shrubs; this process can be accelerated by burning. Under these conditions the heath is increasingly reduced to small islands within a grassy matrix, and eventually to scattered bushes of *Ulex gallii* among stands of *Festuca-Agrostis-Galium* grassland U4. The free-draining soils make the community susceptible to agricultural reclamation and much *Calluna-Ulex gallii* heath now occurs as fragmented stands on marginal land.

# H9 *Calluna vulgaris-Deschampsia flexuosa* heath

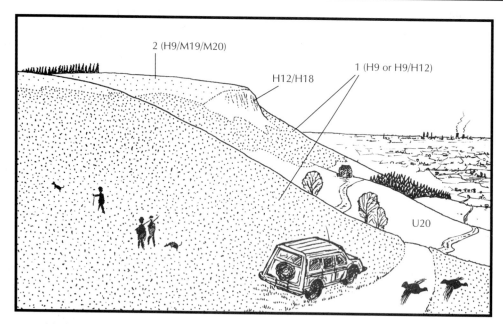

Typical habitats of H9 *Calluna-Deschampsia* heath (stippled):

1   Well-drained hill slopes where repeated or severe moor-burning has led to floristic impoverishment of the heath vegetation; H9 very extensive or in mosaics with other vegetation types including U20 *Pteridium-Galium* community (on lower slopes), H12 *Calluna-Vaccinium* heath and H18 *Vaccinium-Deschampsia* heath (on rocky, less severely burned slopes).
2   Level to gently sloping deep peat, dried out by moor-burning; H9 occupying the most severely burnt and dried-out areas, in mosaics with M19 *Calluna-Eriophorum* bog and M20 *Eriophorum vaginatum* bog.

## Description

Stands of *Calluna-Deschampsia* heath typically consist of a dark even-aged monoculture of *Calluna vulgaris* growing through the crusted surface of dried-out humus or peat. The sward is usually spiked with a little *Deschampsia flexuosa* with its silky panicles of flowers overtopping the heather in summer, and there is generally a green speckling of the moss *Pohlia nutans* over the ground.

The vegetation beneath the heather varies according to how recently the heath has been burnt. The five sub-communities described in the NVC encompass this variation. The *Hypnum cupressiforme* sub-community H9a takes in degenerate stands with an open canopy of old heather plants. Much of the ground is covered with a thick yellow-green weft of *Hypnum cupressiforme s.l.* (probably all *H. jutlandicum*), and there can be numerous clumps and cushions of *Dicranum scoparium*. *Pteridium aquilinum* grows in some stands. The *Vaccinium myrtillus-Cladonia* species sub-community H9b occurs in the early stages of recovery from fire when the heather is short and there is much bare peat. *Vaccinium myrtillus* recovers more quickly from fire than *Calluna* and can thicken up in the canopy. *Empetrum nigrum* or *V. vitis-idaea* commonly occur in small quantity, and over the soil there is a mixed and in some places dense mat of bryophytes including *Campylopus flexuosus* and *Gymnocolea inflata*, and a few encrusting lichens such as *Cladonia chlorophaea*, *C. squamosa* and *C. diversa*. These persist until the heather

242

canopy closes or until the heath is burnt again. The Species-poor sub-community H9c occurs when heather is in the building phase and is usually completely dominant. There can be a meagre scattering of *Deschampsia flexuosa*, *Vaccinium myrtillus* and the mosses *Pohlia nutans* and *Campylopus flexuosus*. The *Galium saxatile* sub-community H9d has a greener sward with more forbs and grasses, generally because the heather has been suppressed by grazing. The most common species are *Potentilla erecta*, *Galium saxatile*, *Festuca rubra* and *Rumex acetosella*. The *Molinia caerulea* sub-community H9e has little to distinguish it except scattered tufts of *Molinia*.

### Differentiation from other communities
Other forms of dry heath are rarely as totally dominated by *Calluna* as is *Calluna-Deschampsia* heath. The most similar community is *Calluna-Vaccinium* heath H12, which generally has a more mixed canopy of dwarf shrubs, including *Vaccinium myrtillus*, and a more continuous underlayer of large pleurocarpous mosses such as *Hylocomium splendens*, *Pleurozium schreberi* and *Rhytidiadelphus loreus*. Conversely, the small moss *Pohlia nutans* is commoner in *Calluna-Deschampsia* heath. Stands of the *Molinia* sub-community of *Calluna-Deschampsia* heath may superficially resemble *Trichophorum-Erica* wet heath M15 and *Erica-Sphagnum compactum* wet heath M16, but lack *Erica tetralix*, *Trichophorum cespitosum*, *Potentilla erecta* and *Sphagnum* species. The lowland *Calluna vulgaris-Festuca ovina* heath H1 can resemble *Calluna-Deschampsia* heath but has little or no *Deschampsia flexuosa*. Where *Vaccinium myrtillus*, *V. vitis-idaea* and *Empetrum nigrum* become temporarily common in stands of *Calluna-Deschampsia* heath as a result of burning and grazing, the community can be separated from *Vaccinium-Deschampsia* heath H18 by the scarcity of *Hypnum jutlandicum*, *Hylocomium splendens*, *Pleurozium schreberi*, *Dicranum scoparium* and other large mosses.

### Ecology
This is a heath of acid podsolised mineral soils and sandy soils in the lowlands and upland fringes, occurring from low altitudes to almost 600 m. In the North York Moors, the Pennines and southern Scotland it is also common on dried-out peat, replacing more natural bog vegetation. The community clothes slopes and rolling moorland, and is common around the moorland edge where its lower limit coincides with the upper edge of enclosed grassland. At its upper limit it usually peters out into blanket mire in the cooler and wetter climate of the higher plateaux.

### Conservation interest
All *Calluna* heaths are internationally scarce vegetation types that are restricted to the western fringes of Europe. However, *Calluna-Deschampsia* heath is produced and maintained by intensive management and is the least natural form of upland heath in Great Britain. It has a poor flora and is generally less valuable for nature conservation than more natural types of heathland. No rare plant species grow in this community. Red Grouse *Lagopus lagopus* and Merlin *Falco columbarius* breed here, but few other birds occur apart from the ubiquitous Meadow Pipit *Anthus pratensis*. It is only in late summer that there is anything particularly attractive about heaths of this type, when the heather flowers in vast sweet-smelling purple sheets. At this time the heaths are attractive to insects too, and in some places beekeepers transport their hives to the moorlands so that the bees can make heather honey.

### Management
In the uplands most *Calluna-Deschampsia* heath is managed as grouse moor by frequent and repeated burning. The heaths are burned in small patches every 10–15 years to

## H9 *Calluna vulgaris-Deschampsia flexuosa* heath

*Calluna-Deschampsia* heath is most common on grouse moors in the southern Pennines northwards to lower Wharfedale and on the North York Moors. It is extensive on the eastern dip slope of the Pennines as far north as the Tyne Gap. It also occurs in parts of eastern Scotland, on the Cheviot (not shown on the distribution map), on the Clwydian range in north Wales, on the Shropshire hills, and in the lowlands of Lancashire, Cheshire and the English Midlands.

There is related vegetation elsewhere in western Europe, especially where there is some atmospheric pollution.

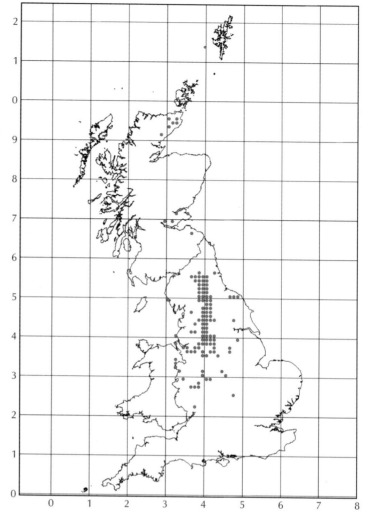

destroy the old leggy heather and to encourage more nutritious young growth. This treatment tends to eliminate most bryophytes and to suppress the growth of shrubs apart from heather. Most stands of *Calluna-Deschampsia* heath are also grazed by sheep. To manage the heaths for nature conservation it is best for any burning to be carried out according to the guidelines of good practice (MAFF 1992, Phillips *et al.* 1993, Moorland Working Group 1998, Scotland's Moorland Forum 2003), and to avoid heavy grazing. Many upland examples of *Calluna-Deschampsia* heath cover ground where there would once have been a more mixed array of dwarf shrubs, bryophytes and other species. By limiting burning and grazing the heath may become more diverse, possibly developing into less impoverished types of moorland vegetation such as *Calluna-Vaccinium* heath, *Trichophorum-Erica* wet heath, *Erica-Sphagnum compactum* wet heath or *Calluna-Eriophorum* blanket mire M19. Excessive grazing will eventually eliminate *Calluna*, and the heathland will be replaced by *Deschampsia flexuosa* grassland U2, *Festuca-Agrostis-Galium* grassland U4 or *Nardus-Galium* grassland U5, depending on the soils.

*Calluna-Deschampsia* heath occurs within the altitudinal range of woodland, and would revert to woodland or scrub in the absence of grazing and burning, provided there is a source of seed. In lowland situations *Calluna-Deschampsia* heath can be the direct result of woodland clearance. The flora is similar to that of *Quercus-Betula-*

*Deschampsia* woodland W16, and there are instances where this form of woodland has replaced *Calluna-Deschampsia* heath in the absence of management, with the heath vegetation persisting on similar soils on adjacent open ground. In some places where *Calluna-Deschampsia* heath exists today, there might once have been *Quercus-Betula-Dicranum* woodland W17, with a richer bryophyte flora.

In some localities, especially in the southern Pennines, atmospheric pollution as well as severe burning is believed to have played a part in impoverishing the bryophyte and lichen flora of these heaths (Rodwell 1991b). In Scotland, where pollution levels are low, *Calluna-Deschampsia* heath is a successional stage that develops after other forms of heath are burnt, and larger mosses eventually recolonise, but in more polluted areas further south impoverished *Calluna-Deschampsia* heath can persist for a much longer time.

# H10 *Calluna vulgaris-Erica cinerea* heath

## Synonyms

*Callunetum vulgaris* (McVean and Ratcliffe 1962 *p.p.*; Birks 1973); *Calluna vulgaris-Sieglingia decumbens* Association (Birks 1973); Atlantic heather moor (Birse and Robertson 1976); *Calluna* dry heath B1a *p.p.* and *Calluna vulgaris-Danthonia decumbens* heath B1d (Birks and Ratcliffe 1980).

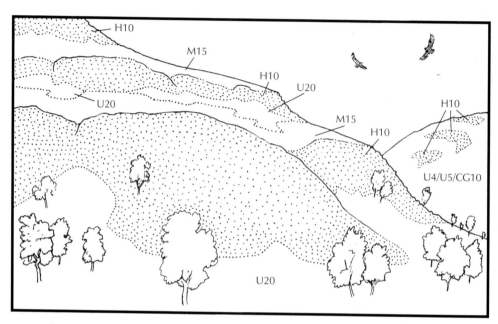

H10 *Calluna-Erica* heath (stippled): on steep, well-drained slopes with thin acidic soils, giving way to M15 *Trichophorum-Erica* wet heath on moist peat on gentle slopes, and to U20 *Pteridium-Galium* community on deeper and richer well-drained soils. This kind of pattern is common on lightly grazed to moderately grazed slopes in the western Highlands.

On the more heavily grazed hill at the right of the picture there are smaller patches of H10 among U4 *Festuca-Agrostis-Galium*, U5 *Nardus-Galium* and CG10 *Festuca-Agrostis-Thymus* grasslands.

H10 is also shown in the pictures for H8, M15, U4, CG10 and MG5.

## Description

These are dry heaths with a low, dark-coloured canopy of *Calluna vulgaris* and *Erica cinerea*. The dwarf shrubs are typically overtopped by the long, deep-green leaves and drooping brownish flowers of *Carex binervis*; *Potentilla erecta* and *Galium saxatile* scramble over the ground below the shrubs. There is usually a thick carpet of mosses such as *Pleurozium schreberi*, *Rhytidiadelphus loreus*, *Hylocomium splendens* and *Hypnum jutlandicum*. *E. cinerea* begins to flower before the *Calluna*, enlivening the heaths with patches of rosy-pink before the purple heather blossoms open in late summer. When both shrubs are in flower the effect is stunning, and whole hillsides can be transformed into richly coloured, almost glowing, fragrant spreads of flowers.

There are four sub-communities, all well-defined and generally easy to recognise. The Typical sub-community H10a has no special distinguishing species, and usually has a thick, continuous canopy of dwarf shrubs. The *Racomitrium lanuginosum* sub-community H10b occurs mainly at higher elevations, and has a shorter, sparser and more open canopy that reveals the thick silvery mats of *Racomitrium lanuginosum* clothing the ground beneath the dwarf shrubs. There can be a speckling of lichens such as *Cladonia portentosa* and *C. uncialis* ssp. *biuncialis*, and there may be scattered plants of *Trichophorum cespitosum*, *Huperzia selago* or *Antennaria dioica*. The *Festuca ovina-Anthoxanthum odoratum* sub-community H10c is a paler-coloured, grassier assemblage of species in which the dwarf shrubs are interleaved with *Festuca ovina*, *F. rubra*, *Anthoxanthum odoratum* and *Agrostis capillaris*. The *Thymus polytrichus-Carex pulicaris* sub-community H10d is species-rich, usually with a profusion of *Thymus polytrichus* and small herbs under a short and open sward of dwarf shrubs. Characteristic species include *Lotus corniculatus*, *Linum catharticum*, *Prunella vulgaris*, *Viola riviniana*, *Primula vulgaris* and *Pilosella officinarum*.

The growth of *E. cinerea* appears to be particularly favoured by a mild, oceanic climate, and in the far western Highlands and on Skye, and more commonly in Ireland, there are *Calluna-Erica* heaths with more *E. cinerea* than *Calluna* (e.g. Hobbs and Averis 1991b). In the northern Highlands and on Orkney and Shetland there are stands of *Calluna-Erica* heath in which the ground is blanketed with white *Cladonia* lichens; some of these are at low altitudes close to the coast (e.g. Coppins and Coppins 1999).

## Differentiation from other communities

This is a dry *Calluna* heath, distinguished from the damper *Calluna-Vaccinium-Sphagnum* H21 and *Vaccinium-Rubus* H22 heaths by the scarcity of Sphagna, and from *Trichophorum-Erica* M15 and *Erica-Sphagnum compactum* M16 wet heaths by the scarcity of *Erica tetralix*, *Molinia caerulea*, *Trichophorum cespitosum* and Sphagna. In contrast to *Calluna-Deschampsia* heath H9 and *Calluna-Vaccinium* heath H12, this community contains much *Erica cinerea* and little or no *Vaccinium myrtillus*. *Ulex gallii-Agrostis* heath H4 and *Calluna-Ulex gallii* heath H8 contain both *Calluna* and *E. cinerea*, but also have much *Ulex gallii*, which is rare in *Calluna-Erica* heath.

The *Racomitrium* sub-community may be confused with *Calluna-Racomitrium* heath H14, the *Racomitrium* sub-community of *Nardus-Galium* grassland U5e, or the *Cladonia* sub-community of *Trichophorum-Erica* wet heath M15c, all of which share *Calluna*, *E. cinerea*, *Racomitrium lanuginosum* and *Potentilla erecta*. However, the heather in *Calluna-Racomitrium* heath is prostrate or severely dwarfed by the wind, and the vegetation usually contains montane species such as *Festuca vivipara*, *Diphasiastrum alpinum*, *Carex bigelowii*, *Ochrolechia frigida* and *Cetraria islandica*. The *Racomitrium* sub-community of *Nardus-Galium* grassland has more *Nardus stricta*, *Trichophorum* and *Galium saxatile* than similar forms of *Calluna-Erica* heath. The *Cladonia* sub-community of *Trichophorum-Erica* wet heath has more *E. tetralix*, *Molinia* and *Trichophorum*.

The herb-rich *Thymus-Carex* sub-community could be confused with heathy stands of the *Calluna-Primula* sub-community of *Luzula-Geum* tall-herb vegetation U17d, but this has more *Luzula sylvatica*, *Angelica sylvestris*, *Geum rivale*, *Sedum rosea* and *Filipendula ulmaria*. Some forms of the *Thymus-Carex* sub-community resemble the coastal *Calluna vulgaris-Scilla verna* heath H7, but they generally lack maritime species, such as *Scilla verna* and *Plantago maritima*, as well as having less *Holcus lanatus*, *Plantago lanceolata* and *Hypochaeris radicata*.

## Ecology

This is a heath of well-drained mineral soils, and is common on steep, stony slopes. Many patches of *Calluna-Erica* heath stand out clearly as patches of dark vegetation on steeper ground, contrasting with the paler tones of the surrounding bogs and wet heaths or the varied greens of upland grasslands. The soils are generally acid, although those

*Calluna-Erica* heath is wide-spread in western and northern Great Britain from Devon to Shetland, and is most common in the western Highlands and the Hebrides.

There is similar vegetation in Ireland (Hobbs and Averis 1991b; Horsfield *et al.* 1991). Heaths of this type also occur rather locally on the Faroes (Hobbs and Averis 1991a) and in western Norway.

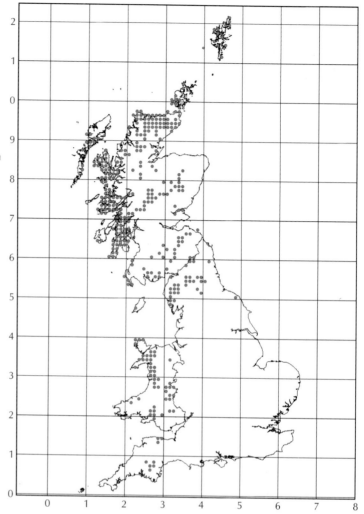

with the herb-rich *Thymus-Carex* sub-community are evidently more basic; this sub-community is especially common on the basalts of Skye and Mull. *Calluna-Erica* heath rarely occurs above about 400 m except in the mild, oceanic climate of the western Highlands and the Hebrides where it can ascend to almost 800 m. Further north and east it is usually associated with warm south-facing or west-facing slopes at low altitudes.

## Conservation interest

*Calluna-Erica* heaths comprise part of the range of variation within the internationally important heather moors of Great Britain. They are a good habitat for upland birds, including Twite *Carduelis flavirostris*, Merlin *Falco columbarius*, Short-eared Owl *Asio flammeus*, Hen Harrier *Circus cyaneus* and Ring Ouzel *Turdus torquatus*. Red Grouse *Lagopus lagopus* also occur, although they are generally rather scarce in the west where these heaths are so common. Most forms of the community are not noted for rare plants, but the herb-rich *Thymus-Carex* sub-community can include a few uncommon species such as *Silene acaulis*, *Rubus saxatilis*, *Galium boreale*, *Gentianella campestris*, *Anthyllis vulneraria*, *Selaginella selaginoides*, *Lathyrus linifolius*, *Pyrola media*, *Filipendula vulgaris*, *Orobanche alba*, *Trifolium medium*, *Saxifraga hypnoides* and *Geranium sylvaticum*. South-facing stands of *Calluna-Erica* heath along the coasts of some of the Inner Hebrides are the habitat of the rare Marsh Fritillary butterfly *Euphydryas aurinia* and the Transparent Burnet moth *Zygaena purpuralis*.

**Management**

Some stands of the *Racomitrium* sub-community of *Calluna-Erica* heath at high elevations and in exposed situations may be near-natural, as may some stands of the *Thymus-Carex* sub-community on ungrazed cliffs. However, most *Calluna-Erica* heath in the British uplands lies within the altitudinal range of woodland, and would become colonised by trees in the absence of burning and grazing. The heathy *Quercus-Betula-Dicranum* woodland W17 would be the most likely type of woodland to become established, at least at first. More herb-rich woodlands might develop after a few years when the fertility of the soils would have increased.

*Calluna-Erica* heaths are not generally managed by burning in small patches as are the grouse moors further east, but are typically burnt occasionally in large patches to provide nutritious new growth for sheep, cattle and deer. In many parts of the west Highlands the heather is simply set on fire and allowed to burn in an uncontrolled way over a large area. The ideal practice is to burn the heather before it becomes old, leggy and degenerate, and to burn lightly enough to remove the old woody growth but not to destroy the herbs, lichens and bryophytes (e.g. Phillips *et al.* 1993; Moorland Working Group 1998). Severe fires on rocky slopes with thin soil can strip the humic layers, leaving an acid crust of poor soil and an impoverished flora. Following burning, the heaths should be lightly grazed so that the dwarf shrubs persist in a short canopy over a diverse array of herbs and lower plants.

Intensive grazing can weaken the dwarf shrubs and may lead to their eventual disappearance. Small stands of *Calluna-Erica* heath within a matrix of wet heaths and bogs can be preferentially grazed because the soils are drier and more fertile and the vegetation more productive. Under heavy grazing, patches of *Calluna-Erica* heath develop a distinctive appearance, with dark, browsed hummocks of *Calluna* and *E. cinerea* set in a green, tightly grazed turf. Where grazing is excessively heavy the dwarf shrubs are eventually eliminated, and the heath is replaced by *Festuca-Agrostis-Galium* grassland U4 or *Nardus-Galium* grassland U5 on acid soils, or by *Festuca-Agrostis-Thymus* grassland CG10 on more basic soils.

Bracken can spread into stands of *Calluna-Erica* heath, especially where the heath covers deep brown soils adjacent to stands of bracken and there is heavy grazing. This is less likely to happen if the heaths are burnt and grazed lightly.

# H12 *Calluna vulgaris-Vaccinium myrtillus* heath

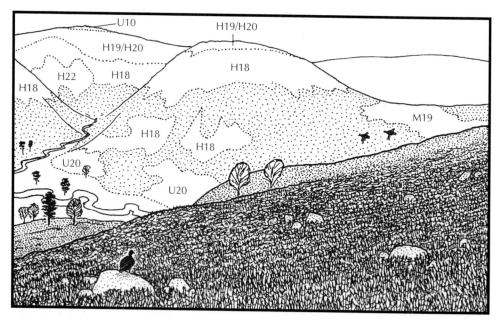

Extensive H12 *Calluna vulgaris-Vaccinium myrtillus* heath (foreground and stippled): on well-drained, gentle to steep slopes, mostly downslope of H18 *Vaccinium-Deschampsia* heath and upslope of U20 *Pteridium-Galium* community. In the distance at the right of the picture, H12 has replaced previous M19 *Calluna-Eriophorum* blanket mire on deep peat which has become dried out by repeated moor-burning.

Other vegetation types in this picture:

H19 *Vaccinium-Cladonia* heath
H20 *Vaccinium-Racomitrium* heath
H22 *Vaccinium-Rubus* heath
U10 *Carex-Racomitrium* moss-heath

H12 is also shown in the pictures for H4, H8, H9, H16, H19–H22, U3, CG10 and U4.

## Synonyms

*Callunetum vulgaris p.p.* (McVean and Ratcliffe 1962); Boreal heather moor (Birse and Robertson 1976); *Calluna* dry heath B1a *p.p.* (Birks and Ratcliffe 1980).

## Description

*Calluna-Vaccinium* heaths are usually extensive and visually uniform, typically covering whole landscapes with a vast, unbroken, red-brown sward. When the heather comes into flower in the late summer, the dull dark moorlands become a blaze of vivid purple. The vegetation consists of a springy, deep canopy of *Calluna vulgaris*, entwined with the bright-green leafy shoots of *Vaccinium myrtillus*. There can also be much *Empetrum nigrum* and *V. vitis-idaea*. After a fire, when the heather is recovering, *V. myrtillus* may be temporarily dominant. Other vascular plants are usually inconspicuous below the canopy of dwarf shrubs, but the fine, dark-green leaves of *Deschampsia flexuosa* can be seen above the heather. Most stands of *Calluna-Vaccinium* heath have a rich golden-red underlay of large pleurocarpous mosses, including *Hylocomium splendens*, *Pleurozium schreberi*, *Rhytidiadelphus loreus* and *Hypnum jutlandicum*.

There are three sub-communities; these are rather less distinct than the sub-communities of *Calluna-Erica* heath H10 but represent similar ecological gradients. The *Calluna vulgaris* sub-community H12a may be regarded as the typical form of dry heather moor, with no distinguishing species other than those that define the community as a whole. The *Vaccinium vitis-idaea-Cladonia portentosa* sub-community H12b is characteristic of higher altitudes; as well as more *V. vitis-idaea*, it has more *Empetrum nigrum*, and usually a few lichens such as *Cladonia portentosa* and *C. uncialis* ssp. *biuncialis*. The *Galium saxatile-Festuca ovina* sub-community H12c is a more heavily grazed and grassy form of heath, with *Potentilla erecta*, *Galium saxatile*, *Festuca ovina* and *Nardus stricta*. Stands tend to be less mossy than the other sub-communities, and the canopy is paler, more variegated and less continuous, with green spreads of grassland showing between the heather bushes.

In the eastern Highlands, the Cairngorm foothills, the lower hills north of Ballater and on Orkney, there are *Calluna-Vaccinium* heaths in which the ground under the dwarf shrubs is white with bushy lichens, including *Cladonia arbuscula*, *C. portentosa* and *C. uncialis* ssp. *biuncialis*. These have some similarities to the *Vaccinium-Cladonia* sub-community, but do not appear to be represented in the NVC.

## Differentiation from other communities

*Calluna-Vaccinium* heath may be confused with any of the wet, damp or dry heaths in which *Calluna vulgaris* is the dominant species.

The general rarity of *Erica tetralix*, *Trichophorum cespitosum*, *Molinia caerulea* and *Sphagnum* species distinguishes the community from *Trichophorum-Erica* M15 and *Erica-Sphagnum compactum* M16 wet heaths. *Sphagnum capillifolium* is typically rare or absent, in contrast to *Calluna-Vaccinium-Sphagnum* damp heath H21. *Vaccinium-Rubus* damp heath H22 usually contains *S. capillifolium*, and also northern or montane species such as *Empetrum nigrum* ssp. *hermaphroditum*, *Carex bigelowii*, *Vaccinium uliginosum*, *Rubus chamaemorus* and *Cornus suecica*.

*Calluna-Vaccinium* heath differs from *Calluna-Arctostaphylos uva-ursi* heath H16 in the scarcity of *Arctostaphylos uva-ursi*. The community can be distinguished from the montane *Calluna* heaths (*Calluna-Cladonia* H13, *Calluna-Racomitrium* H14, *Calluna-Juniperus* H15 and *Calluna-Arctostaphylos alpinus* H17) because the heather is never prostrate or severely dwarfed by the wind, and montane species are generally absent.

Some forms of *Calluna-Vaccinium* heath closely resemble *Calluna-Erica* heath, and transitional stands can occur. The two communities can usually be distinguished by the relative amounts of *Vaccinium myrtillus* and *Erica cinerea*: *V. myrtillus* is the commoner species in *Calluna-Vaccinium* heath, and the reverse is true of *Calluna-Erica* heath. Among other differences, pleurocarpous mosses are generally more common in *Calluna-Vaccinium* heath, and *Potentilla erecta* is more common in *Calluna-Erica* heath. *Calluna-Vaccinium* heath differs from *Calluna-Ulex gallii* heath H8 in the absence or scarcity of *Ulex gallii*, and from *Ulex gallii-Agrostis* heath H4 in the absence or scarcity of both *U. gallii* and *Agrostis curtisii*. It generally has more *V. myrtillus* and less *Pohlia nutans* than *Calluna-Deschampsia* heath H9; the *Vaccinium-Cladonia* sub-community H9b may contain much *V. myrtillus* but it lacks the dense underlay of pleurocarpous mosses typical of *Calluna-Vaccinium* heath.

## Ecology

*Calluna-Vaccinium* heath occurs over a wide variety of siliceous rocks, including sandstone, gritstone, meta-sediments and granite, or on drift and gravel derived from acid rock. The soils are mostly moist but free-draining, nutrient-poor, acid podsols, but the community can also occur on rankers, brown earths and brown podsolic soils. *Calluna-*

*Vaccinium* heath occurs on hillsides, on crags and ledges, among scree and boulders on sun-exposed slopes, and on the sides of ravines. Most stands are between 200 m and 600 m, and on the heathery hills of the eastern Highlands *Calluna-Vaccinium* heath commonly forms a well-marked band of dark heathy vegetation at moderate altitudes, giving way upslope to montane *Calluna* and *Vaccinium* heaths.

## Conservation interest

*Calluna-Vaccinium* heath forms a large proportion of the total extent of heather moorland in the British uplands, and as such it is of international significance for nature conservation as well as for its contribution to the visual appeal of the upland landscape. This type of heath is the prime breeding and feeding habitat for Red Grouse *Lagopus lagopus*, and other notable nesting birds include Black Grouse *Tetrao tetrix*, Merlin *Falco columbarius*, Hen Harrier *Circus cyaneus*, Short-eared Owl *Asio flammeus* and Twite *Carduelis flavirostris*. The uncommon *Listera cordata*, *Lycopodium annotinum* and *Diphasiastrum complanatum* ssp. *issleri* can occur in *Calluna-Vaccinium* heath. In some places there is a rich fauna of plant-eating insects, including the large and impressive Emperor Moth *Pavonia pavonia*. In Scotland, Mountain Hares *Lepus timidus* live in this community and are a common prey of Golden Eagles *Aquila chrysaetos*.

This form of heath occurs widely in the British uplands from south-west England to Shetland, and has been recorded from most upland areas apart from parts of the Hebrides. It accounts for most of the *Calluna* heathland in the central and eastern Highlands, the eastern parts of the Southern Uplands, the northern Pennines, the eastern Lake District and eastern Wales. It is most common in the northern Pennines and eastern Scotland. It is more widespread than *Calluna-Erica* heath in the uplands of eastern Great Britain, where the climate is cooler and less oceanic than in more westerly regions; it also tends to occur at higher altitudes. However, the two communities form mosaics in many areas, with *Calluna-Vaccinium* heath usually on deeper soils than *Calluna-Erica* heath.

*Calluna-Vaccinium* heaths resembling this community have been recorded in eastern Ireland (Hobbs and Averis 1991b) and in western Norway.

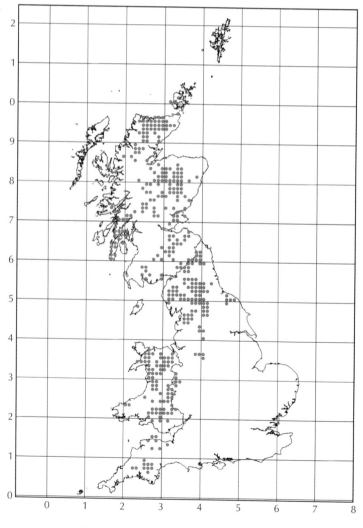

## Management

Almost all *Calluna-Vaccinium* heaths are the result of woodland clearance, although some high-altitude stands may be near-natural. The original woodlands may have been felled or may have died out because grazing and burning prevented the establishment of young trees. In many places there is likely to have been a direct succession from heathy woodland to heather moor. In other places the woodland might have been less heathy, and heather would have become commoner once the habitat became more open and the soils became increasingly leached and impoverished. There are obvious similarities between *Calluna-Vaccinium* heath and the ground flora of dry *Pinus-Hylocomium* woodland W18 and heathy *Quercus-Betula-Dicranum* woodland W17; these communities comprise the natural vegetation of many of the places where *Calluna-Vaccinium* heaths are now so common.

*Calluna-Vaccinium* heaths are the backbone of the grouse moors of northern England, the Southern Uplands and the eastern Highlands. In the Highlands they also form substantial tracts of deer forest. Most *Calluna-Vaccinium* heaths have been maintained as open moorland for decades by controlled burning in small patches. This produces a patchwork of heather of different ages, with a good supply of young nutritious growth and also tall stands that provide shelter for nesting birds. Management of this sort provides an ideal habitat for red grouse and enables large populations to build up. However, too frequent burning can lead to loss of dwarf shrubs, and the replacement of heaths by grasslands (Thompson *et al.* 1995). This has been done deliberately in many parts of Wales where grassland has been considered more valuable than heathland for agricultural purposes (Ratcliffe 1959). Recommendations for good burning practice are given by MAFF (1992), Phillips *et al.* (1993), the Moorland Working Group (1998), and Scotland's Moorland Forum (2003). Burning is beneficial for nature conservation provided that steep, rocky slopes are excluded, and that the heaths are not burned too frequently nor at too high a temperature. Richer examples of *Calluna-Vaccinium* heath tend to be on ground where burning is on a long, rather than a short, rotation. A rich array of herbs, bryophytes and lichens can flourish when the dense canopy of shrubs is opened up, especially on more basic soils, although the larger mosses recover slowly from burning and can easily be lost.

Most stands of *Calluna-Vaccinium* heath (including those that are managed as grouse moors) are grazed by sheep and deer, and locally by cattle, goats and ponies. Intensive grazing, especially when combined with injudicious burning, can cause a gradual change from heath to grassland dominated by *Festuca ovina*, *Agrostis* species, *Nardus stricta* or *Juncus squarrosus*, through intermediate stages during which the heather becomes increasingly contorted and degenerate, and grasses gain a firmer foothold in the sward. Large areas of heather moorland in southern Scotland, northern England and Wales have been converted to grassland because of unsustainable sheep grazing over many years. This can happen in a fairly short time: for example, during the 1980s at least one stand of *Calluna-Vaccinium* heath in the eastern Lake District was converted to species-poor *Agrostis-Festuca* grassland purely by grazing sheep.

*Calluna-Vaccinium* heaths can be regarded as a seral stage in the progression from grassland to woodland (Legg 1995), and reversion to woodland (e.g. *Quercus-Betula-Dicranum*) may occur if heaths are not burned and the number of grazing animals is sufficiently low. This process may occur rapidly at the margins of existing woodland, but elsewhere can take a long time. A site in the North York Moors some 1–2 miles from existing woodland has been unburnt and ungrazed for 60 years and is only just developing a little regeneration in the form of short, young birch and rowan (Mick Rebane, pers. comm.).

# H13 *Calluna vulgaris-Cladonia arbuscula* heath

## Synonyms

*Cladineto-Callunetum* (McVean and Ratcliffe 1962); *Alectorio-Callunetum vulgaris* (Birse and Robertson 1976); Species-poor dwarf *Calluna* heath B2a and *Calluna*-lichen heath B2c (Birks and Ratcliffe 1980).

Typical habitats of H13 *Calluna-Cladonia* heath (stippled):

1 Exposed ridges and shoulders on middle to upper slopes
2 In mixed heath/grassland mosaics on exposed undulating plateaux
3 On tops of moraines in exposed situations
4 On raised areas among high, exposed blanket bogs

H13 is also shown in the picture for H19–H22.

## Description

*Calluna-Cladonia* heaths are thin, tightly woven mats of vegetation in which the creeping, prostrate shoots of heather spread out over dry stony soils, usually with a thick white frosting of lichens. Other typical species include montane plants that can endure frost and biting winds, such as *Carex bigelowii*, *Empetrum nigrum* ssp. *hermaphroditum*, *Vaccinium uliginosum* and the clubmosses *Diphasiastrum alpinum* and *Huperzia selago*. There can be a sprinkling of *Nardus stricta*, *Agrostis canina* and other grasses. The crisp carpet of lichens is composed of *Cladonia arbuscula*, *C. portentosa*, *C. uncialis* ssp. *biuncialis*, *C. rangiferina*, *Coelocaulon aculeatum*, *Alectoria nigricans*, *Cetraria islandica* and other robust species. Bryophytes are less common than lichens, but there can be a thin weft of *Hypnum jutlandicum*, a little *Racomitrium lanuginosum*, and a few tufts of *Polytrichum alpinum* or *P. juniperinum*. In some stands, for example on solifluction lobes on gently sloping ground, the vegetation forms distinctive patterns in which the heather grows in parallel lines or rows of crescents with the shoot tips pointing away from the prevailing wind.

254

There are three sub-communities. The *Cladonia arbuscula-Cladonia rangiferina* sub-community H13a includes the less montane stands. *Erica cinerea, E. tetralix, Molinia caerulea, Nardus* and *Festuca ovina* can occur here, although they are not common. Although Rodwell (1991b) states that lichens are more common in the *Cladonia* sub-community than in the other sub-communities and that they tend to cover more ground than the dwarf shrubs, this difference is not apparent from the published data tables. In the more montane *Empetrum nigrum* ssp. *hermaphroditum-Cetraria nivalis* sub-community H13b, the montane lichen *Cetraria nivalis* grows in conspicuous creamy-yellow patches among the dwarf shrubs; *Deschampsia flexuosa* is also common. The *Cladonia crispata-Loiseleuria procumbens* sub-community H13c is distinguished by the dwarf shrubs *Vaccinium myrtillus* and *Loiseleuria procumbens*, and the lichens *Cladonia crispata, Ochrolechia frigida* and *Thamnolia vermicularis*.

Throughout the Highlands and in the Lake District and north Wales there are prostrate *Calluna* heaths with very few lichens and little or no *Racomitrium lanuginosum*. These heaths typically have few montane species. They have similarities to both *Calluna-Cladonia* heath and its western counterpart *Calluna-Racomitrium* heath H14, but have no clear place within the NVC. They correspond to Birks and Ratcliffe's (1980) Species-poor dwarf *Calluna* heath B2a.

### Differentiation from other communities

*Calluna-Cladonia* heath is distinguished from the sub-montane *Calluna-Erica* H10, *Calluna-Vaccinium* H12 and *Calluna-Arctostaphylos uva-ursi* H16 dry heaths by the extremely short and typically prostrate growth-form of the heather and other dwarf shrubs. There are also more montane species. Some stands of the *Cladonia* sub-community of *Trichophorum-Erica* wet heath M15c have a conspicuous underlayer of lichens, but again the heather is not prostrate, and wet heath species such as *Erica tetralix, Trichophorum cespitosum, Molinia* and Sphagna are common. Among the montane heaths, *Calluna-Cladonia* heath can be confused only with *Calluna-Racomitrium* heath, but it has more lichens and correspondingly less *Racomitrium lanuginosum*.

### Ecology

*Calluna-Cladonia* heath is a montane community of high, exposed, windswept spurs, shoulders and summits: places that catch the bitter winds of winter, are blown clear of snow, and are afflicted by frosts. It usually occupies convex slopes with acid, podsolised, free-draining, stony soils. *Calluna vulgaris* reaches its highest altitudes in Great Britain and Ireland in this type of vegetation, at over 1100 m in the Cairngorms (Ratcliffe 1977; Nethersole-Thompson and Watson 1981). Most stands lie above the altitudinal limit of woodland and so are truly montane, but some small stands occur on the exposed tops of moraines below 600 m. *Calluna-Cladonia* heath is the eastern, boreal counterpart of the more oceanic montane *Calluna-Racomitrium* heath.

### Conservation interest

*Calluna-Cladonia* heath is one of a series of near-natural, high-altitude heath communities that comprise an internationally important element of the British uplands. The effects of a severe montane climate on the vegetation can be clearly seen in this community, and it forms a valuable link with the lichen-rich montane heaths of Scandinavia. The rare montane lichens *Alectoria ochroleuca, A. sarmentosa* ssp. *vexillifera* and *Pertusaria xanthostoma* are almost confined to this type of vegetation, and it is one of the main British habitats of *Cetraria nivalis* (Fryday 1997). Heaths of this type are used as nesting habitat by Dotterel *Charadrius morinellus* and Ptarmigan *Lagopus mutus*.

## H13 *Calluna vulgaris-Cladonia arbuscula* heath

The distribution of *Calluna-Cladonia* heath is centred on the east and central Highlands: the great plateaux of the Cairngorms, Lochnagar and the Drumochter hills. It extends west to Beinn Dearg at the head of Loch Broom and other hills in Wester Ross, and to Ben Dubhchraig in western Perthshire. There are a few fragmentary stands in the Southern Uplands, the Lake District and the northern Pennines. In north Wales, semi-prostrate *Calluna* heaths with few lichens and no *Racomitrium lanuginosum* occur on the Carneddau and Rhinog ranges and on Cadair Idris (Averis and Averis 2000b).

There is vegetation similar to *Calluna-Cladonia* heath in western Norway (McVean and Ratcliffe 1962) and Ireland, but it is not recorded in the Faroes.

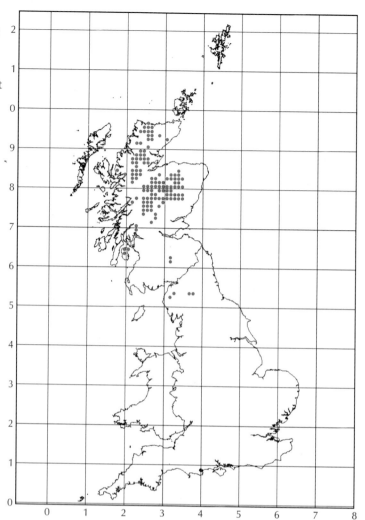

### Management

Virtually all stands of *Calluna-Cladonia* heath above the tree-line are evidently climax vegetation and are not maintained by management. This is a fragile vegetation type, which can be damaged by heavy or even moderate grazing in the montane zone. The trampling feet of sheep and deer break up the carpet of lichens in dry weather, and nutrients in animal dung and urine enrich the soils and change the species composition (e.g. Thompson *et al.* 1987, Thompson and Brown 1992). The lichens can also be damaged by vehicle tracks, notably where vehicles rather than ponies are used for carrying deer carcasses down from the hills. Localised damage may also be caused by human trampling on hills that are popular with walkers, not least because people like to ascend the higher parts of hills by way of their spurs and outlying ridges. Some heaths have been damaged or even destroyed by fires that have run out of control and swept up into the montane zone.

# H14 *Calluna vulgaris-Racomitrium lanuginosum* heath

## Synonyms
*Rhacomitreto-Callunetum* (McVean and Ratcliffe 1962; Birks 1973); *Calluna-Racomitrium lanuginosum* heath B2b (Birks and Ratcliffe 1980).

Typical habitats of H14 *Calluna-Racomitrium* heath (stippled):

1  Exposed ridges and shoulders on middle to upper slopes
2  In mixed heath/grassland mosaics on exposed undulating plateaux
3  On tops of moraines in exposed situations
4  On raised areas among high, exposed blanket bogs

H14 is also shown in the picture for H19–H22.

## Description
Like *Calluna-Cladonia* heath H13, this is a montane heath in which the stems of heather either lie flat along the ground or are more erect but severely pruned by the wind to a height of only about 5 cm. The heather is surrounded by a dense, silvery-grey carpet of *Racomitrium lanuginosum*. Pricking up through this low mat of moss and heather are a few small vascular plants including *Carex bigelowii*, *Erica cinerea*, *Diphasiastrum alpinum*, *Dactylorhiza maculata* and, in some stands, *Loiseleuria procumbens*, *Arctostaphylos alpinus* or *Juniperus communis* ssp. *nana*. In many places the vegetation is discontinuous and distributed in patches over bare stony ground. The patches may be aligned in noticeable stripes parallel to the direction of the prevailing wind; these can occur on solifluction lobes and terraces.

There are three sub-communities. The *Festuca ovina* sub-community H14a is a grassy form of heath with *Potentilla erecta*, *Huperzia selago*, *Carex pilulifera*, *Festuca vivipara* or *F. ovina* and *Agrostis canina*. The *Empetrum nigrum* ssp. *hermaphroditum* sub-community H14b is the most montane form of the community, and is characterised by *Empetrum nigrum* ssp. *hermaphroditum* and the lichens *Cetraria islandica*, *Cladonia gracilis* and *Ochrolechia frigida*. The *Arctostaphylos uva-ursi* sub-community H14c has a more mixed carpet of dwarf shrubs including *Arctostaphylos uva-ursi, A. alpinus, Erica cinerea* and *Empetrum nigrum* ssp. *nigrum*. In practice it can be difficult to separate the sub-communities; many stands have an intermediate flora.

Throughout the Highlands and in the Lake District and north Wales there are prostrate *Calluna* heaths with little or no *R. lanuginosum* and few lichens. They resemble both *Calluna-Racomitrium* heath and its eastern counterpart *Calluna-Cladonia* heath, but have no obvious place in the current NVC scheme.

## Differentiation from other communities

*Calluna-Racomitrium* heath may be confused with other forms of prostrate montane heath. *Calluna-Cladonia* heath has less *Racomitrium lanuginosum* and more lichens. *Calluna-Juniperus* heath H15 has more *Juniperus communis* ssp. *nana*, and the heather is usually sparser. *Calluna-Arctostaphylos alpinus* heath H17 has more *Arctostaphylos alpinus* together with other montane dwarf shrubs such as *Loiselurea procumbens*.

Some forms of *Calluna-Racomitrium* heath resemble *Carex-Racomitrium* moss-heath U10 but have more *Calluna*. Western stands of the *Cladonia* sub-community of *Trichophorum-Erica* wet heath M15c can have a similar flora to *Calluna-Racomitrium* heath, including much *R. lanuginosum*, but have more *Trichophorum cespitosum*,

*Racomitrium lanuginosum* is a common plant in upland vegetation only in the west, and *Calluna-Racomitrium* heath has a markedly western distribution in Great Britain. The community is known mainly from the north-west Highlands, Skye, the Outer Hebrides, Orkney and Shetland. It also occurs sparingly in the eastern Highlands, on Mull and Islay, and in north Wales.

Montane heath vegetation of this type also occurs in Ireland (Hobbs and Averis 1991b) but has not been described in upland areas elsewhere in the world.

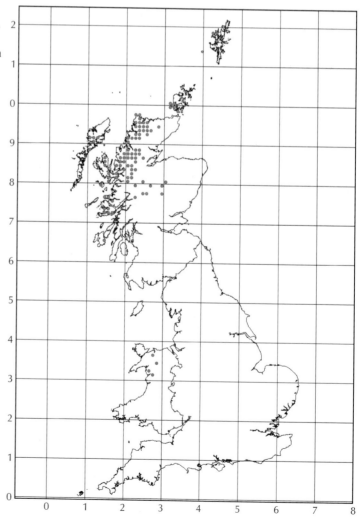

*Molinia caerulea* and *Erica tetralix*. *Calluna-Erica* heath H10 also has a similar flora to *Calluna-Racomitrium* heath but the canopy is never as short and tight, is less obviously wind-pruned, and has less *R. lanuginosum* and no montane species.

## Ecology
*Calluna-Racomitrium* heath is a community of exposed open ground on ridges and high shoulders that are blown clear of snow. It generally occurs on flat or gently sloping ground. It develops mainly over acid rocks, on free-draining rankers or podsols. Many stands are on gravelly moraines. The community appears to be natural climax vegetation, with its structure and flora determined by climate and topography, although in some places it is grazed. Although most common at higher altitudes, it descends to below 200 m in the far north-west – lower than other montane *Calluna* heaths (McVean and Ratcliffe 1962).

## Conservation interest
*Calluna-Racomitrium* heath is notable as a near-natural vegetation type that is very rare globally. Consequently, it is of great value for nature conservation. Several scarce plants can occur in this community. These include *Arctostaphylos alpinus*, *Juniperus communis* ssp. *nana*, *Loiseleuria procumbens* and *Pseudorchis albida*, oceanic liverworts such as *Herbertus aduncus* ssp. *hutchinsiae*, *Plagiochila carringtonii*, *Scapania nimbosa*, *S. ornithopodioides*, *Anastrophyllum donnianum* and the Red Data Book species *A. joergensenii*, and montane lichens such as *Ochrolechia frigida*, *Thamnolia vermicularis* and *Siphula ceratites*. Dotterel *Charadrius morinellus* and, more widely, Ptarmigan *Lagopus mutus* are often encountered first in this type of heath as one ascends the high mountain tops in the Highlands.

## Management
As is the case with other near-natural montane heaths, *Calluna-Racomitrium* heath is highly susceptible to management treatments. Uncontrolled fires can seriously damage it; regrowth is slow in the severe montane environment. The community can withstand light grazing but may have been lost from some places as a result of heavy grazing and associated loss of dwarf shrubs (Thompson and Brown 1992); this may have reduced its distribution in southern Great Britain. However, it appears that the community may be less susceptible to grazing damage than *Calluna-Cladonia* heath.

# H15 *Calluna vulgaris-Juniperus communis* ssp. *nana* heath

## Synonyms
*Juniperetum nanae* (McVean and Ratcliffe 1962); *Juniperus nana* nodum (Birks 1973); Mixed prostrate dwarf shrub heath B2d (Birks and Ratcliffe 1980).

Typical habitats of H15 *Calluna-Juniperus* heath (stippled):

1   Exposed ridges and shoulders on middle to upper slopes
2   In mixed heath/grassland mosaics on exposed undulating plateaux
3   On tops of moraines in exposed situations

## Description
This is one of our more striking upland heaths, with the silvery-green bushes of *Juniperus communis* ssp. *nana* sprawling over bare or stony ground, surrounded by prostrate *Calluna vulgaris* and with a sparse open mat of bryophytes such as *Racomitrium lanuginosum* and *Hypnum jutlandicum*. The plants grow in a habitat that looks bleak and desolate, yet beneath the low and prickly mats of juniper there is enough shelter for an array of small vascular plants and bryophytes to grow, including *Erica cinerea*, *Huperzia selago*, *Potentilla erecta*, *Pleurozia purpurea*, *Mylia taylorii* and *Scapania gracilis*. Notable oceanic liverworts such as *Herbertus aduncus* ssp. *hutchinsiae* and *Plagiochila carringtonii* occur locally, and testify to the humid and sheltered microclimate beneath the dwarf shrubs in an otherwise exposed environment. The dwarf shrubs *Arctostaphylos uva-ursi* and *A. alpinus* are common in some stands.

## Differentiation from other communities
This community has more *Juniperus communis* ssp. *nana* than any of the other prostrate *Calluna* heaths (*Calluna-Cladonia* heath H13, *Calluna-Racomitrium* heath H14 and *Calluna-Arctostaphylos alpinus* heath H17). In the west Highlands, *J. communis* ssp. *nana* occurs in some stands of the *Cladonia* sub-community of *Trichophorum-Erica* wet

heath M15c, but the wet heath has *Erica tetralix*, *Trichophorum cespitosum* and *Molinia caerulea*, all of which are scarce in *Calluna-Juniperus* heath. In a few places, prostrate juniper grows in *Calluna-Erica* H10 and *Calluna-Vaccinium-Sphagnum* H21 heaths, but both of these communities have taller and thicker canopies of dwarf shrubs than *Calluna-Juniperus* heath, and *Calluna-Vaccinium-Sphagnum* heath also contains *Sphagnum capillifolium*.

## Ecology

*Calluna-Juniperus* heath occurs on shattered rock debris, cliff ledges and stony plateaux, generally where the parent rocks are hard and acid. Some of the largest examples are on quartzite, but there are also good stands on Torridonian sandstone, Lewisian gneiss, gabbro and granite. The soils are thin humic rankers lying directly over the bedrock.

The community occurs at moderate to high altitudes in the western Highlands, descending almost to sea-level in the extreme north-west. It is most common around the former tree-line. It is a near-natural climax vegetation type which was probably once much more extensive, and which may at one time have formed a zone of scrubby heath on acid soils above and within the tree-line throughout the western Highlands and Islands, the Lake District and north Wales.

## Conservation interest

*Calluna-Juniperus* heath is a rare and fragile community, known only in a few parts of Great Britain and Ireland and nowhere else in the world. It provides an important habitat for some scarce oceanic liverworts, including *Herbertus aduncus* ssp. *hutchinsiae* and *Plagiochila carringtonii*. The stands on Beinn Eighe are most famous for the very rare *H. borealis*, which forms large rich-orange-red cushions contrasting spectacularly with the grey-green juniper leaves and the shining white background of quartzite. This is the only known British locality for *H. borealis*, a species that is known elsewhere in the world only from a few sites in south-west Norway. It is extraordinary that it does not appear to occur anywhere else in Scotland.

*Calluna-Juniperus* heath and certain forms of the *Cladonia* sub-community of *Trichophorum-Erica* wet heath are the principal habitats for *Juniperus communis* ssp. *nana* in Great Britain. The differences between *J. communis* ssp. *communis* and *J. communis* ssp. *nana* do not seem to be consistent. Ssp. *communis* is defined as erect or spreading, prickly, with narrow, gradually tapering leaves that spread almost at right angles to the stem; ssp. *nana* is prostrate and less prickly, with ascending or loosely appressed, broader leaves that taper suddenly to a short point (Clapham *et al.* 1987). However, it is quite easy to find prostrate plants with the leaves and prickles of ssp. *communis*, erect bushes with the leaves of ssp. *nana*, and even individual plants in which some shoots have the characteristics of ssp. *communis* and others those of ssp. *nana*. McVean (1961) has speculated that there was once continuous variation in the stature and growth-form of juniper, ranging from prostrate forms in the west (ssp. *nana*) to upright forms in the east (ssp. *communis*), and that fragmentation of the original population has led to isolated populations of distinct forms that are now given the status of sub-species. The juniper in the superb heaths on Beinn Eighe appears to be closer to ssp. *communis* than to ssp. *nana* (Donald McVean, pers. comm.). *Calluna-Juniperus* heaths are important as a source of material for detailed biosystematic studies on juniper, and as places where the remaining genetic diversity is probably high. Current research is investigating the past status of juniper in Great Britain and its present-day phenotypic and genotypic variation.

## H15 *Calluna vulgaris-Juniperus communis* ssp. *nana* heath

The community is almost confined to the north-west Highlands of Scotland, the Outer Hebrides, Skye, Jura, Islay and Arran; records from Jura and Islay are insufficiently localised to be shown on the distribution map. There are fragmentary outliers on the Borrowdale Fells in the Lake District and in Snowdonia in north Wales. The largest stands are on the Cambrian quartzite hills of Foinaven and Cranstackie in Sutherland (McVean and Ratcliffe 1962; Ratcliffe 1977; Averis and Averis 1997a), on Beinn Eighe in Wester Ross (McVean and Ratcliffe 1962; Averis and Averis 1998a), and on the Cuillins of Skye (Averis and Averis 1997c, 1998b, 1999c). There are also some fine stands in South Harris (A B G Averis 1994).

There is similar vegetation in western Ireland (Horsfield *et al.* 1991, Averis *et al.* 2000).

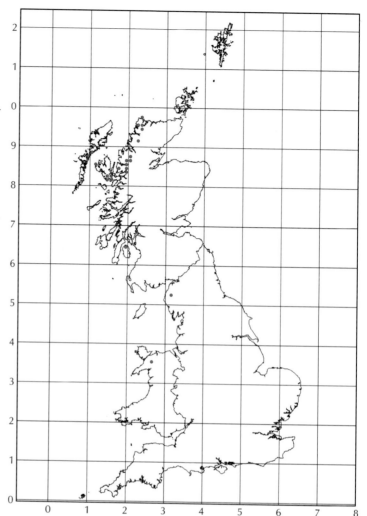

## Management

This is a climax vegetation type and is easily destroyed by burning. Dwarf juniper and the characteristic oceanic bryophytes are particularly sensitive to fire. Juniper remains in peat have been reported widely across Scotland, and it seems that this species may once have been more common throughout the Highlands (McVean 1961). McVean and Ratcliffe (1962) considered that *Calluna-Juniperus* heath was probably once widespread on a variety of acid rocks in Scotland. All remnant stands of *Calluna-Juniperus* heath should be protected from burning.

# H16 *Calluna vulgaris-Arctostaphylos uva-ursi* heath

## Synonyms
*Arctostaphyleto-Callunetum* (McVean and Ratcliffe 1962); *Vaccinio-Ericetum cinereae* (Birse and Robertson 1976); *Arctostaphylos uva-ursi-Calluna vulgaris* heath B1e (Birks and Ratcliffe 1980).

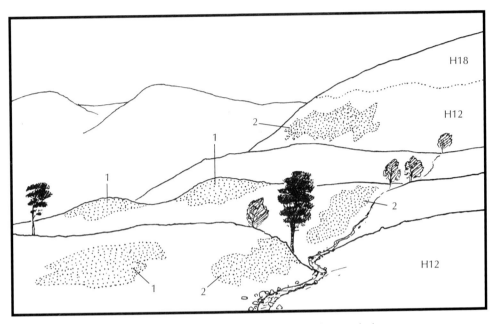

Typical habitats of H16 *Calluna-Arctostaphylos uva-ursi* heath (stippled):

1 Burned areas among H12 *Calluna-Vaccinium* heath, where H16 can be a temporary phase before H12 becomes re-established
2 More stable stands on thin soils on steep or rocky slopes

H16 heath is also shown in the picture for M16

## Description
These are short, dense to rather open dwarf-shrub heaths in which the dark and sombre tones of *Calluna vulgaris*, *Erica cinerea* and *Vaccinium vitis-idaea* are enlivened by the shining, bright-green, net-veined leaves of *Arctostaphylos uva-ursi* and in some places by scattered plants of *Genista anglica*, with its narrow green shoots and golden flowers. The shrubs are overtopped by the thin dark leaves of *Deschampsia flexuosa*, and at the feet of the vascular plants there is a discontinuous thin weft of mosses and lichens, such as *Pleurozium schreberi*, *Hylocomium splendens*, *Hypnum jutlandicum* and *Cladonia portentosa*.

There are three sub-communities: one herb-rich, one heathy, and one species-poor with many lichens. The *Pyrola media-Lathyrus linifolius* sub-community H16a is the herb-rich form, and has mesotrophic species such as *Pyrola media*, *Lathyrus linifolius*, *Viola riviniana*, *Lotus corniculatus*, *Anemone nemorosa* and *Rhytidiadelphus triquetrus* among the dwarf shrubs. The heathy *Vaccinium myrtillus-Vaccinium vitis-idaea* sub-community H16b has a less varied sward containing more *Vaccinium myrtillus* and commonly some *Empetrum nigrum* ssp. *nigrum*. In the *Cladonia* species sub-community H16c there is much *Trichophorum cespitosum* and *Carex pilulifera* together with a thin

263

frosting of *Cladonia uncialis* ssp. *biuncialis, C. floerkeana, C. coccifera, Coelocaulon aculeatum* and other lichens.

In western Scotland, there is a related type of heath composed mainly of *Calluna* and *A. uva-ursi*, with much *E. cinerea* but little or no *V. vitis-idaea* or *V. myrtillus*, which is not described in the NVC. *Salix repens* is common in some of these stands.

On Dalradian limestone in a few places in the south-eastern Highlands there is a very different, calcicolous form of *Arctostaphylos uva-ursi* heath with species including *Thymus polytrichus, Helianthemum nummularium, Festuca ovina, Briza media, Helictotrichon pratense, Carex flacca, Lotus corniculatus* and *Saxifraga aizoides*.

## Differentiation from other communities

The flora of *Calluna-Arctostaphylos uva-ursi* heath is similar to that of *Calluna-Vaccinium* heath H12, from which it differs in being co-dominated by *Calluna* and *Arctostaphylos uva-ursi*. *Erica cinerea* is common in both *Calluna-Arctostaphylos uva-ursi* heath and *Calluna-Erica* heath H10, but *Vaccinium* species are rare in *Calluna-Erica* heath and *A. uva-ursi* is not co-dominant with *Calluna*. *Genista anglica* is more common in *Calluna-Arctostaphylos uva-ursi* heath than in any other type of British heathland.

The western heaths dominated by *Calluna* and *Arctostaphylos uva-ursi* and with much *E. cinerea* differ from typical *Calluna-Arctostaphylos uva-ursi* heath mainly in the absence of *Vaccinium* species. The co-dominance of *Calluna* and *A. uva-ursi* distinguishes them from *Calluna-Erica* heath. They can be regarded as a counterpart of *Calluna-Erica* heath with much *A. uva-ursi*, just as typical *Calluna-Arctostaphylos uva-ursi* heath is a counterpart of *Calluna-Vaccinium* heath. In both cases the vegetation is worthy of recognition as a separate heathland type in which *A. uva-ursi* has gained a strong competitive edge as a result of some kind of disturbance or stress to the habitat.

## Ecology

This is a community of stony, acid, free-draining brown soils and podsols at altitudes up to about 750 m; the species-rich *Pyrola-Lathyrus* sub-community occurs on richer brown-earth soils. In the sub-montane zone of the eastern Highlands *Calluna-Arctostaphylos uva-ursi* heath is generally maintained by fire. *Arctostaphylos uva-ursi* can compete well with *Calluna vulgaris* when heaths are burned, but can be overwhelmed when the heather becomes taller and denser. Stands above 600–700 m may be natural vegetation maintained by the severe climate. In such situations the shrubs can be dwarfed; the heaths clinging to thin stony soils on the upper slopes and grading into prostrate *Calluna-Cladonia* heaths H13 on windswept summits and ridges.

The western form of *Calluna-Arctostaphylos uva-ursi* heath with much *Erica cinerea* seems to be a more or less natural and stable vegetation type. Here *A. uva-ursi* evidently gains a competitive edge where the growth of *Calluna* is restricted by thin, patchy soils on very rocky ground, or suppressed by strong westerly winds and salt spray.

## Conservation interest

*Calluna-Arctostaphylos uva-ursi* heath is one of the distinctive sub-montane communities of heather moorland in Great Britain, and is of international importance for nature conservation. It includes the eastern and boreal end of the range of variation within British heathland vegetation, characterised by northern or boreal species such as *Listera cordata, Trientalis europaea* and *Antennaria dioica*. The scarce *Lycopodium annotinum, Orthilia secunda* and *Pyrola media* are occasionally recorded in the community. Red grouse *Lagopus lagopus* feed in this type of vegetation.

*Calluna-Arctostaphylos uva-ursi* heath occurs mostly in the eastern and central Highlands, with outliers in the Southern Uplands and fragmentary stands in the Lake District, in Upper Teesdale (not shown on the distribution map) and on Orkney (Birks and Ratcliffe 1980). The western *Calluna-Arctostaphylos uva-ursi* heaths (mapped as open circles) have a scattered distribution through the western Highlands and Hebrides, from Colonsay and Mull north to Sutherland.

There are communities related to *Calluna-Arctostaphylos uva-ursi* heath in Norway, Sweden and Denmark.

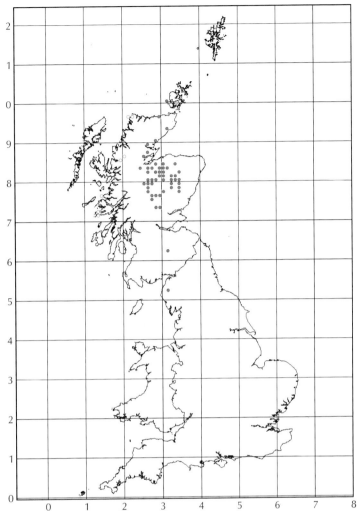

## Management

The typical forms of *Calluna-Arctostaphylos uva-ursi* heath in eastern Scotland are within grouse moors that are managed by rotational burning. The ability of the associated heathland species to re-establish after burning depends on the growth-phase of the *Calluna* at the time of the burn. In order to manage the community for nature conservation, it is important that heaths are burnt at the most favourable stage, when the heather is actively growing but not yet mature. It is also important that heaths are not excessively grazed after burning, as this can prevent the regeneration of heather and result in a change from heath to grassland (Rodwell 1991b). If the community is not managed at all it is most likely to revert to secondary birch woodland.

The related western form of heath with much *Erica cinerea* seems to be more natural. It occurs on cliffs and exposed rocky slopes where the vegetation is neither repeatedly burnt nor heavily grazed, and is in situations that might not naturally be wooded.

# H17 *Calluna vulgaris-Arctostaphylos alpinus* heath

## Synonyms

*Arctoeto-Callunetum* (McVean and Ratcliffe 1962); Mixed prostrate dwarf shrub heath B2d (Birks and Ratcliffe 1980).

Typical habitats of H17 *Calluna-Arctostaphylos alpinus* heath (stippled):

1   Exposed ridges and shoulders on middle to upper slopes
2   In mixed heath/grassland mosaics on exposed undulating plateaux
3   On tops of moraines in exposed situations
4   On raised areas among high, exposed blanket bogs

## Description

This community comprises tight, prostrate carpets of dwarf shrubs, with the shoot tips growing away from the prevailing wind. The canopy of dwarf shrubs is more varied than that of other montane heaths. There is a flattened layer of *Calluna vulgaris* hugging the ground in a tight, purple-brown mat, but this is enlivened by many other species: *Arctostaphylos alpinus* and *A. uva-ursi* with their shiny, bright-green, oval leaves, the small shoots of *Loiseleuria procumbens*, the yellow-green leaf-whorls of *Empetrum nigrum* (both ssp. *nigrum* and ssp. *hermaphroditum*), and in many places the prickly stems of *Juniperus communis* ssp. *nana*. Plants piercing through the mat of shrubs include *Huperzia selago* and *Deschampsia flexuosa*. The vascular plants are underlain by a patchy silvery layer of the moss *Racomitrium lanuginosum*, and interspersed with lichens, including the crisp white shoots of *Cladonia* species and the dark shiny thalli of *Cetraria islandica* and *Coelocaulon aculeatum*.

The two sub-communities differentiate between the more montane stands of the community and those that occur at lower altitudes. The *Loiseleuria procumbens-Platismatia glauca* sub-community H17a is the more montane form, with plants such as *Carex bigelowii*, *Diphasiastrum alpinum*, *Empetrum nigrum* ssp. *hermaphroditum* and *Vaccinium uliginosum* in the sward, together with *Coelocaulon aculeatum*, *Alectoria nigricans*, *Sphaerophorus globosus*, *Cetraria islandica* and other lichens. In the

*Empetrum nigrum* ssp. *nigrum* sub-community H17b, *E. nigrum* ssp. *nigrum*, *Potentilla erecta*, *Trichophorum cespitosum*, *Erica cinerea* and *Hypnum jutlandicum* take the place of the more montane species of the *Loiseleuria-Platismatia* sub-community, and there are fewer lichens.

## Differentiation from other communities

*Calluna-Arctostaphylos alpinus* heath is similar to the other montane prostrate dwarf-shrub heaths: *Calluna-Cladonia* heath H13, *Calluna-Racomitrium* heath H14 and *Calluna-Juniperus* heath H15. It can be distinguished from all of these communities by its more mixed canopy of dwarf shrubs which includes *Arctostaphylos alpinus*.

## Ecology

This form of heath occurs on high, bleak, windswept spurs and shoulders and where stony moraines break through blanket peat. The carpet of dwarf shrubs spreads over fine-grained, well-drained, usually acid soils and rock debris. Most stands are above 500 m, although the community occurs below 200 m in the extreme north-west Highlands. Winters are long and bitter at these altitudes and latitudes, and the dwarf shrubs are pruned by exposure to frost and wind.

The distribution of *Calluna-Arctostaphylos alpinus* heath is defined largely by that of *A. alpinus*, which is rare south of the Great Glen. All records of the community are from the north-western Highlands and Orkney.

There is similar vegetation in Scandinavia, although it generally has little *Calluna*.

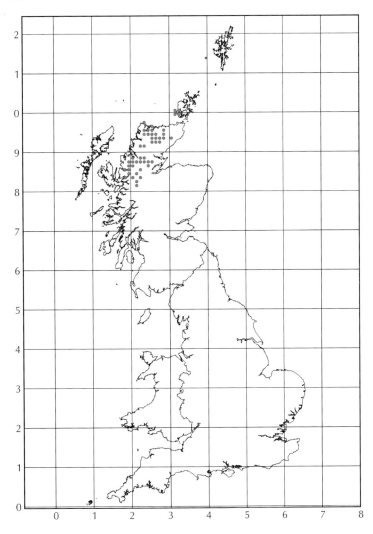

## Conservation interest

*Calluna-Arctostaphylos alpinus* heath is an important oceanic counterpart of the wind-swept lichen-rich montane heaths of eastern Scotland and Scandinavia. The quantity and diversity of lichens decreases with increasing distance west, and bryophytes become more common. Some western stands have a rich flora of large oceanic liverworts: *Pleurozia purpurea, Herbertus aduncus* ssp. *hutchinsiae, Scapania gracilis* and *Plagiochila carringtonii* have been recorded in the community in Wester Ross and Sutherland.

## Management

This form of montane heath is a climax community. It is probably little affected by grazing as long as the density of animals is not too great, but burning is severely damaging. One fire may be enough to eliminate the heath in places where regeneration is prevented by severe weather and soil erosion. All lichen-rich heaths are susceptible to damage by trampling and by the use of vehicles on high ground, and some stands of *Calluna-Arctostaphylos alpinus* heath could easily be damaged as a result of increased recreation in the Highlands.

# H18 *Vaccinium myrtillus-Deschampsia flexuosa* heath

## Synonyms

*Vaccineto-Empetretum* and *Festuceto-Vaccinetum* (McVean and Ratcliffe 1962); Southern *Vaccinium myrtillus* heath B3a, Snow-bed *Vaccinium myrtillus* heath B3b, Species-rich *V. myrtillus* heath B3c, *Vaccinium myrtillus-Nardus stricta* heath B3d and *Vaccinium vitis-idaea* heaths B3g (Birks and Ratcliffe 1980).

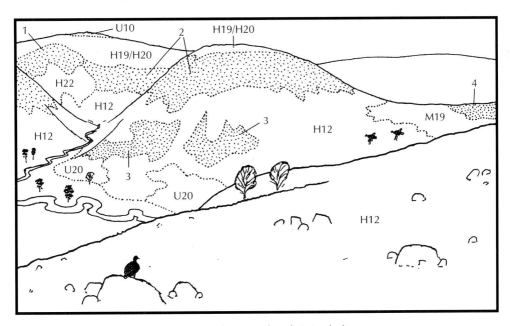

Typical habitats of H18 *Vaccinium-Deschampsia* heath (stippled):

1   High slopes where snow lies late into the year
2   Well-drained middle slopes, above the upper limit of *Calluna* heaths
3   Patches of H18 in heavily burned or grazed places among H12 *Calluna-Vaccinium* heath at lower altitudes
4   On deep peat, H18 having replaced M19 *Calluna-Eriophorum* blanket mire because of moor-burning combined with grazing

Other vegetation types in this picture:

H19 *Vaccinium-Cladonia* heath
H20 *Vaccinium-Racomitrium* heath
H22 *Vaccinium-Rubus* heath
U10 *Carex-Racomitrium* moss-heath
U20 *Pteridium-Galium* community

H18 is also shown in the pictures for M20, H8, H9, H16, H19–H22, U4–U6 and CG10–CG12.

## Description

Many hectares of hillsides in the uplands are clothed with stands of this rich-green, mossy heathland. It usually has a thick vigorous canopy of *Vaccinium myrtillus*, pierced through by the long narrow leaves and delicate flowers of *Deschampsia flexuosa*. *Galium saxatile* trails its dark-green shoots over the ground, and there is a deep carpet of mosses such as *Pleurozium schreberi*, *Dicranum scoparium*, *Hylocomium splendens* and *Hypnum jutlandicum*.

Three sub-communities are described in the NVC, although they are rather tenuously separated from each other. The *Hylocomium splendens-Rhytidiadelphus loreus* sub-community H18a is the typical and most common form of the community. The thick canopy of dwarf shrubs covers a dense soft layer of mosses including *Hylocomium splendens* and *Rhytidiadelphus loreus*, and is studded with short rosettes of *Blechnum spicant*. The *Alchemilla alpina-Carex pilulifera* sub-community H18b is grassier, with short sprigs of *V. myrtillus* set in a matrix of *Festuca ovina*, *Agrostis capillaris* and *Anthoxanthum odoratum*. There is usually a loose mat of *Alchemilla alpina* with its yellow-green flowers in summer; *Potentilla erecta* is common too. The common grassland moss *Rhytidiadelphus squarrosus* also grows here. Some grassy stands without *A. alpina* have been included in this sub-community, which confuses what is otherwise a well-defined unit of vegetation; such stands might be better classed as intermediate between the *Hylocomium-Rhytidiadelphus* sub-community and the *Vaccinium* sub-community of *Festuca-Agrostis-Galium* grassland U4e. The *Racomitrium lanuginosum-Cladonia* species sub-community H18c is a relatively species-poor heath, generally with a rather short and open sward. The species that define the other two sub-communities may occur here but are not so common. There can be a conspicuous pale frosting of the lichens *Cladonia arbuscula*, *C. portentosa*, *C. gracilis* and *C. uncialis* ssp. *biuncialis* encrusting the ground, and *Hypogymnia physodes* may grow in thick greyish clots on the stems of the dwarf shrubs. Patches of the moss *Racomitrium lanuginosum* can be common.

In northern England and eastern Wales there are heaths that closely resemble the *Hylocomium-Rhytidiadelphus* sub-community except that *V. vitis-idaea* takes the place of *V. myrtillus*. These heaths are common on rocky slopes just below the Millstone Grit 'edges' in the southern Pennines.

The current classification of *Vaccinium-Deschampsia* heaths does not adequately show the relationships between flora, environment and management. This is probably the result of inadequate sampling, and the fact that both near-natural and anthropogenic forms of *Vaccinium-Deschampsia* heath have been grouped together because of their floristic similarity. It may be more satisfactory to separate the more natural montane heaths containing *Empetrum nigrum* ssp. *hermaphroditum*, *Vaccinium uliginosum* and other montane plants from the less natural sub-montane heaths with *E. nigrum* ssp. *nigrum* or no *Empetrum* at all. The *Rhytidiadelphus-Hylocomium* sub-community of *Vaccinium-Racomitrium* heath H20d would probably be better incorporated with the montane form of the *Hylocomium-Rhytidiadelphus* sub-community of *Vaccinium-Deschampsia* heath. It seems sensible to separate off the sub-montane lichen-rich and *Racomitrium*-rich *Vaccinium-Deschampsia* heaths. The form of the *Alchemilla-Carex* sub-community with much *Alchemilla alpina* is a well-defined type.

### Differentiation from other communities
*Vaccinium-Deschampsia* heath can be confused only with other heath communities in which *V. myrtillus* is dominant.

The lichen-rich *Racomitrium-Cladonia* sub-community can resemble *Vaccinium-Cladonia* heath H19, although it occupies different habitats. It occurs mainly at moderate altitudes on dried-out peat within eroding blanket bogs, on grazed slopes and in snow-beds, whereas *Vaccinium-Cladonia* heath is characteristic of upper slopes and ridges at high altitudes. *Vaccinium-Cladonia* heath usually has a more continuous underlay of lichens, and has more montane species such as *Carex bigelowii*, *Empetrum nigrum* ssp. *hermaphroditum*, *Salix herbacea*, *Loiseleuria procumbens* and the lichens *Cetraria islandica*, *Ochrolechia frigida* and *Thamnolia vermicularis*. *Vaccinium-Deschampsia* heath has more acid grassland species, including *Potentilla erecta*, *Galium saxatile*, *Agrostis capillaris* and *Anthoxanthum odoratum*.

The *Rhytidiadelphus-Hylocomium* sub-community of *Vaccinium-Racomitrium* heath H20d can be similar to the *Hylocomium-Rhytidiadelphus* sub-community of *Vaccinium-Deschampsia* heath, and indeed it is closer floristically to this vegetation type than to the other sub-communities of *Vaccinium-Racomitrium* heath. However, montane species such as *C. bigelowii* and *E. nigrum* ssp. *hermaphroditum* are generally more common in *Vaccinium-Racomitrium* heath. The other three sub-communities of *Vaccinium-Racomitrium* heath are more obviously distinct from *Vaccinium-Deschampsia* heath, as they have much more *Racomitrium lanuginosum* and fewer pleurocarpous mosses.

The *Alchemilla-Carex* sub-community can be confused with the Typical sub-community of *Festuca-Agrostis-Alchemilla* grassland CG11a, but has more *Vaccinium myrtillus*, *Deschampsia flexuosa* and *Pleurozium schreberi*, and very little *Thymus polytrichus*. *Alchemilla alpina* and *V. myrtillus* are common in a local form of mossy *Festuca-Agrostis-Galium* grassland U4e, which also has much *D. flexuosa*. This grassland differs from *Vaccinium-Deschampsia* heath in that *V. myrtillus* is clearly subordinate to grasses and *A. alpina* (it is at least co-dominant in the heath), there is more *R. lanuginosum* and a less continuous layer of pleurocarpous mosses, and there may be some montane species, such as *Gnaphalium supinum* and *Luzula spicata*.

Grazed heath resembling *Calluna-Vaccinium-Sphagnum* heath H21 but without *Calluna* can be distinguished from *Vaccinium-Deschampsia* heath because it has much *Sphagnum capillifolium*.

## Ecology

This vegetation type takes in *Vaccinium* heaths that range from near-natural to quite clearly anthropogenic. They occur within the altitudinal range of *Calluna vulgaris* in situations where *Calluna* has been excluded by prolonged late snow-lie or by grazing. The *Hylocomium-Rhytidiadelphus* sub-community occurs on high slopes in the Highlands and more locally in the Southern Uplands, the Lake District and north Wales where snow lies moderately late. However, it is equally characteristic of slopes at low to moderate altitudes in the uplands of southern Scotland, England and Wales as a derivative of *Calluna-Vaccinium* heath H12. This happens when *Calluna-Vaccinium* heaths are burned and then grazed too hard to allow *Calluna* to re-establish itself. The *Alchemilla-Carex* sub-community includes grassy heaths on middle slopes which are obviously grazed derivatives of more natural vegetation – perhaps sub-alpine birch woods or scrub. It also includes shallow snow-beds on steep unstable slopes. The *Racomitrium-Cladonia* sub-community can cover eroded peats and grazed slopes at moderate altitudes, or occur around *Nardus-Carex* snow-beds U7 in the Highlands.

Despite the wide variation in habitat, the community is consistently associated with well-drained, acid to neutral mineral soils, humic rankers and dry peats. Stands can occur on slopes of all aspects, but the more natural examples are confined to cold, shaded slopes facing north or east where snow lies late. The community has a wide altitudinal range, from near sea-level to over 900 m; most stands lie between 400 m and 800 m.

## Conservation interest

*Vaccinium-Deschampsia* heath is an important part of the natural altitudinal sequences of vegetation in the British uplands, replacing *Calluna* heaths where snow lies moderately late in spring. These snow-beds can be home to a few rare species. The most notable of these is *Phyllodoce caerulea*, which grows in grassy *Vaccinium* heath on the Sow of Atholl and on Ben Alder. *Salix lapponum*, *Barbilophozia lycopodioides* and oceanic liverworts such as *Scapania ornithopodioides* may also be hidden away in these montane forms of

*Vaccinium-Deschampsia* heaths occur throughout the British uplands from south-west England to northern Scotland. The largest stands are in the central and eastern Highlands, but the community is also very common in Wales. It becomes less common towards the north and west, and is not recorded from Orkney, Shetland and the Outer Hebrides.

There is similar vegetation in western Norway.

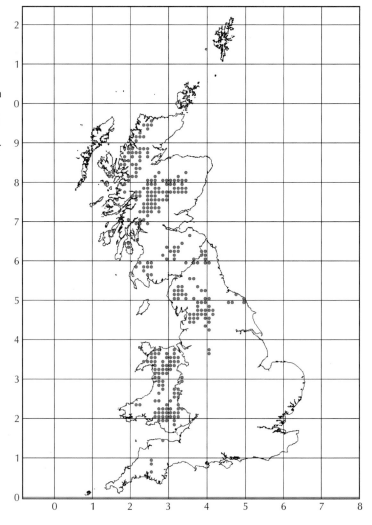

*Vaccinium-Deschampsia* heath. The anthropogenic heaths of southern Scotland, England and Wales are generally of less interest for nature conservation. They have few rare species, and are probably derived from *Calluna-Vaccinium* heath by grazing or burning.

## Management

*Vaccinium myrtillus* is heavily grazed by sheep, deer, hares and voles. The berries of *Vaccinium* and *Empetrum* are eaten by a range of birds including Red Grouse *Lagopus lagopus*, Ptarmigan *L. mutus* and Ring Ouzel *Turdus torquatus*, and the purple-stained droppings are a common sight on rocks and bare ground in the hills. The thalli of crustose lichens on rocks can take up the purple colour and then stand out from their normally coloured neighbours (Gilbert 2000). Before they were exterminated in Great Britain, bears ate the berries too, as they still do in parts of mainland Europe.

At higher altitudes *Vaccinium-Deschampsia* heath constitutes climax vegetation. *V. myrtillus* can tolerate repeated burning but the heath is burnt only where it occurs in mosaic with *Calluna*-dominated heaths and more rarely, grassland. *Vaccinium-Deschampsia* heath can be maintained by moderate levels of grazing. Moderate grazing can open up the canopy and allow a wide range of herbs and bryophytes to flourish, but if too intensive it can weaken the dwarf shrubs and eventually grasses become more common and convert the heath into grassland. On steep, rocky slopes heavy grazing may cause *Vaccinium-Deschampsia* heath to be replaced by bare and unstable scree. Lightly grazed stands of sub-montane *Vaccinium-Deschampsia* heath may be invaded by *Calluna*; this is often the desired aim when the heaths are managed for nature conservation. If they are not grazed at all, they may, in the long term, develop into oak-birch (e.g. *Quercus-Betula-Dicranum*) or pine (*Pinus-Hylocomium*) woodland provided there is a source of seed.

# H19 *Vaccinium myrtillus-Cladonia arbuscula* heath

## Synonyms

*Cladineto-Vaccinetum* and *Festuceto-Vaccinetum rhacomitrosum p.p.* (McVean and Ratcliffe 1962); *Vaccinium*-lichen heath B3f (Birks and Ratcliffe 1980).

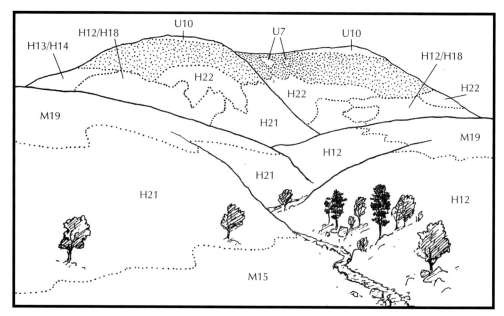

Typical habitats of H19 *Vaccinium-Cladonia* heath (stippled): high, exposed, gentle to steep slopes and plateaux with thin, stony soils.

Other vegetation types in this picture:

H12 *Calluna-Vaccinium* heath
H13 *Calluna-Cladonia* heath
H14 *Calluna-Racomitrium* heath
H18 *Vaccinium-Deschampsia* heath
H20 *Vaccinium-Racomitrium* heath
H21 *Calluna-Vaccinium-Sphagnum* heath
H22 *Vaccinium-Rubus* heath
M15 *Trichophorum-Erica* wet heath
M19 *Calluna-Eriophorum* blanket mire
U7 *Nardus-Carex* grass heath
U10 *Carex-Racomitrium* moss heath

H19 is also shown in the pictures for H12, H18, U7–U8, U11–U14 and U18.

## Description

This is a strikingly attractive montane heath, in which the rich green shoots of *Vaccinium myrtillus* and *Empetrum nigrum* ssp. *hermaphroditum* are set in a crisp, white, tangled matrix of lichens. Growing through the low canopy of dwarf shrubs (up to about 15 cm high but usually much less) there are a few other species including the montane *Carex bigelowii*, *Diphasiastrum alpinum*, *Salix herbacea* and *Loiseleuria procumbens*, and less demanding plants such as *Huperzia selago*, *Deschampsia flexuosa* and *Vaccinium vitis-idaea*. The layer of lichens is composed of *Cladonia arbuscula*, *C. portentosa*, *Coelocaulon aculeatum*, *Alectoria nigricans* and other widespread species, but the montane *Cetraria islandica* and *Ochrolechia frigida* are also common here. The lichens are crowded so thickly and deeply that the vegetation often feels springy or spongy to walk on. A few mosses may also occur, especially large pleurocarpous species such as *Pleurozium schreberi* and *Hypnum jutlandicum*, but these are never as obvious as the lichens.

There are three sub-communities. The *Festuca ovina-Galium saxatile* sub-community H19a is a grassy assemblage with much *Festuca ovina* or *F. vivipara* and *Galium saxatile* among the dwarf shrubs. The *Racomitrium lanuginosum* sub-community H19b has slightly more *Racomitrium lanuginosum*, and there are a few more montane plants such as *Alchemilla alpina*, *Salix herbacea* and *Silene acaulis*. It is typical of shallow snowbeds in the eastern Highlands as well as high slopes in the wetter west of the country. The *Empetrum nigrum* ssp. *hermaphroditum-Cladonia* species sub-community H19c is the most lichen-rich form; it is further distinguished by *Empetrum nigrum* ssp. *hermaphroditum*, *Dicranum scoparium*, *Pleurozium schreberi*, *Cladonia rangiferina* and *C. gracilis*.

*Vaccinium-Cladonia* heath is fairly widely distributed in the higher uplands from north Wales and the Lake District north to Sutherland. It has not been recorded in the Hebrides or the Northern Isles. The community is most characteristic of the more continental parts of the country: the largest stands are in the eastern Breadalbanes, the Cairngorms, on Lochnagar and on Ben Wyvis. Stands in the west are smaller and less distinctive.

Lichen-rich vegetation such as this is one of the nearest approaches in Great Britain to the continental heaths of Scandinavia with their deep, species-rich mats of lichens. Vegetation similar to *Vaccinium-Cladonia* heath occurs locally in central and eastern Norway.

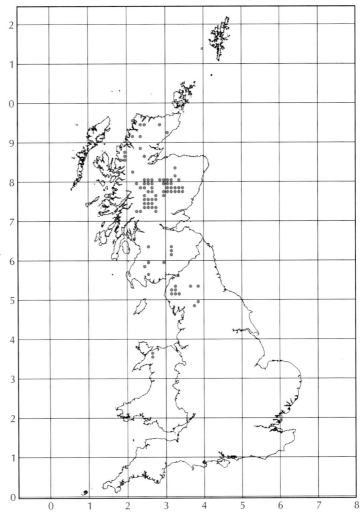

## Differentiation from other communities

This distinctive type of heath is most similar to *Vaccinium-Deschampsia* heath H18 and *Vaccinium-Racomitrium* heath H20. Compared to *Vaccinium-Deschampsia* heath, *Vaccinium-Cladonia* heath has a more continuous underlayer of lichens, and more montane species such as *Empetrum nigrum* ssp. *hermaphroditum*, *Carex bigelowii*, *Salix herbacea* and *Loiseleuria procumbens*. It has more lichens and fewer mosses than *Vaccinium-Racomitrium* heath; *Racomitrium lanuginosum* is generally rare in *Vaccinium-Cladonia* heath.

## Ecology

*Vaccinium-Cladonia* heath is a community of high slopes, including places where snow lies moderately late. It forms a zone above the montane *Calluna-Cladonia* H13 and *Calluna-Racomitrium* H14 heaths, and below the *Carex-Racomitrium* moss-heaths U10 of the highest ground. Most stands are at altitudes above 650 m. The soils are strongly leached and base-poor, even over calcareous parent rocks such as the mica-schist of the Breadalbanes.

## Conservation interest

This is a near-natural plant community with a well-defined position in the altitudinal sequence of vegetation in the British uplands. Although it hosts few rare species, there are usually fine assemblages of upland lichens including *Cetraria nivalis* and *Ochrolechia frigida* (Fryday 1997). Ptarmigan *Lagopus mutus* and Dotterel *Charadrius morinellus* nest here, and Snow Buntings *Plectrophenax nivalis* feed on the many insects that breed in the layer of lichens.

## Management

*Vaccinium-Cladonia* heath is a climax vegetation type and does not require management for its maintenance. The slopes where it occurs are often favoured by sheep and deer, whose grazing can blur the distinction between this community and sub-montane *Vaccinium-Deschampsia* heath. This seems to have happened particularly in the Lake District, southern Scotland and the eastern Breadalbanes, all of which have a long history of grazing by sheep and deer. Like the *Calluna-Cladonia* heath, the lichens can be damaged through trampling by people and animals, by vehicles being driven over the upper hill slopes, and by fire outbreaks.

# H20 *Vaccinium myrtillus-Racomitrium lanuginosum* heath

## Synonyms

*Rhacomitreto-Empetretum* (McVean and Ratcliffe 1962; Birks 1973); *Empetrum*-hypnaceous moss heath (McVean and Ratcliffe 1962); *Alchemilla alpina-Vaccinium myrtillus* nodum (Birks 1973); *Vaccinium myrtillus-Empetrum nigrum* ssp. *hermaphroditum* heath B3e and *Empetrum nigrum* ssp. *hermaphroditum-Racomitrium* heath E1d (Birks and Ratcliffe 1980).

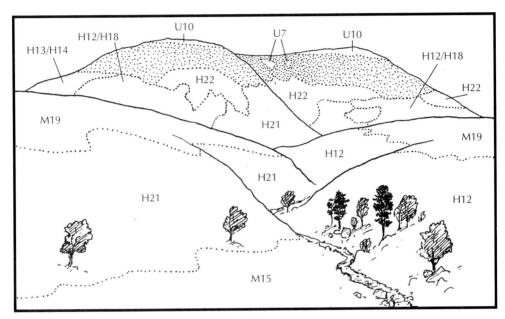

Typical habitats of H20 *Vaccinium-Racomitrium* heath (stippled): high, exposed, gentle to steep slopes and plateaux with thin, stony soils.

Other vegetation types in this picture:

H12 *Calluna-Vaccinium* heath
H13 *Calluna-Cladonia* heath
H14 *Calluna-Racomitrium* heath
H18 *Vaccinium-Deschampsia* heath
H19 *Vaccinium-Cladonia* heath
H21 *Calluna-Vaccinium-Sphagnum* heath
H22 *Vaccinium-Rubus* heath
M15 *Trichophorum-Erica* wet heath
M19 *Calluna-Eriophorum* blanket mire
U7 *Nardus-Carex* grass heath
U10 *Carex-Racomitrium* moss heath

H20 is also shown in the pictures for H12, H18, U7–8, U11–14 and U18.

## Description

This is an attractive, rich-bright-green montane heath. It is composed of a short, even mixture of *Vaccinium myrtillus* and *Empetrum nigrum* ssp. *hermaphroditum* bound together by a deep lush turf of mosses, of which the most common are the silvery-green *Racomitrium lanuginosum*, the golden *Rhytidiadelphus loreus*, *Hylocomium splendens* and *Pleurozium schreberi*, and the pale green *Hypnum jutlandicum*. The sward is speckled with small vascular plants such as *Nardus stricta, Deschampsia flexuosa, Carex bigelowii, Alchemilla alpina, Diphasiastrum alpinum* and *Galium saxatile.*

There are four sub-communities. The *Viola riviniana-Thymus polytrichus* sub-community H20a is a rather herb-rich heath in which the dwarf shrubs are mixed with *Potentilla erecta, Huperzia selago, Alchemilla alpina, Thymus polytrichus* and *Viola riviniana*. The *Cetraria islandica* sub-community H20b is the most lichen-rich form of the community, with *Cetraria islandica, Cladonia gracilis* and *Coelocaulon aculeatum* among the mat of *Racomitrium lanuginosum*. The *Bazzania tricrenata-Mylia taylorii* sub-community H20c is perhaps the most distinctive form; it is a damp heath with a rich-red underfelt of *Sphagnum capillifolium* interleaved with liverworts including *Bazzania tricrenata, Mylia taylorii* and the oceanic *Pleurozia purpurea, Bazzania pearsonii, Scapania ornithopodioides, S. nimbosa, Plagiochila carringtonii* and *Anastrophyllum donnianum*. The *Rhytidiadelphus loreus-Hylocomium splendens* sub-community H20d includes less distinctive stands with few distinguishing species apart from *Vaccinium vitis-idaea, Hylocomium splendens* and *Rhytidiadelphus loreus*; pleurocarpous mosses are commoner than *R. lanuginosum*.

## Differentiation from other communities

*Vaccinium-Racomitrium* heath may be confused with other forms of bilberry heath. It has more montane species, fewer grasses and more *Racomitrium lanuginosum* than *Vaccinium-Deschampsia* heath H18. It is mossier than *Vaccinium-Cladonia* heath H19, with more *R. lanuginosum* and fewer lichens. Sparser stands of *Vaccinium-Racomitrium* heath could be mistaken for *Carex-Racomitrium* moss-heath U10, but generally the sward of *Vaccinium myrtillus* is thicker and more continuous.

## Ecology

*Vaccinium-Racomitrium* heath is typical of rocky slopes, boulder fields and exposed ridges at high altitudes in areas with a cold, very wet, oceanic climate. The underlying rocks are usually acid although the community does extend onto richer basalt and Moine schist. Soils are damp humic rankers lying directly on broken rocks (McVean and Ratcliffe 1962). The *Bazzania-Mylia* sub-community is virtually confined to slopes facing north or east, but the other sub-communities can occur on slopes of any aspect. The community occurs between the *Calluna-Erica* H10, *Calluna-Vaccinium* H12, *Calluna-Vaccinium-Sphagnum* H21 and *Vaccinium-Rubus* H22 heaths of lower altitudes, and the more montane *Carex-Nardus* grasslands U7 and *Carex-Racomitrium* moss-heaths U10. It is generally above the tree-line; most examples are above 500 m, although there are dwarfed rowans in some of the more sheltered stands.

## Conservation interest

*Vaccinium-Racomitrium* heath is a part of the internationally important series of oceanic heaths in the north-western uplands of Great Britain. Ptarmigan *Lagopus mutus*, and sometimes Dotterel *Charadrius morinellus*, nest in this vegetation. The *Bazzania-Mylia* sub-community is a valuable habitat for scarce oceanic liverworts, which flourish in this type of vegetation in the northern and central Highlands where they are protected from the severe winter temperatures by a moderate cover of snow. Notable species include *Anastrophyllum donnianum, A. joergensenii, Bazzania pearsonii, Mastigophora woodsii, Plagiochila carringtonii, Scapania ornithopodioides, S. nimbosa* and *Herbertus aduncus* ssp. *hutchinsiae*. The montane *A. donnianum, A. joergensenii* and *S. nimbosa* are generally more common in this form of heath than in any other type of vegetation (A M Averis 1994).

This community is recorded almost entirely from the Scottish Highlands and the Hebrides, and is especially widespread in the north-west. It also occurs in the hills of north-west Wales and may yet be found in the Lake District. To some extent it is an oceanic replacement for the *Vaccinium-Cladonia* heaths of the eastern Highlands, although the two types have overlapping distributions. The *Bazzania-Mylia* sub-community is almost confined to the north-west Highlands, extending as far east as the Affric-Cannich hills, Creag Meagaidh and Ben Hope, but there are outlying and species-poor fragments in north Wales (A M Averis and A Turner, pers. obs.).

There seems to be no comparable vegetation elsewhere in Europe, but structurally and floristically similar communities have been noted in Iceland and Tristan da Cunha (McVean and Ratcliffe 1962), and in western Norway.

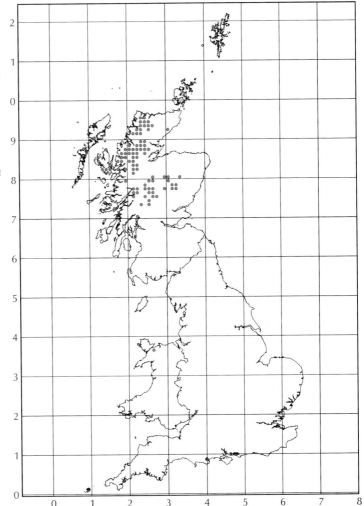

## Management

Many stands of *Vaccinium-Racomitrium* heath appear to be near-natural and ideally require no management. Grazing may convert the heath to grassland with a more impoverished structure and flora. The *Bazzania-Mylia* sub-community is particularly vulnerable to burning and excessive grazing, both of which damage the bryophyte layer.

# H21 *Calluna vulgaris-Vaccinium myrtillus-Sphagnum capillifolium* heath

## Synonyms

*Vaccineto-Callunetum hepaticosum* (McVean and Ratcliffe 1962; Birks 1973); *Calluna vulgaris-Sphagnum* damp heath B1b and *Calluna vulgaris*-hepatic heath B1c (Birks and Ratcliffe 1980).

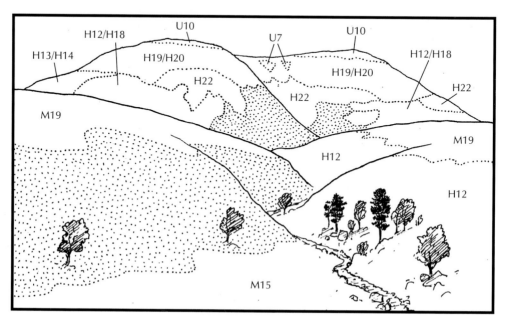

H21 *Calluna-Vaccinium-Sphagnum* heath (stippled): on steep, damp middle slopes facing between north-west and east, both on open hillsides and in more sheltered stream valleys and ravines.

Other vegetation types in this picture:

H12 *Calluna-Vaccinium* heath
H13 *Calluna-Cladonia* heath
H14 *Calluna-Racomitrium* heath
H18 *Vaccinium-Deschampsia* heath
H19 *Vaccinium-Cladonia* heath
H20 *Vaccinium-Racomitrium* heath
H22 *Vaccinium-Rubus* heath
M15 *Trichophorum-Erica* wet heath
M19 *Calluna-Eriophorum* blanket mire
U7 *Nardus-Carex* grass heath
U10 *Carex-Racomitrium* moss heath

H21 is also shown in the picture for M15.

## Description

This is a dark, shaggy damp heath in which the rough, purple-green canopy of *Calluna vulgaris* is mixed with clumps of *Erica cinerea* and light-green shoots of *Vaccinium myrtillus*, and speckled in summer with the yellow flowers of *Potentilla erecta*. The vascular plants grow out of a deep, richly coloured quilt of bryophytes, the most common of which is *Sphagnum capillifolium*. Other Sphagna, such as *S. subnitens*, may also occur, and there can be much *Racomitrium lanuginosum*.

There are two sub-communities. The *Calluna vulgaris-Pteridium aquilinum* sub-community H21a has a red-green carpet of Sphagna below a canopy of dwarf shrubs and in some places a sprinkling of *Pteridium aquilinum*. In the *Mastigophora woodsii-Herbertus aduncus* ssp. *hutchinsiae* sub-community H21b the red hummocks of *Sphagnum* are mixed with a great profusion of elegant oceanic and other western liverworts. The bright orange-red patches of *Herbertus aduncus* ssp. *hutchinsiae*, the pale gold *Mastigophora woodsii* and *Bazzania tricrenata*, the crisp yellow-green tufts of *B. pearsonii*, the ochre shoots of *Scapania gracilis*, the purplish shoots of *S. nimbosa*, *S. ornithopodioides* and *Pleurozia purpurea*, the variegated clumps of *Mylia taylorii*, the pale green upright shoots of *Plagiochila carringtonii*, and the dark hooked stems of *Anastrophyllum donnianum* make a bright tapestry under the sombre canopy of dwarf shrubs. The oceanic filmy fern *Hymenophyllum wilsonii* can grow among boulders and in the mats of liverworts.

In Wales, north-west England and parts of Scotland there are damp heaths with a dense layer of *Sphagnum* under a thick canopy of *V. myrtillus*, but with no *Calluna*. These are obviously derived from *Calluna-Vaccinium-Sphagnum* heath by heavy grazing.

## Differentiation from other communities

The conspicuous cushions of Sphagna in the bryophyte layer distinguish *Calluna-Vaccinium-Sphagnum* heath from the drier *Calluna-Erica* H10 and *Calluna-Vaccinium* H12 heaths. Some forms of *Vaccinium-Rubus* heath H22 have much heather and resemble *Calluna-Vaccinium-Sphagnum* heath, but they have *Rubus chamaemorus* or *Cornus suecica*. However, the definition of *Vaccinium-Rubus* heath as a vegetation type that can be co-dominated by *Calluna* and *Vaccinium* presents problems for its separation from *Calluna-Vaccinium-Sphagnum* heath (see account for *Vaccinium-Rubus* heath).

## Ecology

Most stands of *Calluna-Vaccinium-Sphagnum* heath occupy steep, damp, cool and shaded places: rocky precipitous slopes facing north or east and sheltered ledges on the sides of ravines. However, both sub-communities also occur more rarely on shallow open slopes and on level ground. The soils are thin, humic, greasy, peaty rankers, in many places lying precariously over barely stable scree. Most stands are on hard, coarse-grained and acid bedrock, but the community occurs on richer substrates too. The community can occur from a few metres above sea-level to over 700 m (A M Averis 1994), but most stands are between 300 m and 600 m.

## Conservation interest

Among the various forms of *Calluna* heath in the British uplands, this community (and especially the *Mastigophora-Herbertus* sub-community) is of great value for nature conservation because it is so rare in Europe and because the bryophyte flora is so notable. The *Mastigophora-Herbertus* sub-community of *Calluna-Vaccinium-Sphagnum* heath and the *Bazzania-Mylia* sub-community of *Vaccinium-Racomitrium* heath H20c are the main habitats in Great Britain for the scarce oceanic liverworts *Anastrophyllum donnianum*, *Bazzania pearsonii*, *Mastigophora woodsii*, *Plagiochila carringtonii*, *Scapania ornithopodioides* and *S. nimbosa*. The Red Data Book liverwort *Adelanthus lindenbergianus* grows only in this type of vegetation in Great Britain. *Anastrophyllum joergensenii*, another nationally rare plant, can occur in the *Mastigophora-Herbertus* sub-community, although its main habitat is heaths and snow-beds at higher altitudes.

# H21 *Calluna vulgaris-Vaccinium myrtillus-Sphagnum capillifolium* heath

The *Calluna-Pteridium* sub-community is common in the Highlands, and extends as far east as the Cheviot Hills and even the North York Moors on shaded slopes that are moist enough for Sphagna to grow. Its distribution runs south through Wales to a few steep banks on Dartmoor (not shown on the map). The *Mastigophora-Herbertus* sub-community, with its more exacting oceanic bryophytes, is virtually confined to hills in the western Highlands and Hebrides, with fragmentary outliers in north Wales and the Lake District.

The liverwort-rich *Mastigophora-Herbertus* sub-community is a type of vegetation unique to Great Britain and Ireland. The most similar related vegetation is in the Faroes and perhaps western Norway (Hobbs and Averis 1991a; A M Averis 1994). Although there are good stands in Kerry, Galway, Mayo and Donegal in western Ireland, nowhere else does the vegetation attain the luxuriance of stands in the north-west Highlands.

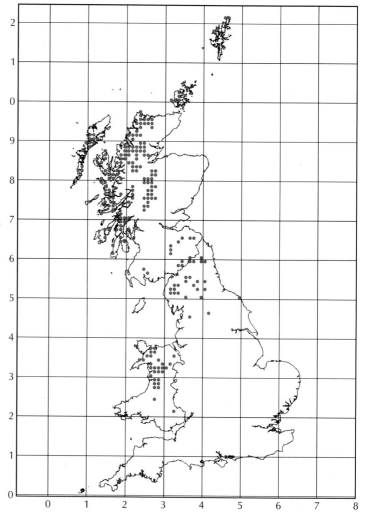

The filmy fern *Hymenophyllum wilsonii* is common in the *Mastigophora-Herbertus* sub-community. Other scarce species in both sub-communities include *Listera cordata* and the mosses *Campylopus setifolius*, *Dicranodontium uncinatum* and *Myurium hochstetteri*. In some places in the extreme west, the diminutive oceanic liverworts *Colura calyptrifolia*, *Drepanolejeunea hamatifolia* and *Leptoscyphus cuneifolius* grow as epiphytes on old heather in *Calluna-Vaccinium-Sphagnum* heath.

## Management

*Calluna-Vaccinium-Sphagnum* heath is highly susceptible to damage by burning and grazing. Burning destroys the bryophytes, opens up the dwarf-shrub canopy, and gives the soil a dry, hard surface which makes it difficult for the scarce leafy liverworts to recolonise. The community can withstand light grazing, but excessive numbers of grazing animals disrupt the canopy of dwarf shrubs, encouraging the growth of grasses, and break up the bryophyte carpets by trampling and wallowing. The *Mastigophora-Herbertus* sub-community in particular has been much reduced in extent by grazing and burning, and there are many places in the west Highlands where a few tufts of the characteristic liverworts linger on in grassy, degenerate heath on rocky slopes. The *Calluna-Pteridium* sub-community is more resilient but here too the characteristic Sphagna can be lost if the heaths are burnt too frequently or grazed too hard. Stands of *Calluna-Vaccinium-Sphagnum* heath at higher elevations are probably near-natural, but at lower altitudes it is possible that the community has been derived from woodland.

# H22 *Vaccinium myrtillus-Rubus chamaemorus* heath

## Synonyms

*Vaccineto-Callunetum suecicosum* and *Vaccinetum chionophilum p.p.* (McVean and Ratcliffe 1962); *Calluna vulgaris-Sphagnum* damp heath B1b (Birks and Ratcliffe 1980).

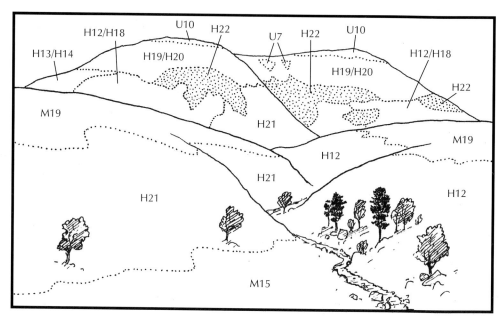

H22 *Vaccinium myrtillus-Rubus chamaemorus* heath (stippled): on upper slopes, especially where facing between north west and east and where snow lies late into the year. The community may be extensive, or in smaller patches marking out areas of more prolonged snow-lie among other heaths.

Other vegetation types in this picture:

H12 *Calluna-Vaccinium* heath
H13 *Calluna-Cladonia* heath
H14 *Calluna-Racomitrium* heath
H18 *Vaccinium-Deschampsia* heath
H19 *Vaccinium-Cladonia* heath
H20 *Vaccinium-Racomitrium* heath
H21 *Calluna-Vaccinium-Sphagnum* heath
M15 *Trichophorum-Erica* wet heath
M19 *Calluna-Eriophorum* blanket mire
U7 *Nardus-Carex* grass heath
U10 *Carex -Racomitrium* moss-heath

H22 is also shown in the picture for H12 and H18.

## Description

These are damp, richly coloured heaths with a canopy of *Calluna vulgaris, Vaccinium myrtillus, V. vitis-idaea* and *Empetrum nigrum* ssp. *hermaphroditum* growing through a deep, spongy and continuous carpet of large mosses such as *Hylocomium splendens, Pleurozium schreberi, Rhytidiadelphus loreus* and *Sphagnum capillifolium*. *V. myrtillus* is usually sufficiently common to give a rich green tone to the vegetation, making it conspicuous from a distance, although the dark tones of heather predominate in many stands, especially in the east. *Potentilla erecta* is usually common, and two especially distinctive and characteristic species are the northern *Rubus chamaemorus* and *Cornus suecica*. *Sphagnum fuscum* occurs in some of the more montane stands of the community.

There are two sub-communities, one a mild snow-bed with more *V. myrtillus* than *Calluna*, and the other with more equal quantities of *Calluna* and *V. myrtillus*. The *Polytrichum commune-Galium saxatile* sub-community H22a has a canopy dominated by *V. myrtillus*; *Polytrichum commune* and *Galium saxatile* are also very common here. The *Plagiothecium undulatum-Anastrepta orcadensis* sub-community H22b has more *Calluna* in a mossy heath with *Plagiothecium undulatum*, *Hypnum cupressiforme*, *H. jutlandicum*, *Racomitrium lanuginosum* and *Anastrepta orcadensis*.

## Differentiation from other communities

The *Vaccinium*-dominated *Polytrichum-Galium* sub-community resembles the *Hylocomium-Rhytidiadelphus* sub-community of *Vaccinium-Deschampsia* heath H18a, but has more *Empetrum nigrum* ssp. *hermaphroditum*, *Rubus chamaemorus*, *Cornus suecica*, *Sphagnum capillifolium* and *S. fuscum*, as well as montane species such as *Carex bigelowii*, *Vaccinium uliginosum* and the lichen *Cetraria islandica*.

The *Plagiothecium-Anastrepta* sub-community contains much heather and is very similar to *Calluna-Vaccinium-Sphagnum* heath H21. However, it has more *E. nigrum* ssp. *hermaphroditum*, *Vaccinium vitis-idaea*, *R. chamaemorus* and *C. suecica*, and the canopy of dwarf shrubs is shorter and more open. The mosses *Hylocomium splendens* and *Rhytidiadelphus loreus* are also generally more common in *Vaccinium-Rubus* heath, giving a rich golden tinge to the vegetation, but the rich assemblages of oceanic liverworts that distinguish the *Mastigophora-Herbertus* sub-community of *Calluna-Vaccinium-Sphagnum* heath H21b are absent. The grazed form of *Calluna-Vaccinium-Sphagnum* heath with little or no *Calluna* may resemble *Vaccinium-Rubus* heath, but has fewer pleurocarpous mosses and no montane species. The *Plagiothecium-Anastrepta* sub-community can also resemble *Calluna-Vaccinium* heath H12, but has more *Sphagnum capillifolium* and more of the northern species *E. nigrum* ssp. *hermaphroditum*, *Carex bigelowii*, *Vaccinium uliginosum*, *R. chamaemorus*, *C. suecica* and *Sphagnum fuscum*.

For the most part, the NVC defines *Calluna*-dominated heaths separately from *Vaccinium*-dominated heaths, so that vegetation co-dominated by both species can be understood quite clearly as being floristically intermediate. This pattern is confused by having one heathland community (*Vaccinium-Rubus* heath) that can be either dominated by *V. myrtillus* or co-dominated by *Calluna* and *V. myrtillus*. The *Vaccinium-Rubus* community appears to have been defined to include a heterogeneous range of heathland quadrat samples that contain one or both of *R. chamaemorus* and *C. suecica*. An alternative approach would be to include only the *Vaccinium*-dominated examples of the *Polytrichum-Galium* sub-community within the definition of *Vaccinium-Rubus* heath, leaving the remainder of the vegetation currently assigned to this community to be regarded as a form of *Calluna-Vaccinium-Sphagnum* heath containing scattered plants of *R. chamaemorus* and *C. suecica*. Apart from *Vaccinium-Rubus* heath, all upland heathland NVC types are defined by their dwarf shrubs and bryophyte or lichen carpets: the species that make up the bulk of the vegetation. For the *Vaccinium-Rubus* heath to follow suit, *R. chamaemorus* and *C. suecica* should not be important defining species, and the community should include *Vaccinium* heaths that lack *R. chamaemorus* or *C. suecica*, but have pleurocarpous mosses, *Sphagnum capillifolium*, and montane species such as *E. nigrum* ssp. *hermaphroditum* and *V. uliginosum*.

## Ecology

This is a montane community of moist, acid, peaty soils on damp slopes above 400 m. In the north-west and especially in the eastern Highlands it can be extensive on slopes of almost any aspect, as long as there is a moderately deep layer of peat or humus. Towards the south-eastern part of its range the community becomes increasingly restricted to shady slopes facing north or east (McVean and Ratcliffe 1962). Another habitat is around the edges of stands of *Calluna-Eriophorum* blanket bog M19, where the slope begins to get steeper and *Vaccinium-Rubus* heath forms a transitional zone between the bog and *Calluna-Vaccinium* heath.

*Vaccinium-Rubus* heath is evidently a near-natural climax community, maintained by low temperatures or late-lying snow. The *Polytrichum-Galium* sub-community occurs on shaded slopes and in sheltered hollows where a mildly prolonged cover of snow suppresses *Calluna* in favour of *V. myrtillus*. The *Plagiothecium-Anastrepta* sub-community is less associated with long-lying snow and is co-dominated by *Calluna*.

## Conservation interest

*Vaccinium-Rubus* heath is an important component of the suite of dwarf-shrub heaths that are so much more widespread in the British uplands than they are in other parts of

*Vaccinium-Rubus* heath is widely distributed in the Highlands with outliers in the Southern Uplands, Cheviots and north Wales (A B G Averis, pers. obs.). In the milder Hebrides and south-western Highlands it is generally replaced by the more oceanic *Calluna-Vaccinium-Sphagnum* heath. The geographical ranges of the two heaths overlap and both can occur on the same hill.

There is similar vegetation to *Vaccinium-Rubus* heath in western Norway.

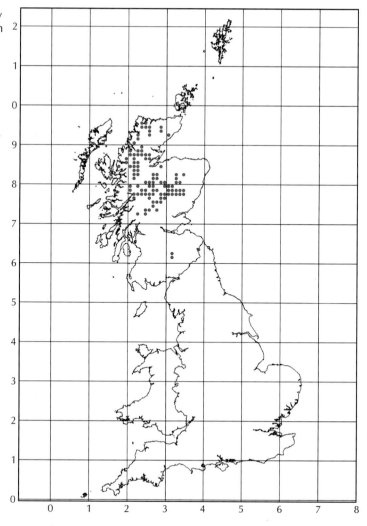

Europe. It is one of the most important habitats in Great Britain for the uncommon species *Rubus chamaemorus*, *Cornus suecica* and *Lycopodium annotinum*. The rare Atlantic liverworts *Scapania ornithopodioides*, *S. nimbosa* and *Plagiochila carringtonii* grow in some western stands transitional to the *Mastigophora-Herbertus* sub-community of *Calluna-Vaccinium-Sphagnum* heath (e.g. Averis and Averis 1998a).

## Management

Management by burning is probably not necessary to maintain the flora of these near-natural heaths. The canopy is kept open by the harsh upland climate, and *Calluna* is not able to suppress the distinctive herbs and bryophytes. Burning may not always be deleterious, but the bryophytes and the peaty soils are easily damaged by fire, and observations suggest that *Vaccinium-Rubus* heath can be converted to *Calluna-Vaccinium* heath by frequent or severe fires. If moors are managed in accordance with recommended good practice, following Phillips *et al.* (1993), the Moorland Working Group (1998) and Scotland's Moorland Forum (2003), *Vaccinium-Rubus* heaths should escape burning because they are on steep, rocky, damp and sheltered slopes. Most stands of *Vaccinium-Rubus* heath are lightly grazed, and can merge into the surrounding anthropogenic *Vaccinium* heaths (McVean and Ratcliffe 1962). The best way of conserving the community and its ecological relationships with other vegetation types is to control grazing and burning so that these natural sequences are maintained.

# MG2 *Arrhenatherum elatius-Filipendula ulmaria* tall-herb grassland

MG2 *Arrhenatherum-Filipendula* tall-herb grassland (ungrazed slope in foreground) with ash woodland on steep Carboniferous limestone slope.

## Description

This is a lush, species-rich grassland in which large grasses such as *Arrhenatherum elatius*, *Dactylis glomerata* and *Festuca rubra* form a rank, tussocky sward. They are interspersed with tall forbs such as *Angelica sylvestris*, *Filipendula ulmaria* and *Valeriana officinalis*, and woodland species including *Mercurialis perennis*, *Dryopteris filix-mas*, *Silene dioica* and *Heracleum sphondylium*. There may be a profuse underlay of mosses including *Plagiothecium denticulatum*, *Plagiomnium undulatum*, *Eurhynchium hians* and *Lophocolea bidentata*.

   Two sub-communities are described in the NVC, but they are rather poorly separated. This is partly because of biased sampling in the NVC, which has focused on the habitat of the rare *Polemonium caeruleum*. The *Filipendula ulmaria* sub-community MG2a has more *Angelica sylvestris* and *Trisetum flavescens*. The *Polemonium caeruleum* sub-community MG2b has *P. caeruleum*, more *Festuca rubra*, and a few lowland species such as *Stellaria holostea*, *Geranium robertianum*, *Scabiosa columbaria* and *Galium aparine*.

## Differentiation from other communities

*Arrhenatherum-Filipendula* tall-herb grassland can be confused with the *Filipendula ulmaria* sub-community of *Arrhenatherum elatius* grassland MG1c, but it is more species-rich and has more woodland species such as *Mercurialis perennis*, *Silene dioica*, *Stellaria holostea* and *Oxalis acetosella*. It might also be confused with *Filipendula-Angelica* tall-herb fen M27, which is a wetter community with more *Filipendula ulmaria*, fewer grasses and fewer woodland species. Flushed forms of *Festuca-Agrostis-Galium* grassland U4 with *F. ulmaria* and other mesotrophic forbs can bear some resemblance to *Arrhenatherum-Filipendula* grassland, but the swards are shorter, have less

*Arrhenatherum* and *Dactylis glomerata*, and more mosses typical of acid grassland such as *Rhytidiadelphus squarrosus* and *Hylocomium splendens*.

## Ecology
*Arrhenatherum-Filipendula* grassland occurs on steep, lightly grazed or ungrazed rocky slopes, usually where there is some shade. The most characteristic habitat is near the margins of woodland. Soils are damp, base-rich rendzinas, derived mainly from calcareous rocks such as limestone. This sub-montane type of vegetation has been found mainly between 200 m and 400 m, but occurs just above sea-level in western Scotland.

## Conservation interest
These are among our most species-rich *Arrhenatherum* grasslands, and are uncommon in Great Britain. Stands with the rare *Polemonium caeruleum* are especially important for nature conservation.

## Management
Most of the tall herbs in *Arrhenatherum-Filipendula* grassland are susceptible to heavy grazing. Many stands are on slopes that are too steep to be grazed regularly by large

This is a scarce community recorded mainly in northern England on the Carboniferous limestone of Derbyshire and the Craven Pennines, but also known from the Cheviot, southeast Scotland and Mull.

There is similar vegetation in mainland Europe.

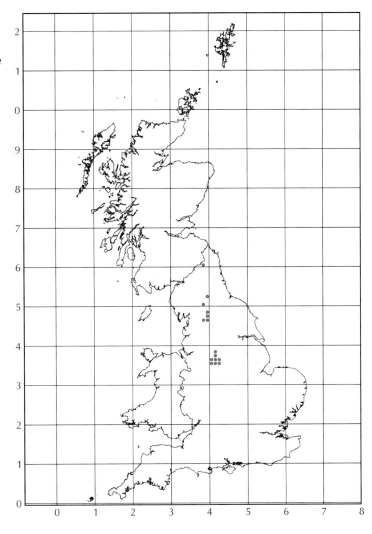

herbivores. However, light or sporadic grazing is probably necessary to prevent succession to scrub or woodland where the community occurs on shallower slopes with deeper soils where trees or shrubs could grow. Since many of the species in *Arrhenatherum-Filipendula* grassland can tolerate neither heavy grazing nor deep shade, it has probably always been a scarce community of woodland glades and margins and steep ground where trees could not form a continuous canopy.

# MG3 *Anthoxanthum odoratum-Geranium sylvaticum* grassland

Typical habitats of MG3 *Anthoxanthum-Geranium* grassland (stippled):

1  Unimproved or little-improved enclosed meadow
2  On banks of river
3  On roadside verges

## Description

The thick swards of this grassland are generally made up of mixtures of the grasses *Anthoxanthum odoratum, Agrostis capillaris, Festuca rubra, Holcus lanatus, Poa trivialis, Dactylis glomerata* and *Cynosurus cristatus*. They are densely crowded with large, vigorous forbs such as *Geranium sylvaticum, Plantago lanceolata, Conopodium majus, Rumex acetosa, Ranunculus acris, R. bulbosus, Alchemilla glabra* and *Sanguisorba officinalis*. The sward can reach a height of 60–80 cm by midsummer. Under the tall plants there may be smaller species such as *Cerastium fontanum, Trifolium repens* and *Bellis perennis*. Bryophytes can grow in dense wefts among the grasses; the commonest species are usually *Rhytidiadelphus squarrosus, Brachythecium rutabulum* and *Eurhynchium praelongum*.

There are three sub-communities. The *Bromus hordeaceus* ssp. *hordeaceus* sub-community MG3a is the most species-poor form, containing few forbs but with much *Bromus hordeaceus* ssp. *hordeaceus* and *Lolium perenne*. The *Briza media* sub-community MG3b is the most species-rich form, with *Rhinanthus minor, Trifolium pratense, Lotus corniculatus, Hypochaeris radicata, Centaurea nigra, Briza media, Luzula campestris* and other herbs. The *Arrhenatherum elatius* sub-community MG3c has a tall sward with much *Arrhenatherum elatius, Dactylis glomerata* and *Helictotrichon pubescens*.

## Differentiation from other communities

The flora of *Anthoxanthum-Geranium* grassland is generally similar to that of *Cynosurus-Centaurea* grassland MG5. The main difference between these two communities is that *Geranium sylvaticum, Geum rivale, Trollius europaeus, Cirsium*

291

*heterophyllum* and *Alchemilla* species are commoner in *Anthoxanthum-Geranium* grassland.

The tall swards of the *Arrhenatherum* sub-community can resemble *Arrhenatherum elatius* grassland MG1 and *Arrhenatherum-Filipendula* grassland MG2, but differ in that they contain *Geranium sylvaticum* and *Sanguisorba officinalis*. This sub-community might equally well have been placed within the NVC as a *G. sylvaticum-S. officinalis* sub-community of *Arrhenatherum* grassland.

Some herb-rich examples of *Festuca-Agrostis-Galium* grassland U4 can look very like *Anthoxanthum-Geranium* grassland, but are generally mossier and have more *Potentilla erecta* and *Galium saxatile*. Northern flushed forms of *Festuca-Agrostis-Galium* grassland, with species including *Filipendula ulmaria* in the west or *Helianthemum nummularium* and *Persicaria vivipara* in the east, can have *G. sylvaticum*, but are mossier than *Anthoxanthum-Geranium* grassland, have more *P. erecta* and lack *S. officinalis*. The *Lathyrus-Stachys* sub-community U4c can contain *S. officinalis*, but also has *P. erecta*, *G. saxatile* and *Stachys officinalis*, and lacks *G. sylvaticum*. Stands transitional between the two communities occur locally, for example on hay-meadow banks in northern England (Richard Jefferson, pers. comm.).

*Anthoxanthum-Geranium* grassland is a scarce community that is most common in upland valleys in the mid- to north Pennines. It also occurs less commonly in the Lake District, the Southern Uplands, and the eastern and southern Highlands, where there are patches of it along roadside verges and on river banks.

There is similar vegetation on the Faroes (Hobbs and Averis 1991b) and in mainland Europe.

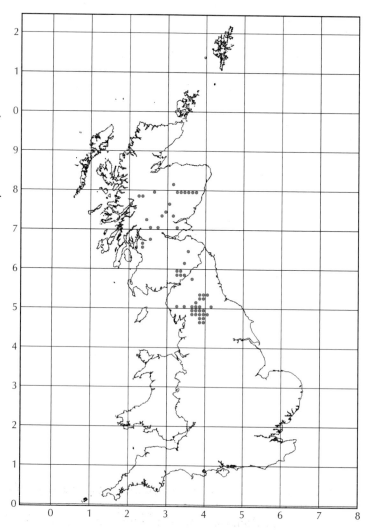

## Ecology

*Anthoxanthum-Geranium* grassland generally occurs on moist brown earth soils in enclosed farmland on level to sloping ground at around 200–400 m. It is usually grazed by sheep or cattle in autumn, winter and early spring before being shut up for hay in May. The tall hay crop is mown in late summer, after which the subsequent young growth (aftermath) may be grazed. The community also occurs on lightly grazed or mown roadside verges and on river banks.

## Conservation interest

This colourful, herb-rich grassland is of great value for nature conservation. It is home to the uncommon *Geranium sylvaticum* and *Sanguisorba officinalis*, and is the main habitat in Great Britain of the scarce *Alchemilla acutiloba*, *A. monticola* and *A. subcrenata*. Over 700 ha of *Anthoxanthum-Geranium* grassland are protected within candidate SACs; this is estimated to be at least 70% of the total extent of the community in Great Britain.

## Management

*Anthoxanthum-Geranium* grasslands are maintained by less intensive management than that of most enclosed farmland in the British uplands. The more species-rich *Briza* sub-community is associated with hay-meadows that traditionally receive a light application of farmyard manure and some lime. Chemical fertilisation or ploughing of stands of this sub-community, followed by abandonment or re-seeding, leads to the development of the more impoverished *Bromus* sub-community. If the sward continues to be artificially fertilised it will eventually become semi-improved *Lolium perenne-Cynosurus cristatus* grassland MG6 of little value for nature conservation. The *Arrhenatherum* sub-community occurs where there is very light grazing which allows taller grasses to become more common and oust some of the smaller plants. Most *Anthoxanthum-Geranium* grassland has probably been derived from woodland of the *Alnus-Fraxinus-Lysimachia* W7 and *Fraxinus-Sorbus-Mercurialis* W9 types, and might revert to such woodland if left ungrazed for many years.

# MG5 *Cynosurus cristatus-Centaurea nigra* grassland

## Synonyms

*Centaureo-Cynosuretum* (Birks 1973); *Lolio-Cynosuretum luzuletosum p.p.* (Birse and Robertson 1976).

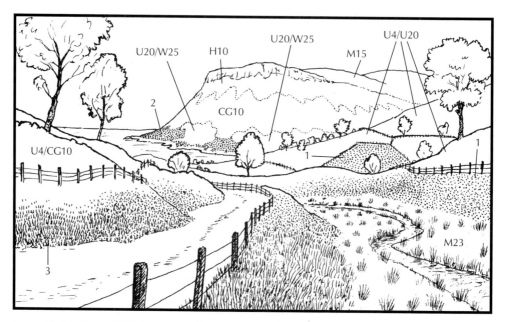

Typical habitats of MG5 *Cynosurus-Centaurea* grassland (stippled):

1  Unimproved or little-improved enclosed pasture and meadow, in mixtures with bracken, rush-dominated mire and other types of grassland
2  In varied mosaics of grassland, mire and bracken in low-altitude unenclosed land
3  On roadside verges

Other vegetation types shown in this picture:

U4 *Festuca-Agrostis-Galium* grassland
U20 *Pteridium-Galium* community
M15 *Trichophorum-Erica* wet heath
M23 *Juncus-Galium* rush-pasture
CG10 *Festuca-Agrostis-Thymus* grassland
H10 *Calluna-Erica* heath
W25 *Pteridium-Rubus* underscrub

## Description

This herb-rich pasture or meadow grassland has a dense and lush sward that is bright with flowers in summer. The bulk of the sward is composed of the grasses *Cynosurus cristatus, Festuca rubra, Agrostis capillaris, Holcus lanatus, Anthoxanthum odoratum* and *Dactylis glomerata*. The grasses are entwined with forbs such as *Plantago lanceolata, Trifolium repens, T. pratense, Lotus corniculatus, Ranunculus acris, R. bulbosus, Prunella vulgaris, Rumex acetosa, Centaurea nigra, Taraxacum officinale* agg., *Rhinanthus minor, Euphrasia officinalis* and *Hypochaeris radicata*. There is little space for bryophytes in the dense turf, but there is usually at least some *Brachythecium rutabulum, Scleropodium purum, Eurhynchium praelongum* and *Rhytidiadelphus squarrosus*.

There are three sub-communities. The *Lathyrus pratensis* sub-community MG5a has *Lolium perenne*, *Bellis perennis* and *Lathyrus pratensis*. The *Galium verum* sub-community MG5b tends to occur on more base-rich soils, and includes plants such as *Galium verum*, *Trisetum flavescens*, *Carex flacca*, *Sanguisorba minor* and *Koeleria macrantha*, but of these species only *G. verum* is common in upland stands of *Cynosurus-Centaurea* grassland. The *Danthonia decumbens* sub-community MG5c has *Danthonia decumbens*, *Potentilla erecta* and *Succisa pratensis*.

### Differentiation from other communities

The swards of forbs and grasses in *Cynosurus-Centaurea* grassland resemble those of several other grassland types. *Arrhenatherum elatius* grassland MG1 has a coarser sward with more *Arrhenatherum* and *Heracleum sphondylium*, and less *Cynosurus cristatus*. *Arrhenatherum-Filipendula* grassland MG2 also comprises tall swards with much *Arrhenatherum*, and is further distinguished by woodland plants such as *Mercurialis perennis*, *Urtica dioica*, *Dryopteris filix-mas* and *Silene dioica*, and species of damper soils such as *Geum rivale* and *Filipendula ulmaria*. *Anthoxanthum-Geranium* grassland MG3 has many similarities with *Cynosurus-Centaurea* grassland, but has *Geranium sylvaticum* and *Alchemilla* species. Species-poor forms of *Cynosurus-Centaurea* grassland can resemble *Lolium perenne-Cynosurus cristatus* grassland MG6, but *Lolium perenne* is generally not dominant, and *Centaurea nigra* and *Lotus corniculatus* are commoner.

The *Lathyrus-Stachys* sub-community of *Festuca-Agrostis-Galium* grassland U4c can look very like *Cynosurus-Centaurea* grassland, but has more *Galium saxatile* and *Potentilla erecta* and more mosses, and *Cynosurus cristatus* and tall mesotrophic forbs are less common. *Festuca-Agrostis-Thymus* grassland CG10 has less *C. cristatus* and other mesotrophic species, and much *Thymus polytrichus*. Grasslands dominated by *Festuca rubra* and *Holcus lanatus* are quite common at low altitudes in some upland areas, such as the western Highlands (Cooper and MacKintosh 1996; Rodwell *et al.* 2000), but are not described in the NVC. They can resemble lush forms of *Cynosurus-Centaurea* grassland but have less *C. cristatus* and are generally less herb-rich.

### Ecology

This is a managed grassland of low-lying, well-drained, neutral mineral soils. It is maintained by manuring, grazing and cutting for hay. In upland Great Britain it occurs mainly within enclosed fields, but also along some unenclosed coastlines in the western Highlands and the Inner Hebrides. It is mostly below 300 m, and in the western Highlands is rare above 100 m.

### Conservation interest

This community, best known as the colourful, herb-rich, unimproved grassland that used to be common in Great Britain, is an important element in the landscape of the upland margins. It is a rich assemblage of common species rather than a habitat of rarities, but some stands provide a home for uncommon plants such as *Carum verticillatum*, *Vicia orobus*, *Ophioglossum vulgatum*, *Orchis morio* and *Dactylorhiza purpurella*. Unimproved grasslands have suffered a catastrophic decline over the last 50 years, and many examples of *Cynosurus-Centaurea* grassland survive as isolated fragments surrounded by intensively managed agricultural land. Remnant stands have received considerable attention from nature conservationists and are well represented within the SSSI series in Great Britain.

## MG5 *Cynosurus cristatus-Centaurea nigra* grassland

*Cynosurus-Centaurea* grassland is widespread at low altitudes throughout Great Britain. Most examples of the community in upland Great Britain belong to the *Danthonia* sub-community, but the other two sub-communities also occur on the upland fringes.

There is similar vegetation in Ireland but nothing comparable has been described from mainland Europe.

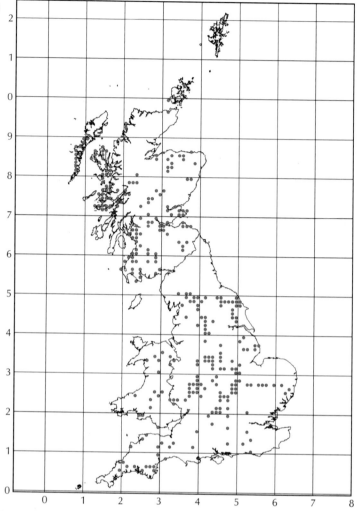

### Management

*Cynosurus-Centaurea* grassland can be managed as either meadow or pasture. Meadows are shut up in spring, given a light dressing of farmyard manure, and left to grow a crop of hay. After the hay is cut, stock are allowed to graze the aftermath and are often left on the land over the winter. The community is dependent on traditional, low-intensity agricultural practices, and stands will persist only if these practices are continued and chemical fertilisers are avoided. In recent decades, re-seeding or the application of artificial fertilisers has converted much *Cynosurus-Centaurea* grassland to species-poor *Lolium-Cynosurus* grassland or *Lolium perenne* grasslands MG7. These grasslands provide a larger yield of grass that matures earlier in the year, but this is obtained at the expense of the forbs characteristic of *Cynosurus-Centaurea* grassland.

If *Cynosurus-Centaurea* grassland is neither grazed nor cut, the swards become taller; rank grasses such as *Arrhenatherum*, *Dactylis glomerata*, *Festuca rubra* or *Holcus lanatus* become more common, and in just a few years the vegetation can develop into *Arrhenatherum* grassland. In the longer term, *Ulex europaeus* and other shrubs may colonise, and eventually the vegetation would revert to some kind of woodland, perhaps of the *Fraxinus-Sorbus-Mercurialis* W9 type.

# MG8 *Cynosurus cristatus-Caltha palustris* grassland

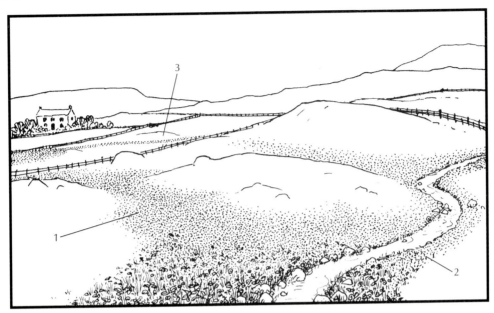

Typical habitats of MG8 *Cynosurus-Caltha* grassland (stippled) in northern Great Britain:

1  damp ground among unimproved or semi-improved grasslands and mires in enclosed, unintensively managed farmland
2  along streams
3  along ditches

## Description

This is a herb-rich community in which the most common plant is usually *Caltha palustris*, its masses of bright yellow flowers forming eye-catching sheets in late spring and early summer. The *Caltha* is set within a green matrix of grasses including *Anthoxanthum odoratum*, *Holcus lanatus*, *Festuca rubra* and smaller quantities of *Agrostis capillaris*, *A. stolonifera*, *Poa trivialis*, *Alopecurus pratensis*, *A. geniculatus* and *Lolium perenne*; *Cynosurus cristatus* is very common in English stands, but scarcer in Scottish examples. Tucked into the deep, lush sward are forbs such as *Ranunculus acris*, *R. repens*, *Rumex acetosa*, *Trifolium repens*, *T. pratense*, *Rhinanthus minor*, *Filipendula ulmaria*, *Cardamine pratensis* and *Montia fontana*. In some examples of *Cynosurus-Caltha* grassland in northern England the sward is enriched with a few northern or upland species such as *Trollius europaeus*, *Crepis paludosa*, *Geum rivale* and *Dactylorhiza purpurella*. There can be much *Carex nigra*; *Juncus acutiflorus* is common in English stands, but its place is taken by *J. articulatus* in northern Scotland. Bryophytes are very sparse in English examples of the community, where the commonest species are the mosses *Calliergonella cuspidata* and *Climacium dendroides*, but are more common and diverse in Scotland, where typical species include the mosses *C. cuspidata*, *Brachythecium rivulare*, *B. rutabulum*, *Eurhynchium praelongum* and *Plagiomnium undulatum*, and the liverworts *Chiloscyphus polyanthos* and *Pellia epiphylla*.

## Differentiation from other communities

There is more *Caltha palustris* in *Cynosurus-Caltha* grassland than in any other vegetation type in upland Great Britain. *Caltha* can be very common in *Juncus-Galium* rush-pasture M23, *Molinia-Crepis* mire M26, *Filipendula-Angelica* fen M27 and *Iris-Filipendula* mire M28, but these communities are dominated by rushes, *Molinia caerulea*, *Filipendula ulmaria* or *Iris pseudacorus*.

*Cynosurus-Caltha* grassland shares many species with *Cynosurus-Centaurea* grassland MG5, but has more *Caltha*, *Ranunculus repens* and *F. ulmaria*. These species, and also the scarcity of *Geranium sylvaticum*, distinguish the community from *Anthoxanthum-Geranium* grassland MG3.

## Ecology

Most *Cynosurus-Caltha* grassland in the uplands occurs as small patches within enclosed hay-meadows or pastures, where it marks out more or less permanently wet, flushed mineral or organic soils on fairly level to gently sloping ground. This is rather different from the community's typical lowland habitat of periodically inundated ground, such as water meadows.

In the north Pennines, the community typically occurs among *Anthoxanthum-Geranium* and *Cynosurus-Centaurea* grasslands, *Juncus-Galium* rush-pastures, and *Molinia-Crepis* mires in valley-side hay-meadows, ascending to over 400 m. At higher altitudes it can be more extensive, covering entire fields. In Scotland it occurs in enclosed farmland

Rodwell (1992) describes *Cynosurus-Caltha* grassland from widely scattered places in lowland Great Britain but does not mention it as an upland vegetation type. However, some of the NVC samples seem to have been recorded in upland pastures, as the floristic tables include *Trollius europaeus*, *Geum rivale* and *Crepis paludosa*. The vegetation described here as an upland form of *Cynosurus-Caltha* grassland has been found since the NVC analysis. It is a scarce vegetation type, which occurs in some of the north Pennine valleys, especially Upper Teesdale (Prosser 1990a, 1990b; Richard Jefferson, pers. comm.), and in widely scattered places in the Highlands, the Hebrides and Shetland (Roper-Lindsay and Say 1986; Cooper and MacKintosh 1996; Ben and Alison Averis, pers. obs.). Rodwell (1992) did not provide a map showing the distribution of the NVC samples for *Cynosurus-Caltha* grassland; consequently, lowland English records of the community are not shown on our map.

Vegetation similar to *Cynosurus-Caltha* grassland occurs in Germany and the Netherlands (Rodwell 1992), and is widespread in hay-meadows and ditches in the Faroe Islands (Hobbs and Averis 1991a).

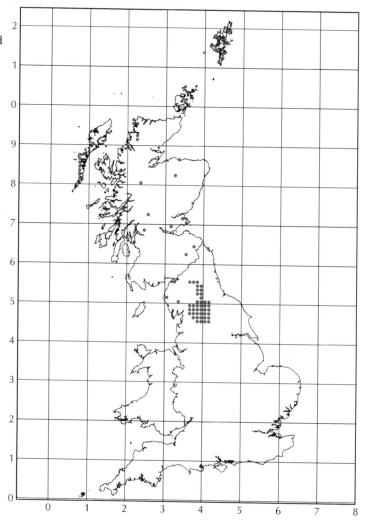

at low altitudes, forming varied mosaics with vegetation types such as *Cynosurus-Centaurea* grassland, *Lolium perenne-Cynosurus cristatus* grassland MG6, *Juncus-Galium* rush-pasture and *Iris-Filipendula* mire. Here it can also be associated with mixed swards of *Agrostis stolonifera*, *Carex nigra*, *Holcus lanatus*, *Anthoxanthum odoratum*, *Ranunculus repens* and *Rumex acetosa*. These grassy or sedgy swards are not described in the NVC, but are among the *Carex nigra-Agrostis stolonifera-Senecio aquaticus* grasslands (*Senecio-Brometum racemosi*) discussed by Rodwell *et al.* (2000).

## Conservation interest

*Cynosurus-Caltha* grassland is a scarce vegetation type associated with traditional or non-intensive forms of land management. It usually occurs among varied mosaics of grasslands and mires, some of which are also uncommon. Several scarce species grow in stands of this community in the north Pennines: *Trollius europaeus*, *Crepis paludosa*, *Geum rivale* and *Dactylorhiza purpurella*.

## Management

In the north Pennines, most *Cynosurus-Caltha* grasslands are managed as traditional hay-meadows (Robertson and Jefferson 2000). The vegetation is grazed by livestock in the spring. The fields are shut up for hay by about mid-May and a hay crop is cut around mid- to late July. The aftermath growth is grazed in late summer and autumn, and there may be some sporadic grazing in winter. The meadows are given periodic dressings of farmyard manure. Scottish stands appear to be subject to periodic light grazing, and some are occasionally cut for hay. In both England and Scotland too many grazing animals might damage the vegetation by trampling, especially when the soil is waterlogged. The main threats to the survival of this form of grassland are probably drainage and associated agricultural improvement. Some *Cynosurus-Caltha* grassland is probably derived from woodland of the *Alnus-Fraxinus-Lysimachia* W7 type or a willow scrub form of this community, and might revert to such woodland or scrub if left ungrazed for a long time.

# MG10 *Holcus lanatus-Juncus effusus* rush-pasture

## Synonyms
*Ranunculus repens-Juncus effusus* community *p.p.* (Birse and Robertson 1976).

Typical habitats of MG10 *Holcus-Juncus* rush-pasture (stippled):

1   Flushes among grassland on low-altitude hill slopes
2   Wet level ground among grassland in valley floor
3   In ditch
4   Along edges of stream
5   Around edges of pool

## Description

This is damp grassland in which tussocks of *Juncus effusus* stand out in species-poor swards of *Holcus lanatus, Agrostis stolonifera, Poa trivialis, Ranunculus repens* and *R. acris.* These plants are interleaved with smaller species such as *Cardamine pratensis* and *Trifolium repens.* The rushes are generally taller than the matrix of grasses and forbs, unless the community is heavily grazed. Bryophytes such as *Eurhynchium praelongum, Brachythecium rutabulum, Calliergonella cuspidata* and *Rhytidiadelphus squarrosus* grow in thin wefts over the damp soil and among the larger plants.

There are three sub-communities. The Typical sub-community MG10a has no distinguishing species of its own. The *Juncus inflexus* sub-community MG10b has more *J. inflexus* than *J. effusus.* The *Iris pseudacorus* sub-community MG10c has *Iris pseudacorus* and in some stands *Alopecurus pratensis, Phalaris arundinacea, Glyceria fluitans, Filipendula ulmaria* and *Lotus uliginosus.*

There is also a form of species-poor vegetation consisting of large tussocks of *J. effusus,* beneath which is a sward of typical acid grassland species including *Agrostis capillaris, Festuca ovina, Anthoxanthum odoratum, Nardus stricta, Galium saxatile, Potentilla erecta* and the moss *Rhytidiadelphus squarrosus.* This does not fit clearly into any NVC type.

## Differentiation from other communities
The *Juncus inflexus* sub-community of *Holcus-Juncus* rush-pasture is the only type of vegetation described in the NVC in which *J. inflexus* is dominant or co-dominant with *J. effusus*. The other two sub-communities can be confused with the *Juncus effusus* sub-community of *Juncus-Galium* rush-pasture M23b, but this is usually rather more species-rich, with more *Galium palustre*, *Cirsium palustre* and *Ranunculus flammula*, and generally with more bryophytes. The *Juncus effusus* sub-community of *Carex echinata-Sphagnum* mire M6c is dominated by *J. effusus* but has Sphagna and *Polytrichum commune*.

## Ecology
This is a vegetation type of damp acid to neutral soils on level to gently sloping ground in enclosed pastures, and in neglected situations such as ditches, pond sides and roadside verges. Most stands are at low altitudes, but the community extends to over 300 m.

*Holcus-Juncus* rush-pasture is widespread in lowland Great Britain. It also occurs at low altitudes in most upland areas, although it is rare in the north-west Highlands, Lewis, Harris, Skye, Orkney and Shetland. Most of the records in the uplands are of the Typical sub-community. The other two sub-communities are more restricted to the lowlands. The *Juncus inflexus* sub-community is recorded mostly from the southern and eastern lowlands, but also occurs on upland limestone in the Craven and northern Pennines. The *Iris* sub-community has been recorded in Wales and south-west England.

There is vegetation similar to *Holcus-Juncus* rush-pasture in mainland Europe.

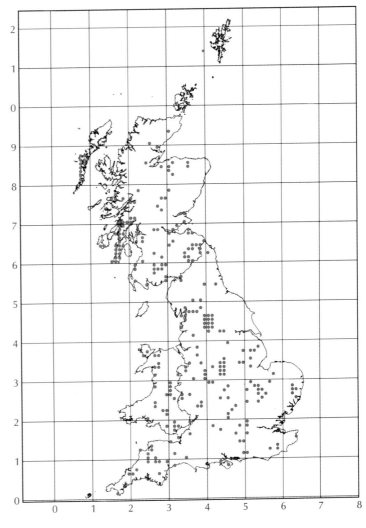

## Conservation interest

This is an impoverished vegetation type and it seldom contains any uncommon species. Although it plays a part in the mixtures of mires and unimproved grasslands around the upland fringes, it is generally regarded as being of less value for nature conservation than other forms of damp grassland such as *Juncus-Galium* rush-pasture. However, vegetation of this type provides important breeding habitats for birds including Curlew *Numenius arquata*, Snipe *Gallinago gallinago*, Lapwing *Vanellus vanellus*, Mallard *Anas platyrhynchos* and Teal *Anas crecca*.

## Management

*Holcus-Juncus* rush-pasture is characteristic of damp pastures, many of which are quite heavily grazed by livestock. Improved pastures on the upland margins can revert to this community if they cease to be fertilised or if drains become ineffective. *Holcus-Juncus* rush-pasture can be reclaimed for agriculture by draining, fertilising and re-seeding to give more productive forms of grassland. If management were to be completely abandoned the vegetation could potentially develop into some form of damp woodland.

# CG9 *Sesleria caerulea-Galium sterneri* grassland

## Synonyms

*Sesleria albicans-Festuca ovina* grassland C1g (Birks and Ratcliffe 1980); *Sesleria albicans-Galium sterneri* grassland (Rodwell 1992).

CG9 *Sesleria-Galium* grassland (stippled): on thin, well-drained rendzina soils on Carboniferous limestone slopes, cliff ledges and scree margins (screes shown as unstippled areas on main slope), and also on similar soils on flatter ground (in foreground).

CG9 is also shown in the picture for M38.

## Description

*Sesleria-Galium* grasslands form extensive pale swards, broken by bone-white tracts of limestone, across the bold isolated hills of the Craven Pennines and over the strange stony moorlands of Upper Teesdale. These species-rich grasslands have a tufted sward of the tough, glaucous leaves of *Sesleria caerulea*. *Koeleria macrantha* and *Festuca ovina* are common here too, and there is much *Briza media* in some stands. *Thymus polytrichus* and *Helianthemum nummularium* trail among the grasses, and there is a rich array of other small herbs such as *Galium sterneri*, *Linum catharticum*, *Viola riviniana* and *Carex flacca*. There is usually some *Ctenidium molluscum*. The turf can be dense, or sparse and thin.

There are five sub-communities. The *Helianthemum oelandicum-Asperula cynanchica* sub-community CG9a is a lowland grassland that usually occurs as fragmented patches over rocky ground and on cliff ledges. Characteristic species include *Helianthemum oelandicum*, *Hippocrepis comosa*, *Scabiosa columbaria*, *Asperula cynanchica*, *Carlina vulgaris* and *Anthyllis vulneraria*. The Typical sub-community CG9b has no distinguishing species other than those which define the community as a whole; there is usually less *H. nummularium* and more *Hypnum cupressiforme s.l.* and *Dicranum scoparium*. It can form continuous swards or more open vegetation on stony ground. The *Carex pulicaris-Carex panicea* sub-community CG9c is a damp upland

303

grassland characterised by the two sedges, *Potentilla erecta, Polygala vulgaris, Hylocomium splendens, Frullania tamarisci* and *Coelocaulon aculeatum*. There are some records here too for *Parnassia palustris, Pinguicula vulgaris, Primula farinosa, Dryas octopetala, Carex hostiana* and *Pimpinella saxifraga*, and in some stands there are a few moorland plants such as *Calluna vulgaris, Molinia caerulea* and *Empetrum nigrum*. The *Carex capillaris-Kobresia simpliciuscula* sub-community CG9d is a damp, sedgy upland grassland that resembles the *Carex pulicaris-Carex panicea* sub-community, but has a suite of rare montane calcicoles including *Kobresia simpliciuscula, Gentiana verna* and *Carex capillaris*, as well as more widespread plants such as *Selaginella selaginoides* and *Persicaria vivipara*. There can be thick mats of bryophytes such as *Racomitrium ericoides, R. lanuginosum, Hylocomium splendens* and *Hypnum cupressiforme s.l.* The *Saxifraga hypnoides-Cochlearia pyrenaica* ssp. *alpina* sub-community CG9e is also a damp grassland with *Carex pulicaris*. Here there is another array of scarce montane calcicoles including *Myosotis alpestris* and *Draba incana*, together with *Saxifraga hypnoides, Cochlearia pyrenaica* ssp. *alpina* and *Plagiochila asplenioides*. The *Carex pulicaris-Carex panicea, Carex-Kobresia* and *Saxifraga-Cochlearia* sub-communities are similar in many ways. Each has a distinctive assemblage of species but in some stands many of the characteristic plants are rare or completely absent. Some vegetation can therefore be hard to assign to a sub-community.

Locally in Wales there are grasslands with swards of *Festuca ovina, Agrostis capillaris, Thymus polytrichus, Linum catharticum, Carex flacca, Galium sterneri* and *Sanguisorba minor*, and with scattered *Carlina vulgaris, Scabiosa columbaria, Carex panicea, Briza media* and *Thalictrum minus*. Despite the absence of *Sesleria caerulea*, this vegetation has some characteristics of *Sesleria-Galium* grassland. It also has affinities with *Festuca ovina-Carlina vulgaris* grassland CG1, *Festuca ovina-Helictotrichon pratense* grassland CG2 and *Festuca-Agrostis-Thymus* grassland CG10.

Given the fact that some calcicolous grasslands outside the peculiarly small British range of *Sesleria* have floras very similar to *Sesleria-Galium* grassland, the various forms of vegetation currently classified as this community could perhaps be regarded as variants of other more widespread grassland types. The *Helianthemum-Asperula* sub-community could be seen as a *Sesleria* form of *Festuca-Carlina* grassland, the Typical sub-community as a *Sesleria* form of *Festuca-Helictotrichon* grassland, and the *Carex pulicaris-Carex panicea, Carex-Kobresia* and *Saxifraga-Cochlearia* sub-communities as a *Sesleria* form of the *Saxifraga-Ditrichum* sub-community of *Festuca-Agrostis-Thymus* grassland. The lowland *Sesleria albicans-Scabiosa columbaria* grassland CG8 could be reclassified in a similar way.

## Differentiation from other communities

The dominance of *Sesleria* separates *Sesleria-Galium* grassland from all other British plant communities except *Sesleria-Scabiosa* grassland. These two *Sesleria* grasslands are very similar. However, *Galium sterneri* is more characteristic of *Sesleria-Galium* grassland, and *Carex panicea* and *C. pulicaris* are common in the damper, more strongly upland sub-communities; these species are all scarce in *Sesleria-Scabiosa* grassland. Conversely, the predominantly lowland species *Plantago media, Galium verum, Primula veris, Leontodon hispidus* and *Sanguisorba minor* are more common in *Sesleria-Scabiosa* grassland.

Stands of the *Carex pulicaris-Carex panicea* sub-community containing much *Dryas octopetala* can be quite similar to *Dryas-Silene* heath CG14 which, however, has more montane species such as *Silene acaulis, Saxifraga aizoides, S. oppositifolia* and *Alchemilla alpina*. The damper *Sesleria-Galium* grasslands can be confused with *Carex-Pinguicula* mire M10. *Sesleria* can grow in *Carex-Pinguicula* flushes, and both communities provide habitats for rare montane calcicoles such as *Kobresia simpliciuscula* and

*Carex capillaris*, and also for the uncommon *Primula farinosa*. *Sesleria-Galium* grasslands have a denser sward, with *Galium sterneri*, *Helianthemum nummularium* and lichens such as *Coelocaulon aculeatum* and *Cetraria islandica*, whereas *Carex-Pinguicula* mires are more open, with *Carex dioica*, *Juncus articulatus*, *Triglochin palustris*, *Drepanocladus revolvens*, *Bryum pseudotriquetrum*, *Campylium stellatum* and other species of wet ground.

## Ecology

*Sesleria-Galium* grassland is almost entirely confined to thin, calcareous, free-draining but moist rendzina soils over drift-free exposures of Carboniferous limestone. It clothes steep slopes below outcrops of limestone, and is also common on cliff ledges, in stable scree, on river shingle and among limestone pavements. The *Carex-Kobresia* sub-community is especially associated with gravelly granular exposures of metamorphosed sugar limestone. The *Helianthemum-Asperula* and *Carex-Kobresia* sub-communities cover thin soils that can become totally parched in summer, and many of the plants that grow in the *Helianthemum-Asperula* sub-community are resistant to drought.

*Sesleria-Galium* grassland comprises both lowland and upland types of vegetation, and has an altitudinal range from near sea-level to over 700 m. The *Helianthemum-Asperula* sub-community has a lowland distribution, the Typical sub-community occurs in both upland and lowland situations, and the *Carex pulicaris-Carex panicea*, *Carex-Kobresia* and *Saxifraga-Cochlearia* sub-communities are exclusively upland. High-altitude stands of *Sesleria-Galium* grassland are probably near-natural. In the harsh upland climate the soils are unstable because of physical and chemical weathering, and although they are highly calcareous they are not rich in plant nutrients. This means that the smaller rare species such as *Viola rupestris*, *Carex capillaris* and *Gentiana verna* are able to flourish because there is not much competition from more vigorous plants.

The soils are derived from limestone and can be rich in lead and other heavy metals. Metallophyte species such as *Minuartia verna* and the lichens *Bacidia bagliettoana*, *B. sabuletorum*, *Cladonia pocillum*, *Collema tenax*, *Diploschistes muscorum*, *Leptogium teretiusculum*, *Peltigera rufescens*, *Polyblastia gelatinosa* and *Vezdaea aestivalis* can be quite common in *Sesleria-Galium* grassland (Gilbert 2000).

## Conservation interest

This is a community in which the dominant and defining species is nationally scarce, and it is a type of grassland that is rare and declining in Europe. It is the main habitat of several rare plants in Great Britain. *Cypripedium calceolus* grows in *Sesleria-Galium* grassland at its sole remaining native British site in the Yorkshire Dales. *Gentiana verna*, *Viola rupestris*, *Kobresia simpliciuscula*, *Myosotis alpestris* and *Carex ornithopoda* are probably more common in this community than in any other type of vegetation in Great Britain, and it is also one of the most important habitats south of the Highlands for *Carex capillaris*, *Persicaria vivipara*, *Dryas octopetala* and *Draba incana*. The uncommon continental species *Carex ericetorum* and *Veronica spicata* grow in some stands. *Sesleria-Galium* grasslands can be nesting-grounds for birds such as Skylark *Alauda arvensis* and Lapwing *Vanellus vanellus*, feeding places for Golden Plover *Pluvialis apricaria*, and home to various invertebrates and small mammals.

## Management

Almost all *Sesleria-Galium* grassland is grazed by sheep and rabbits, and many stands are grazed by cattle as well. In the Craven area of Yorkshire, grassland of this type constitutes a substantial part of the upland hill pastures. Grazing maintains the open sward and prevents succession to woodland, which would otherwise be inevitable in all but

## CG9 *Sesleria caerulea-Galium sterneri* grassland

*Sesleria-Galium* grassland is confined to exposures of Carboniferous limestone in northern England around Morecambe Bay, on the Orton Fells, on the Appleby Fells, in Upper Teesdale, and in the Craven Pennines of Yorkshire. The *Helianthemum-Asperula* subcommunity is confined to the west, occurring in places close to the sea and on limestone scars at the eastern edge of the Lake District. The Typical, *Carex pulicaris-Carex panicea* and *Saxifraga-Cochlearia* sub-communities are widely distributed across the Craven Pennines of north Yorkshire, around Malham and Wharfedale, and around Settle and Whernside. The *Carex-Kobresia* sub-community is known only from Upper Teesdale. There is vegetation similar to *Sesleria-Galium* grassland but without *Sesleria caerulea* on some Carboniferous limestone hills in Wales (see above).

Vegetation almost identical to *Sesleria-Galium* grassland occurs on Carboniferous limestone in western and north-western Ireland (most notably on the Burren), and there are similar grasslands in mainland Europe.

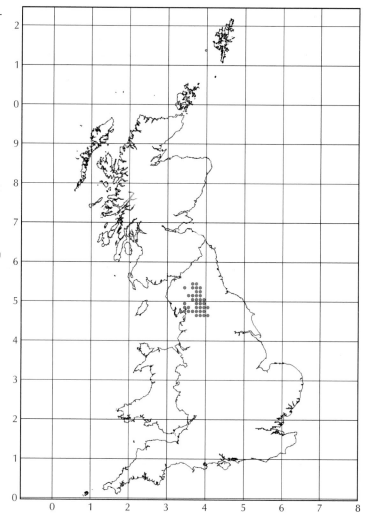

the highest stands where *Sesleria-Galium* grassland is near-natural. The natural vegetation of these places would be herb-rich woodland, such as the *Fraxinus-Sorbus-Mercurialis* community W9 which does indeed occur on rendzina soils adjacent to some stands of *Sesleria-Galium* grassland. In the interests of nature conservation, it is desirable to maintain most of these grasslands as they are, rather than encourage woodland regeneration, because of their importance as a habitat for rare species.

Where stands are inaccessible and ungrazed, *Sesleria* can form tall tussocks that shade out other species, and such stands can have a poor flora. Conversely, too much grazing prevents the herbs from flowering and setting seed. The turf can also be damaged by trampling animals and burrowing rabbits, and the soils readily erode once the continuous layer of vegetation is lost. Where *Sesleria-Galium* grasslands are managed for nature conservation they should be grazed at the lowest intensity that will maintain the swards as grassland rather than as scrub or woodland. The ideal is to maintain a sward of varying height, with short areas where small plants can thrive and taller areas where larger herbs are able to flower. This allows a rich array of species to persist, and provides a fine habitat for invertebrates and small mammals.

# CG10 *Festuca ovina-Agrostis capillaris-Thymus polytrichus* grassland

## Synonyms
Species-rich *Agrosto-Festucetum* and *Saxifrageto-Agrosto-Festucetum* (McVean and Ratcliffe 1962, Birks 1973); Southern species-rich *Agrostis-Festuca* grassland C1e and Northern species-rich *Agrostis-Festuca* grassland C1f (Birks and Ratcliffe 1980); *Festuca ovina-Agrostis capillaris-Thymus praecox* grassland (Rodwell 1992).

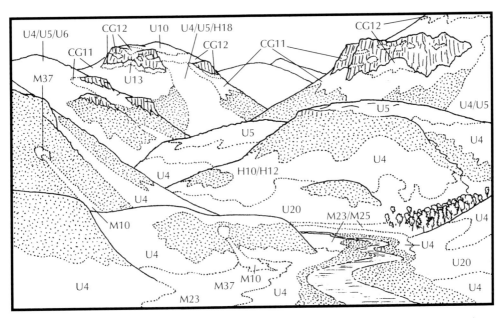

Typical habitats of CG10 *Festuca-Agrostis-Thymus* grassland (stippled): on a wide range of grazed hill slopes at low to high altitude, and also on river shingle (lower right).

Other vegetation types in this picture:

CG11 *Festuca-Agrostis-Alchemilla* grass-heath
CG12 *Festuca-Alchemilla-Silene* dwarf-herb community
U4 *Festuca-Agrostis-Galium* grassland
U5 *Nardus-Galium* grassland
U6 *Juncus-Festuca* grassland
U10 *Carex-Racomitrium* moss-heath
U13 *Deschampsia-Galium* grassland
U20 *Pteridium-Galium* community
H10 *Calluna-Erica* heath
H12 *Calluna-Vaccinium* heath
H18 *Vaccinium-Deschampsia* heath
M10 *Carex-Pinguicula* mire
M23 *Juncus-Galium* rush-pasture
M25 *Molinia-Potentilla* mire
M37 *Palustriella-Festuca* spring

CG10 is also shown in the picture for MG5.

## Description
This is a grassland with a short, green species-rich turf. The grassy matrix of *Festuca ovina, F. rubra, Agrostis capillaris, Anthoxanthum odoratum, Potentilla erecta* and *Viola*

*riviniana* is entangled with the creeping stems of the tiny shrub *Thymus polytrichus*; its dark leaves are fragrant underfoot and its purple-pink flowers are conspicuous in summer. *Festuca vivipara* is common in many Scottish stands. The small mesotrophic forbs *Prunella vulgaris*, *Plantago lanceolata*, *Ranunculus acris* and *Achillea millefolium* are scattered through the sward and bryophytes are common, especially the large, red-gold moss *Hylocomium splendens*.

Three sub-communities are described in the NVC. The *Trifolium repens-Luzula campestris* sub-community CG10a is most common at low to moderate altitudes and on drier soils. It contains much *Anthoxanthum*, *Trifolium repens*, *Galium saxatile*, *Luzula campestris* and *Rhytidiadelphus squarrosus*. There are few sedges in the sward except for some *Carex caryophyllea* and *C. pilulifera*. The *Carex pulicaris-Carex panicea* sub-community CG10b occurs on damper soils. It has a greyer, more sedgy sward than the *Trifolium-Luzula* sub-community, with *Carex panicea*, *C. pulicaris*, *C. flacca* and *Linum catharticum*. *Agrostis canina* can be as common as *A. capillaris*, and in some stands there is a little *Succisa pratensis* and *Selaginella selaginoides*. The *Saxifraga aizoides-Ditrichum flexicaule* sub-community CG10c is the most northern and montane form. It is a grassland of damp soils, and has the characteristic species of the *Carex pulicaris-Carex panicea* sub-community together with an array of small montane herbs such as *Saxifraga aizoides*, *S. oppositifolia*, *Persicaria vivipara*, *Thalictrum alpinum* and *Carex capillaris*, and the calcicole mosses *Ctenidium molluscum*, *Tortella tortuosa*, *Ditrichum flexicaule s.l.* and *Hypnum lacunosum*. In many stands in the Breadalbanes the grass sward of this sub-community is composed mainly of *Festuca ovina*, *Briza media* and *Helictotrichon pratense*, with only a little *Anthoxanthum* and *Agrostis* species, and, rarely, *Sesleria caerulea*. Elsewhere, *Festuca ovina*, *Anthoxanthum* and *Agrostis* species are the most common grasses.

## Differentiation from other communities

*Festuca-Agrostis-Thymus* grassland can be confused with several other short upland grassland communities. It is more species-rich than either *Festuca-Agrostis-Rumex* grassland U1 or *Festuca-Agrostis-Galium* grassland U4. *Thymus polytrichus* is common and there are more small herbs such as *Prunella vulgaris*, *Ranunculus acris* and *Lotus corniculatus*, whereas the acid grasslands have more *Galium saxatile* and generally have a more continuous underlayer of mosses. *Sesleria-Galium* grassland CG9 differs from *Festuca-Agrostis-Thymus* grassland in being dominated by *Sesleria*. *Festuca-Agrostis-Alchemilla* grassland CG11 is distinguished by having much more *Alchemilla alpina*. *Festuca-Alchemilla-Silene* dwarf-herb vegetation CG12 has *Silene acaulis* as well as *A. alpina*, and the turf is more herb-dominated than in *Festuca-Agrostis-Thymus* grassland.

Around the fringes of some Carboniferous limestone hills in Wales, *Festuca-Agrostis-Thymus* grassland can grade into the predominantly lowland *Festuca ovina-Helictotrichon pratense* grassland CG2. *Helictotrichon pratense*, *H. pubescens*, *Briza media*, *Koeleria macrantha*, *Helianthemum nummularium* and lowland forbs such as *Sanguisorba minor*, *Scabiosa columbaria* and *Carlina vulgaris* are more common in *Festuca-Helictotrichon* grassland, whereas the characteristic upland grassland species *Agrostis capillaris*, *Anthoxanthum* and *Potentilla erecta* are scarce.

As noted above, there is marked variation in the composition of the grass swards of the *Saxifraga-Ditrichum* sub-community. This sub-community straddles an important division within British calcicolous grasslands. In the cool, wet climate of the north and west, where soils tend not to be strongly base-rich because of leaching, there are swards of *Festuca ovina*, *F. vivipara*, *Agrostis* species and *Anthoxanthum* with many small forbs and sedges. In the NVC these are accommodated within the *Festuca-Agrostis-Thymus*, *Festuca-Agrostis-Alchemilla* and *Festuca-Alchemilla-Silene* communities. In contrast, in strongly basic habitats in the warmer and drier south and east (but penetrating into cold upland environments in the north Pennines and the Breadalbanes, where the soils are very base-rich), there are grasslands of *F. ovina, Briza media, Helictotrichon pratense, Koeleria macrantha* and, more locally, *Sesleria*. These mostly belong to the *Festuca ovina-Carlina vulgaris* CG1, *Festuca-Helictotrichon* CG2, *Festuca ovina-Pilosella officinarum-Thymus polytrichus/pulegioides* grassland CG7, *Sesleria caerulea-Scabiosa columbaria* CG8, and *Sesleria-Galium* CG9 grassland communities. Some forms of the *Saxifraga-Ditrichum* sub-community of *Festuca-Agrostis-Thymus* grassland belong in the first group of calcicolous grasslands whereas others belong in the second group.

**Ecology**

*Festuca-Agrostis-Thymus* grassland is a community of shallow, brown, silty soils with a pH between 5.3 and 7.2 (McVean and Ratcliffe 1962). The base-status of the soils is maintained either by physical and chemical weathering or by irrigation with base-rich water. *Festuca-Agrostis-Thymus* grassland can clothe large tracts of ground where basic rocks predominate, but can also mark out intrusions of basic rocks or flushed areas in places where the rocks are generally acid. It occurs on a wide range of slopes and

*Festuca-Agrostis-Thymus* grassland is widespread in the British uplands but is not recorded in south-west England or the North York Moors, and is very rare in the southern Pennines. It is especially common on the Dalradian schist of the Breadalbane hills, and there are notably large stands on Mull, in Morvern, and on the Trotternish ridge of Skye.

Vegetation related to *Festuca-Agrostis-Thymus* grassland occurs in mainland Europe.

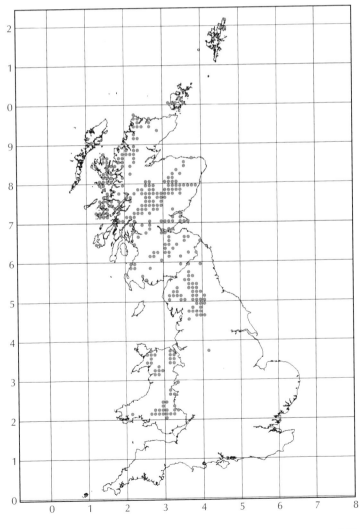

aspects. Most stands are in the sub-montane zone below about 700 m, but it can occur at altitudes of over 1000 m.

## Conservation interest

This species-rich grassland is of great value for nature conservation. Even small stands can add substantially to the species diversity of sub-montane hill slopes, particularly in upland areas with predominantly acidic soils. Rare plants recorded here include *Alchemilla filicaulis* ssp. *filicaulis*, *A. wichurae*, *Carex capillaris*, *C. montana*, *C. rupestris*, *Draba incana*, *Minuartia verna*, *Gentiana verna*, *Myosotis alpestris*, *Potentilla crantzii* and *Sagina saginoides*. There are moles in many stands on deeper soils, even when the grasslands are isolated by miles of blanket bog on deep peat (e.g. Averis and Averis 1995a).

## Management

Most stands of *Festuca-Agrostis-Thymus* grassland are ultimately derived from heath or woodland as a result of grazing and burning. *Fraxinus-Sorbus-Mercurialis* W9 and *Quercus-Betula-Oxalis* W11 woodlands are the most likely precursors of the community, together with *Dryas-Carex* heath CG13 in the far north-west. The grasslands are maintained by sheep and deer grazing; some stands are also grazed by rabbits. Reversion to woodland or calcareous heath can occur if grazing animals are excluded. In comparison with many other upland vegetation types, *Festuca-Agrostis-Thymus* grasslands are preferentially grazed because of their high nutrient content and are consequently enriched by dung and urine. The herbs in most stands rarely flower, and there is little opportunity for the plants to spread by seed dispersal or for the genetic diversity of the populations to increase. At lower altitudes, the community may be converted to more productive agricultural swards following enclosure and intensification of management.

# CG11 *Festuca ovina-Agrostis capillaris-Alchemilla alpina* grass-heath

## Synonyms

*Alchemilleto-Agrosto-Festucetum* and species-rich *Agrosto-Festucetum p.p.* (McVean and Ratcliffe 1962); *Alchemilla-Festuca* grassland C1d (Birks and Ratcliffe 1980).

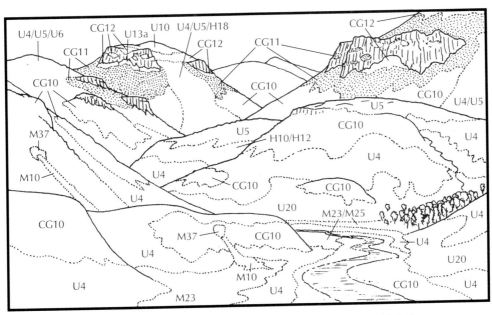

CG11 *Festuca-Agrostis-Alchemilla* grass-heath (stippled) on steep, grazed hill slopes at medium to high altitude, and on cliff ledges (top right). It mostly occurs at higher altitudes than CG10 *Festuca-Agrostis-Thymus* grassland.

Other vegetation types shown in this picture:

CG12 *Festuca-Alchemilla-Silene* dwarf-herb community
U4 *Festuca-Agrostis-Galium* grassland
U5 *Nardus-Galium* grassland
U6 *Juncus-Festuca* grassland
U10 *Carex-Racomitrium* moss-heath
U13 *Deschampsia-Galium* grassland
U20 *Pteridium-Galium* community
H10 *Calluna-Erica* heath
H12 *Calluna-Vaccinium* heath
H18 *Vaccinium-Deschampsia* heath
M10 *Carex-Pinguicula* mire
M23 *Juncus-Galium* rush-pasture
M25 *Molinia-Potentilla* mire
M37 *Palustriella-Festuca* spring

## Description

In this grassland type, a short, species-rich, loosely woven, silvery sward of *Alchemilla alpina*, *Festuca vivipara*, *Anthoxanthum odoratum* and *Agrostis capillaris* forms the background for many small flowering plants such as *Ranunculus acris*, *Persicaria vivipara*, *Thymus polytrichus*, *Thalictrum alpinum* and *Carex pulicaris*. The herbs are entwined with mosses, in particular the red-gold *Hylocomium splendens* and *Rhytidiadelphus loreus* and the large, grey-green *Racomitrium lanuginosum*.

There are two sub-communities. Stands on drier soils with *Viola riviniana*, *Rhytidiadelphus squarrosus*, *Pleurozium schreberi* and *Dicranum scoparium* belong to the Typical sub-community CG11a. In the damper *Carex pulicaris-Carex panicea* sub-community CG11b, *Carex pulicaris* and *C. panicea* are common, together with other plants of damp habitats such as *Selaginella selaginoides*, *Pinguicula vulgaris* and *Viola palustris*.

### Differentiation from other communities

*Festuca-Agrostis-Alchemilla* grass-heath resembles *Festuca-Agrostis-Thymus* grassland CG10 but has much more *Alchemilla alpina*. It is also similar to the *Festuca-Alchemilla-Silene* dwarf-herb community CG12 but lacks the large mats of *Silene acaulis* that are characteristic of that vegetation type, and has less *Deschampsia cespitosa*.

The community might be confused with a distinctive mossy form of the *Vaccinium-Deschampsia* sub-community of *Festuca-Agrostis-Galium* grassland U4e in which there is a thin sward of grasses and *Alchemilla alpina* growing through a silvery carpet of *Racomitrium lanuginosum*. However, this form of *Festuca-Agrostis-Galium* grassland is

*Festuca-Agrostis-Alchemilla* grass-heath is widespread but scarce throughout the Highlands. It is common on the basic Dalradian rocks of the Breadalbanes, the Durness limestone, and basic exposures of the Moine Schist. There are outlying stands on basalt hills in Morvern, on the Ardmeanach in Mull, on the Trotternish ridge on Skye, and on Borrowdale Volcanic rocks in the Lake District.

There is similar vegetation in mainland Europe and on the Faroes.

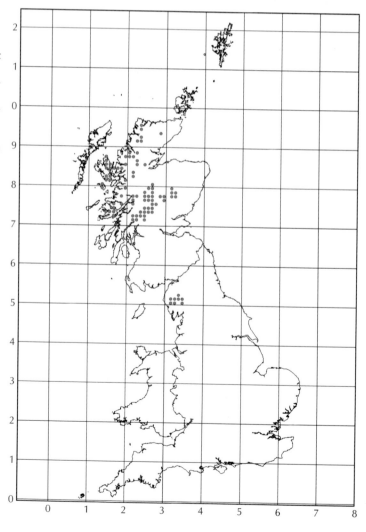

relatively species-poor, and calcicolous species such as *Thymus polytrichus* are very rare or absent; *Vaccinium myrtillus* and *Deschampsia flexuosa* occur in some stands. The floristic table for *Festuca-Agrostis-Alchemilla* grass-heath in the NVC suggests that *V. myrtillus* is common here, but bilberry is a plant of acid habitats and is not characteristic of calcicolous grasslands. It seems possible that the NVC data might include some quadrat samples recorded in vegetation that would be better classed as the *Alchemilla alpina* form of the *Vaccinium-Deschampsia* sub-community of *Festuca-Agrostis-Galium* grassland.

**Ecology**
Most stands of *Festuca-Agrostis-Alchemilla* grass-heath are on high slopes (generally over 600 m and extending to over 1000 m) on moderately basic shallow soils that are flushed but free-draining. The community occurs on slopes of all aspects and on a variety of gradients. Many stands are small, possibly because there is so much leaching at these high altitudes that there are few large spreads of suitably basic soils.

**Conservation interest**
This species-rich community is the habitat for several scarce calcicoles including *Cerastium alpinum*, *Potentilla crantzii* and *Herbertus stramineus*. The oceanic liverwort *Mastigophora woodsii* is locally common, notably on the Trotternish ridge on Skye. The uncommon *Botrychium lunaria* is often recorded in this form of grassland, especially where it covers shallow stony soils along the sides of streams and gullies.

**Management**
*Festuca-Agrostis-Alchemilla* grass-heath is maintained by grazing at most if not all of its sites. In most places, it seems to be stable under current levels of grazing, even on some over-stocked hills in the Lake District. Succession to scrub, woodland or *Dryas* heath might occur in the absence of grazing.

# CG12 *Festuca ovina-Alchemilla alpina-Silene acaulis* dwarf-herb community

## Synonyms

Dwarf Herb nodum (McVean and Ratcliffe 1962); *Silene acaulis-Festuca ovina* sward D3 (Birks and Ratcliffe 1980).

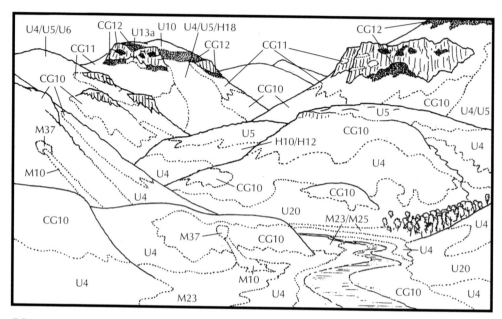

CG12 *Festuca-Alchemilla-Silene* dwarf-herb community (stippled): on high, grazed slopes, on cliff ledges, and below cliffs.

Other vegetation types shown in this picture:

CG10 *Festuca-Agrostis-Thymus* grassland
CG11 *Festuca-Agrostis-Alchemilla* grass-heath
U4 *Festuca-Agrostis-Galium* grassland
U5 *Nardus-Galium* grassland
U6 *Juncus-Festuca* grassland
U10 *Carex-Racomitrium* moss-heath
U13 *Deschampsia-Galium* grassland
U20 *Pteridium-Galium* community
H10 *Calluna-Erica* heath
H12 *Calluna-Vaccinium* heath
H18 *Vaccinium-Deschampsia* heath
M10 *Carex-Pinguicula* mire
M23 *Juncus-Galium* rush-pasture
M25 *Molinia-Potentilla* mire
M37 *Palustriella-Festuca* spring

## Description

These short velvety-green swards of *Silene acaulis* and *Minuartia sedoides*, spiked through with a little *Festuca vivipara*, *Deschampsia cespitosa* and *Anthoxanthum odoratum*, are bright in summer with the pink flowers of *S. acaulis*. They enclose a splendid array of dwarf herbs: *Alchemilla alpina*, *A. glabra*, *Persicaria vivipara*, *Thalictrum alpinum*, *Saxifraga oppositifolia*, *Selaginella selaginoides*, *Luzula spicata*,

*Sibbaldia procumbens, Carex capillaris, Trollius europaeus* and *Geum rivale.* Calcicolous mosses such as *Hypnum lacunosum* and *Ctenidium molluscum* grow through the carpets of vascular plants.

## Differentiation from other communities

This is a distinctive community. It could be confused with some examples of *Festuca-Agrostis-Thymus* grassland CG10 or *Festuca-Agrostis-Alchemilla* grassland CG11, but has more *Silene acaulis*, and generally more forbs and fewer grasses, than either of these communities. It has a similar flora to the *Dryas-Silene* ledge community CG14, but *Dryas octopetala* is much more common in the *Dryas-Silene* community.

Some stands of the *Festuca-Alchemilla-Silene* dwarf-herb community resemble the *Silene* sub-community of *Carex-Racomitrium* moss-heath U10c, but have more *Agrostis capillaris* and *Deschampsia cespitosa*, and less *Racomitrium lanuginosum* and *Carex bigelowii*. The community shares some species with the *Silene-Luzula* sub-community of *Salix-Racomitrium* snow-bed U12a, including *Silene acaulis, Luzula spicata, Thymus polytrichus* and *Persicaria vivipara*, but in the snow-bed these species are set in a matrix of diminutive liverworts and montane mosses, rather than forming a thick, continuous, green sward.

This type of vegetation occurs throughout the Scottish Highlands, and there is an outlier in north Wales. In the Hebrides it has been recorded only on Mull, Skye and Rum. The largest stands are on the calcareous mica-schist of the Breadalbane hills, especially on Ben Lui and Ben Lawers.

There is similar vegetation on the Faroe Islands (Hobbs and Averis 1991a) and in Norway.

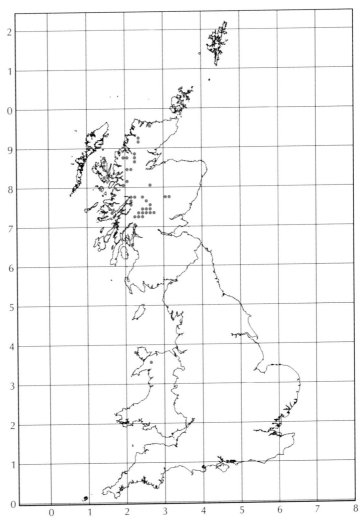

## Ecology

The *Festuca-Alchemilla-Silene* dwarf-herb community occurs on calcareous and damp but free-draining soils at high altitudes: generally above 600 m and extending up above 1000 m. At these altitudes many stands are covered with snow for much of the winter. The community can occur at the foot of cliffs where many of the small herbs must have seeded down from plants on the ungrazed ledges. It occurs on all aspects and a range of gradients.

## Conservation interest

This is one of the most species-rich plant communities in the uplands and many stands include an array of rare species. The most notable are *Cerastium alpinum*, *Draba norvegica*, *Myosotis alpestris*, *Potentilla crantzii*, *Veronica alpina*, *Saxifraga nivalis*, *Gentiana nivalis*, the mosses *Aulacomnium turgidum*, *Hypnum hamulosum*, *Timmia norvegica*, *T. austriaca*, *Pseudoleskea incurvata* and *Ptychodium plicatum*, and the Atlantic liverwort *Herbertus stramineus*.

## Management

*Festuca-Alchemilla-Silene* vegetation is probably a near-natural climax community on montane plateaux and on cliff ledges; in these situations it appears to be little affected by management. However, many Breadalbane stands are grazed by sheep, and some examples of the community may be derived from *Luzula-Geum* tall-herb vegetation U17, *Dryas-Silene* ledge community CG14 (McVean and Ratcliffe 1962) or *Salix-Luzula* scrub W20.

# CG13 *Dryas octopetala-Carex flacca* heath

## Synonyms
*Dryas-Carex flacca* nodum and *Dryas-Carex rupestris* nodum *p.p.* (McVean and Ratcliffe 1962); Low altitude *Dryas* heath B4a (Birks and Ratcliffe 1980).

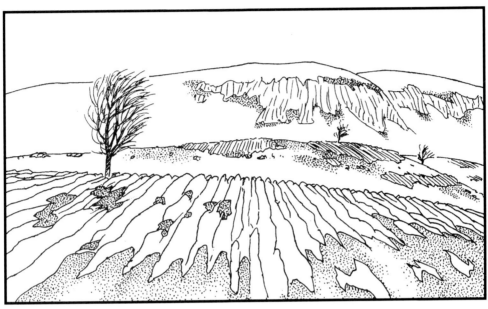

Typical habitat of CG13 *Dryas-Carex* heath (stippled): on thin, well-drained limestone soils, especially around rock outcrops and limestone pavements.

## Description
In this community the small, dark-green, shining leaves of *Dryas octopetala* grow in a tight, low mat over the ground, and the plant is conspicuous only in summer when the large pale-cream delicate flowers unfold. The mat of *Dryas* is interlaced with *Thymus polytrichus*, and pricked through by *Carex flacca, C. panicea, Viola riviniana, Plantago maritima, P. lanceolata, Linum catharticum* and *Festuca ovina*. Under the vascular plants there are tufts and patches of the mosses *Hylocomium splendens, Fissidens dubius, Hypnum cupressiforme s.l.* and, on the rocks that break through the thin skin of soil, *Neckera crispa, Ctenidium molluscum* and *Tortella tortuosa*.

There are two sub-communities: one grassy and one heathy. The grassy form is the *Pilosella officinarum-Ctenidium molluscum* sub-community CG13a, which has an especially rich array of species including *Pilosella officinarum, Potentilla erecta, Carex pulicaris, C. caryophyllea, Prunella vulgaris, Bellis perennis* and *Ctenidium molluscum*; there can be short, stunted shoots of *Calluna vulgaris*. The *Salix repens-Empetrum nigrum* ssp. *nigrum* sub-community CG13b includes more heathy vegetation. The canopy of shrubs is usually a little taller and denser here, and *Dryas* is joined by *Arctostaphylos uva-ursi, Salix repens* (including ssp. *argentea*), *Empetrum nigrum* (both ssp. *nigrum* and ssp. *hermaphroditum*), *Erica cinerea* and, in some stands, *Juniperus communis* ssp. *nana*. There are also a few mosses such as *Scleropodium purum* and *Rhytidiadelphus triquetrus* growing over the ground in loose wefts.

**Differentiation from other communities**

*Dryas-Carex* heath can be confused with its montane counterpart, *Dryas-Silene* heath CG14. *Dryas-Silene* heath has more montane species such as *Silene acaulis, Saxifraga aizoides, S. oppositifolia, Carex capillaris, Alchemilla alpina, Oxyria digyna, Saussurea alpina, Thalictrum alpinum* and *Polystichum lonchitis*; *Sedum rosea, Trollius europaeus, Geum rivale, Geranium sylvaticum* and other tall forbs occur in many stands. Lowland plants such as *Bellis perennis, Koeleria macrantha, Plantago lanceolata* and *Carex flacca* are more common in *Dryas-Carex* heath. The main habitats of the two are also quite different. *Dryas-Carex* heath occurs mainly in grazed mosaics of heath and grassland at low altitudes, whereas *Dryas-Silene* heath is a community of ungrazed cliff ledges at higher altitudes.

*Dryas* can be common in the *Carex pulicaris-Carex panicea* sub-community of *Sesleria-Galium* grassland CG9c, but that vegetation differs clearly from *Dryas-Carex* heath in being dominated by *Sesleria albicans*.

**Ecology**

*Dryas-Carex* heath occurs on shallow calcareous rendzinas and skeletal soils over Durness limestone and dolomite, Dalradian limestone, Jurassic limestone and, in a few places, basalt. The soils are damp but free-draining and are base-rich, either with calcium or magnesium; the pH can be as high as 7.5. The community occurs around limestone pavements, below cliffs and on screes, and also on exposed headlands where calcareous shell-sand has been blown inland over peaty soils. Stands vary in size from a few hectares to patches only a few metres across. It is a sub-montane vegetation type, occurring from just above sea-level along the north coast of Scotland and on Skye to over 430 m in Wester Ross, although most stands are below 100 m. *Dryas* grows slowly and can compete with other, more vigorous plants only where the climate is cool, moist and oceanic and the growing season is short.

**Conservation interest**

During the last glacial period 115,000 to 11,000 years ago, the distribution of *Dryas* extended across Great Britain right down to the south of England. It was probably accompanied by an array of species similar to that with which it grows today. *Dryas-Carex* heaths are valuable surviving fragments of the sort of shrubby, flowery vegetation that must have occurred widely in Great Britain on the wet, base-rich, gravelly, skeletal soils left behind by the retreating ice. As the temperature rose, plants such as *Dryas* would have been ousted by taller herbs and then by trees. When woodland was at its greatest extent the surviving *Dryas* heaths would have been an important refugium for species of open ground at low altitudes.

*Dryas-Carex* heath is one of the rarest types of upland vegetation in Great Britain, with a total extent of a little over 300 ha. Almost all of this is within SSSIs and SACs. Many rare calcicoles grow in this community. One of the most notable is *Carex rupestris*. Other rarities recorded include *Elymus caninus* var. *donianus, Arenaria norvegica* ssp. *norvegica, Carex capillaris, Epipactis atrorubens, Oxytropis halleri* and *Primula scotica*. These heaths are also home to the rare mosses *Conardia compacta, Schistidium robustum, S. trichodon, Seligeria trifaria* and *Syntrichia princeps*. The lichen flora is also noteworthy, although northern limestone outcrops seem to be floristically poorer than those in the south (Gilbert 2000). Rare species include *Ionaspis melanocarpa, Sagiolechia protuberans* and *Staurothele bacilligera; Collema furfuraceum, C. multipartitum, C. polycarpon, Caloplaca arnoldii, Gyalecta jenensis* and *Lecanora crenulata* are unusually common (Gilbert 2000).

In Britain, *Dryas-Carex* heath is confined to the maritime fringes of the north-west Highlands. It extends eastwards along the north coast of Scotland from Durness to Invernaver. From Durness southwards to Inchnadamph, Rassal and Kishorn and onto Skye it follows the outcrops of Durness lime-stone and dolomite along the Moine Thrust Plane. It also occurs on Jurassic limestone on Raasay, on Dalradian limestone on Lismore, and on Tertiary basalt on Skye and Rum.

Similar forms of *Dryas* heath, though with fewer arctic-alpine species, occur at low altitudes on the Burren in western Ire-land. In Scandinavia and other parts of mainland Europe *Dryas octopetala* is more common in montane vegetation.

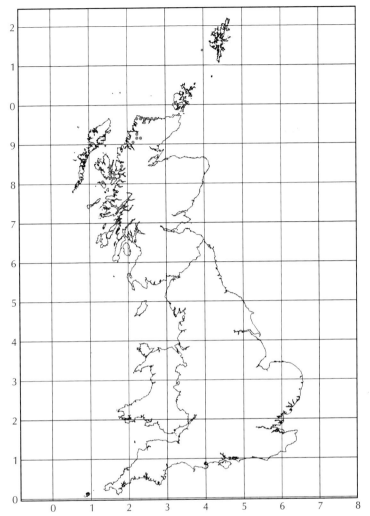

## Management

Most stands of *Dryas-Carex* heath are grazed by sheep, deer and rabbits. The heathy *Salix-Empetrum* sub-community occurs mostly on ground that is less accessible to grazing animals. In contrast, the grassy *Hieracium-Ctenidium* sub-community can be grazed so hard that the spreading mats of *Dryas* can be seen only at close range. The flowers of *Dryas* are often nipped off before they can set seed, but even in hard-grazed heaths, such as those in southern Skye, a proportion of the plants still set seed, and *Dryas* is also able to spread by adventitious rooting from its creeping branches. However, very hard grazing may weaken the *Dryas* and convert the heath to calcicolous swards, such as *Festuca-Agrostis-Thymus* grassland CG10 or *Festuca-Agrostis-Alchemilla* grass-heath CG11. In the current climate, a certain amount of grazing may be necessary to maintain *Dryas-Carex* heaths. Most stands are well within the altitudinal range of wood-land and the ground would be colonised by trees in the absence of grazing animals. Although *Dryas* grows under trees in Strath Suardal in southern Skye (Averis and Averis 2000a) and in Glen Coe (Averis and Averis 2003), it is primarily a plant of open ground. Stands on the exposed northern coasts may be naturally treeless because of the short growing season and constant exposure to wind, salt spray and abrasive wind-blown sand. In more sheltered inland sites *Dryas* heaths would probably revert to hazel scrub with *Salix repens* if grazing was removed. Stands at Inchnadamph and Rassal might also be colonised by *S. myrsinites*. The ideal management is probably moderate grazing, which will prevent scrub encroachment but at the same time allow *Dryas* to grow vigorously and set seed.

Many *Dryas* heaths occur close to human settlements on accessible, low-lying ground. Some stands have been converted to pasture by fertilising and re-seeding with vigorous grasses. Others have been lost to golf courses, and some stands have been eroded by walkers. However, the most serious threat is overgrazing by red deer, domestic stock and rabbits.

# CG14 *Dryas octopetala-Silene acaulis* ledge community

## Synonyms
*Dryas-Salix reticulata* nodum and *Dryas-Carex rupestris* nodum *p.p.* (McVean and Ratcliffe 1962); High altitude *Dryas* heath B4b (Birks and Ratcliffe 1980).

CG14 *Dryas-Silene acaulis* ledge community (stippled) on steep, ungrazed, basic cliff ledges at high altitude.

## Description
These are spreading loose mats of *Dryas octopetala*; its dark-green and silver scallop-edged leaves are decorated with delicate creamy flowers in summer. The low canopy of *Dryas* is interspersed with bright-green cushions of *Silene acaulis* and darker green trailing stems of *Thymus polytrichus*, *Saxifraga aizoides* and *S. oppositifolia*, and is pierced by the neat dark leaves of *Persicaria vivipara* and the pale shoots of *Selaginella selaginoides*. There is generally also an array of taller species such as *Geum rivale*, *Sedum rosea*, *Rubus saxatilis*, *Geranium sylvaticum*, *Ranunculus acris*, *Trollius europaeus* and *Polystichum lonchitis*, and montane willows may occur. There is a deep underlay of bryophytes, comprised of *Ctenidium molluscum*, *Tortella tortuosa*, *Ditrichum flexicaule s.l.*, *Orthothecium rufescens*, *Neckera crispa* and other calcicolous mosses and liverworts.

## Differentiation from other communities
*Dryas-Silene* ledge vegetation might be confused with *Dryas-Carex* heath CG13, but it occurs in quite different situations, and montane species such as *Silene acaulis*, *Carex capillaris*, *Saxifraga oppositifolia* and *Thalictrum alpinum* are more common. *Dryas* can grow in the *Carex pulicaris-Carex panicea* sub-community of *Sesleria-Galium* grassland CG9c, but this is a grassy sward dominated by *Sesleria caerulea*. *Dryas* can also grow in the *Luzula-Geum* tall-herb community U17 but it never forms such extensive mats there as it does in the *Dryas-Silene* ledge community.

## Ecology

This community is confined to calcareous rock ledges with free-draining, base-rich soils; most stands are between 300 m and 900 m. The soils are rendzina-like or moist loams and are irrigated by base-rich water from calcareous parent rocks. Solifluction, freeze-thaw and dry flushing help to keep the soils immature and base-rich with a high pH (McVean and Ratcliffe 1962). *Dryas-Silene* vegetation occurs in places that are either inaccessible or only lightly grazed, including cliff ledges, steep rocky slopes with broken crags, and among boulders. Some stands are on exposed south-facing slopes and are dry in summer.

## Conservation interest

The *Dryas-Silene* ledge community is noteworthy as an example of climax vegetation on calcareous substrates at high altitudes. It is one of the most species-rich plant communities in Great Britain. Many rare species occur here, of which the most notable are *Astragalus alpinus*, *Bartsia alpina*, *Saxifraga nivalis*, *Carex atrata*, *C. capillaris*, *C. rupestris*, *C. vaginata*, *C. norvegica*, *Oxytropis halleri* and *Veronica fruticans*, and the bryophytes *Rhytidium rugosum*, *Ptychodium plicatum*, *Saelania glaucescens*, *Timmia norvegica*, *T. austriaca*, *Syntrichia norvegica*, *Schistidium trichodon*, *Hypnum revolutum*, *Campylophyllum halleri*

The *Dryas-Silene* community is most common on Dalradian mica-schist in the Breadalbane region of Scotland between Ben Lui and Caenlochan. It also occurs in the western and north-western Highlands and the Hebrides, on Durness limestone, basic Moine rocks, basalt and other volcanic outcrops. There are fragmentary outliers in north Wales.

There are similar forms of *Dryas* vegetation in Scandinavia.

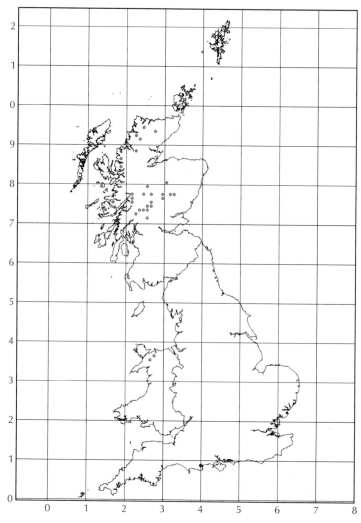

and *Hylocomium pyrenaicum*. The rare montane willows *Salix reticulata*, *S. lanata*, *S. lapponum*, *S. myrsinites* and *S. arbuscula* occur in some stands.

## Management
As the *Dryas-Silene* ledge community is currently almost confined to ungrazed situations, but does occur rarely in remote areas open to grazing animals, it is likely that it was once more extensive in the British uplands. However, it is so dependent on rather peculiar and scarce types of soils that its total range may never have been much greater in recent times than it is today (McVean and Ratcliffe 1962). Hard grazing can lead to the loss of *Dryas* and other grazing-sensitive species, and is likely to have converted some stands into other forms of calcareous vegetation, such as *Festuca-Agrostis-Alchemilla* grass-heath CG11 or *Festuca-Alchemilla-Silene* dwarf-herb community CG12. This has left *Dryas-Silene* vegetation increasingly restricted to inaccessible cliff ledges. It should be encouraged to spread from rock ledges onto adjacent open slopes, preferably by reducing the intensity of grazing. Stockproof fencing may also be useful in suitable places.

# U1 *Festuca ovina-Agrostis capillaris-Rumex acetosella* grassland

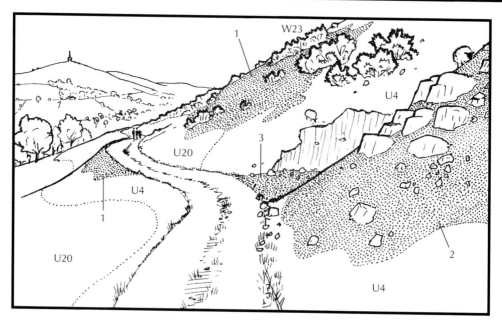

Typical habitats of U1 *Festuca-Agrostis-Rumex* grassland (stippled):

1 Thin, dry, stony soil on steep slopes
2 Thin, dry soil broken by rock outcrops
3 Thin, stony soil on artificially disturbed ground in old quarry

Other vegetation types in this picture:

U4 *Festuca-Agrostis-Galium* grassland
U20 *Pteridium-Galium* community
W23 *Ulex-Rubus* scrub


## Description

This is a low, tufted patchy sward of the grasses *Festuca ovina*, *Aira praecox* and *Agrostis capillaris*, sprinkled with *Rumex acetosella*. There is a sparse layer of bryophytes over the thin dry soil, comprising the mosses *Pleurozium schreberi*, *Dicranum scoparium*, *Brachythecium albicans*, *Polytrichum formosum*, *P. juniperinum* and *P. piliferum*, and in some places the liverwort *Ptilidium ciliare* with its distinctive yellow-gold, fringed leaves. Some stands are scattered with low-grown, heavily browsed bushes of *Ulex gallii* or *U. europaeus*. In many lowland stands there is a distinctive assemblage of annual herbs, but these occur only rarely in the uplands. Although the turf is green in winter it soon dries out in spring and early summer to a pale silver-grey sward.

*Festuca-Agrostis-Rumex* grassland is predominantly a lowland community but at least two of the six sub-communities also occur in the uplands. The *Galium saxatile-Potentilla erecta* sub-community U1e is defined by *Galium saxatile*, *Potentilla erecta* and *Deschampsia flexuosa*. The *Hypochaeris radicata* sub-community U1f has *Hypochaeris radicata*, *Leontodon autumnalis*, *Sedum anglicum* and *Festuca rubra*. Some stands on steep dry rocky slopes in eastern Wales and central Scotland have ephemeral species such as *Aphanes arvensis* and *Erophila verna* and can be moderately herb-rich. These show some affinity to the *Erodium cicutarium-Teesdalia nudicaulis* sub-community U1c.

## Differentiation from other communities

In the uplands this community is likely to be confused only with *Festuca-Agrostis-Galium* grassland U4. *Festuca-Agrostis-Rumex* grassland has a sparser sward in which plants of open ground such as *Rumex acetosella*, *Pilosella officinarum* and *Senecio jacobaea* are common. *Festuca-Agrostis-Galium* grassland has more *Anthoxanthum odoratum*, *Holcus lanatus* and *Viola riviniana* within a more continuous and mossy sward.

## Ecology

*Festuca-Agrostis-Rumex* grassland occurs on thin, dry, acid soils. In the uplands it is most common on south-facing rocky slopes and rocky knolls at low altitudes, and rarely occurs above 400 m. The underlying rocks are generally hard and acid: granite and Old Red Sandstone in south-west England, and Cambrian, Pre-Cambrian, Ordovician, Silurian and Devonian sedimentary and igneous rocks extending from Wales to central Scotland.

*Festuca-Agrostis-Rumex* grassland is widespread in lowland Great Britain, and extends locally into upland areas. The *Galium-Potentilla* sub-community is the commonest form in the uplands, and occurs in scattered localities from south-west England to eastern Scotland. It accounts for much of the grassland vegetation on some of the Shropshire hills. The *Hypochaeris* sub-community is much scarcer in the uplands but has been recorded in the hills of mid-Wales (Woods 1993). Stands showing affinities with the *Erodium-Teesdalia* sub-community are known from eastern Wales and eastern Scotland.

Related forms of grassland occur throughout central Europe as far east as Poland, and also locally in southern and central Scandinavia.

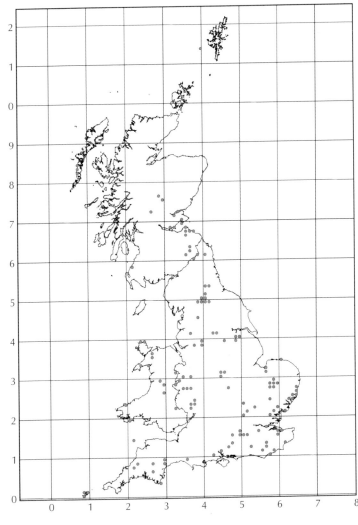

## Conservation interest

Most upland examples of *Festuca-Agrostis-Rumex* grassland are composed of common, widespread plants and are not particularly species-rich, but they are of value for nature conservation because they add to the diversity of vegetation mosaics in the upland fringes in southern and eastern parts of Great Britain. Some stands of the community on steep, rocky slopes in eastern Wales and central Scotland are home to rarities such as *Lychnis viscaria*, *Veronica spicata*, *Dianthus deltoides*, *Scleranthus perennis* and the moss *Bartramia stricta*.

## Management

The community is maintained by sheep and rabbit grazing. If herbivores were removed, *Festuca-Agrostis-Rumex* grassland would probably revert to gorse scrub or dry *Calluna vulgaris* heath: *Ulex-Rubus* scrub W23, *Calluna-Ulex gallii* heath H8, *Calluna-Deschampsia* heath H9 or *Calluna-Vaccinium* heath H12. Some stands, for example in the Shropshire hills, are being invaded by bracken.

# U2 *Deschampsia flexuosa* grassland

## Synonyms
*Deschampsia flexuosa* grassland C1c (Birks and Ratcliffe 1980).

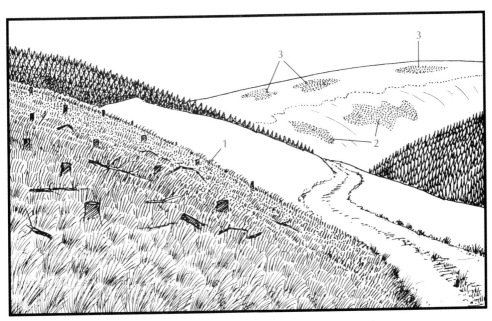

Typical habitats of U2 *Deschampsia* grassland (foreground and stippled):

1  Felled woodland
2  Severely burned areas among heath on hill slopes
3  Severely burned areas among bogs

## Description
This is a tussocky grassland dominated by fine-leaved, dark-green clumps of *Deschampsia flexuosa*. The smooth rounded tufts of this plant can give the sward a characteristic quilted appearance, and the tall, delicate, silvery-pink panicles of flowers are conspicuous in summer. There can be a small amount of *Calluna vulgaris* among the *D. flexuosa*.

There are two sub-communities. The grassy *Festuca ovina-Agrostis capillaris* sub-community U2a includes common plants of upland acid grassland such as *Festuca ovina*, *Agrostis capillaris*, *Potentilla erecta* and *Galium saxatile*. *Pteridium aquilinum* grows in many stands. The sparse mat of bryophytes includes the mosses *Polytrichum piliferum* and *Dicranum scoparium*. The heathy *Vaccinium myrtillus* sub-community U2b has a freckling of *Vaccinium myrtillus* and, less commonly, *Empetrum nigrum*, *Molinia caerulea*, *Eriophorum vaginatum*, *Juncus squarrosus*, *J. effusus* and *Agrostis canina*. *Calluna* is generally commoner here than in the *Festuca-Agrostis* sub-community, and the sparse flora of mosses is augmented by *Pleurozium schreberi*, *Polytrichum commune* and *Hypnum jutlandicum*.

## Differentiation from other communities
*Deschampsia* grassland may be confused with grassy stands of *Calluna-Deschampsia* heath H9 or *Vaccinium-Deschampsia* heath H18, but *Deschampsia flexuosa* is dominant rather than subordinate to dwarf shrubs as it is in heathland vegetation.

In a few high corries of the Cairngorms there are snow-beds dominated by *Deschampsia flexuosa* with a montane flora very different from that of *Deschampsia* grassland, including *Juncus trifidus*, *Huperzia selago* and the mosses *Polytrichum alpinum* and *Dicranum fuscescens*. These are a distinctive form of *Salix-Racomitrium* snow-bed U12 (Rodwell *et al.* 2000).

## Ecology

*Deschampsia* grassland is typical of slopes and valley sides with base-poor, moist, free-draining soils. It is a lowland to sub-montane vegetation type, and rarely, if ever, occurs above 600 m. In many places this community represents the first stage of recolonising vegetation within felled conifer plantations where it forms an untidy grassland among the dead stumps and branches of the felled trees. It can also replace blanket bog vegetation where the peat has dried out as a result of excessive burning.

## Conservation interest

*Deschampsia* grassland has little interest for nature conservation, except as a component of semi-natural mosaics of vegetation around the lower fringes of the uplands.

The community is widespread around the upland margins from Devon to the eastern Highlands, and also occurs in some lowland areas. The *Vaccinium* sub-community is the most common form of *Deschampsia* grassland in the uplands. It does not seem to have been recorded in mainland Europe.

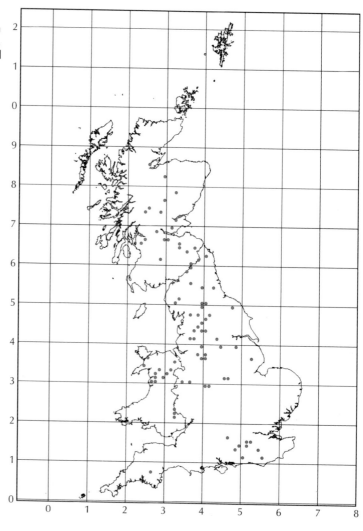

## Management

Much *Deschampsia* grassland has probably been derived from various forms of dwarf-shrub heath as a result of burning and grazing. For example, in eastern Wales and central and northern England many stands of the community occur adjacent to *Calluna-Deschampsia* heaths. Some stands may have developed directly from *Quercus-Betula-Dicranum* woodland W17 or, around the uplands of south-east Scotland, the south Pennines and Wales, from lowland *Quercus* species-*Betula* species-*Deschampsia flexuosa* woodland W16. In a few places, it has replaced blanket bog vegetation on drained peat. The community is maintained by grazing, principally by sheep and rabbits. In general, a reduction in grazing, allowing succession to dwarf-shrub heath or woodland, would be desirable for the purposes of nature conservation.

# U3 *Agrostis curtisii* grassland

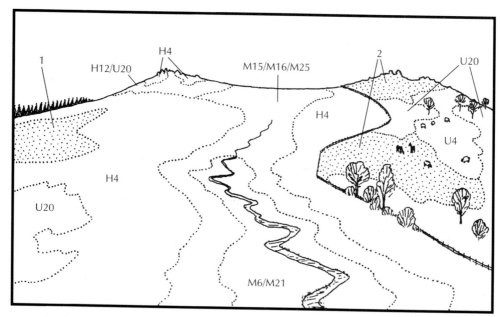

Typical habitats of U3 *Agrostis* grassland (stippled):

1   Among H4 *Ulex gallii-Agrostis* heath, from which U3 is derived by moor-burning. U3 can be a temporary phase after burning, eventually re-developing into heath
2   In less heathy and more heavily grazed places. U3 is more permanent here, and is maintained by grazing

Other vegetation types in this picture:

H12 *Calluna-Vaccinium* heath
M6 *Carex echinata-Sphagnum* mire
M15 *Trichophorum-Erica* wet heath
M16 *Erica-Sphagnum compactum* wet heath
M21 *Narthecium-Sphagnum* valley mire
M25 *Molinia-Potentilla* mire
U4 *Festuca-Agrostis-Galium* grassland
U20 *Pteridium-Galium* community

## Description

From a distance this grassland looks similar to *Festuca-Agrostis-Galium* grassland U4, but here the sward is dominated by the dense, wiry, glaucous tufts of *Agrostis curtisii*. The turf has a blue-grey tinge that is visible from some distance, and is usually short, tightly grazed and species-poor. Associated species include *Agrostis capillaris*, *Festuca ovina*, *Danthonia decumbens*, *Calluna vulgaris*, *Potentilla erecta*, *Galium saxatile* and *Molinia caerulea*. There is no more than a thin and patchy layer of bryophytes comprising species such as *Dicranum scoparium*, *Pleurozium schreberi*, *Hypnum jutlandicum*, *Rhytidiadelphus squarrosus* and *Scleropodium purum*.

## Differentiation from other communities

*Agrostis* grassland is the only British grassland community dominated by *Agrostis curtisii*. It bears a superficial resemblance to *Deschampsia* grassland U2 because *A.*

*curtisii* and *Deschampsia flexuosa* are rather similar in growth-form, though not in colour.

## Ecology

This is one of the most widespread grassland communities in the uplands of south-west England, where it forms mosaics with *Festuca-Agrostis-Galium* grassland and *Ulex gallii-Agrostis* heath H4. It occurs on gentle slopes on moist free-draining acidic soils (Rodwell 1992). It is a community of lower ground, up to about 500 m.

## Conservation interest

This form of acid grassland is of less value for nature conservation than *Ulex gallii-Agrostis* heath H4 with which it is commonly associated and from which it has generally been derived as a result of management treatments. The community does not usually include rare species, apart from *Agrostis curtisii* itself.

## Management

*Agrostis* grassland is probably all derived from dwarf-shrub heaths following grazing and burning. In many places the characteristic dwarf shrubs of the heathland persist as

The distribution of the community is limited by that of *Agrostis curtisii*. It is recorded in south-west England, Dorset, Hampshire, the Isle of Wight, and on Old Castle Down, Mynydd y Gaer and Mynydd Llangeinwyr in south Wales. There are particularly large stands on the grazed commons of Bodmin Moor.

There are similar *Agrostis curtisii* grasslands in western France.

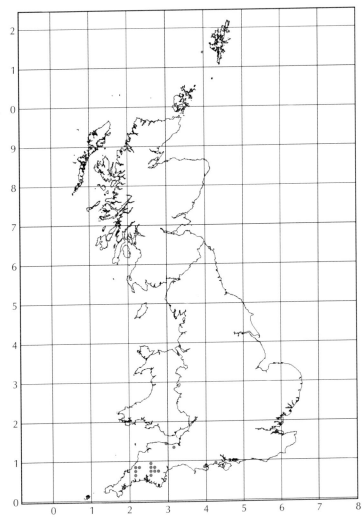

isolated scattered shoots in the grassland. Some intermediate stands show clearly that the community has replaced *Ulex gallii-Agrostis* heath, and the grassland commonly occurs as a seral stage after *Ulex gallii-Agrostis* heaths have been burnt, reverting to heathland as the dwarf shrubs recolonise. Where the intensity of grazing is high or where burning is frequent, *Agrostis* grassland may persist for a long time as an element of the upland landscape.

The commons of south-west England and, to a lesser extent, south Wales, are grazed by many different sorts of animal including sheep, cattle, ponies, donkeys, rabbits and, in some places, geese. It seems to be this mixed grazing that maintains the dominance of *Agrostis curtisii* and *Festuca ovina* on the damp soils and prevents the invasion of *Nardus stricta* (Rodwell 1992). The commons would almost certainly be heathier and less grassy if grazing and burning were better controlled. Bodmin Moor was apparently much more heathy a hundred years ago, but since then the intensity of grazing appears to have increased and many commoners have exercised their grazing rights to the full. The problems of over-grazing have been exacerbated because some areas of common land have been enclosed or planted with conifers. This has left a smaller area of open ground to be grazed by an increasing number of animals.

# U4 *Festuca ovina-Agrostis capillaris-Galium saxatile* grassland

## Synonyms

Species-poor *Agrosto-Festucetum* (McVean and Ratcliffe 1962; Birks 1973); *Achilleo-Festucetum tenuifoliae, Galium saxatile-Poa pratensis* community and *Junco squarrosi-Festucetum tenuifoliae* (Birse and Robertson 1976); *Agrostis canina-A. tenuis* grassland C1a, *Festuca ovina* grassland C1b and *Festuca ovina-Deschampsia flexuosa-Racomitrium* heath E1c *p.p.* (Birks and Ratcliffe 1980).

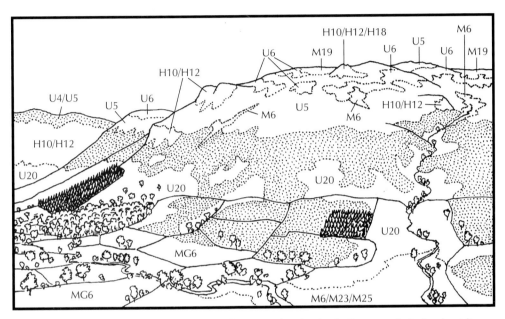

Typical habitats of U4 *Festuca-Agrostis-Galium* grassland (stippled): on well-drained acidic soils from low-altitude enclosed pasture to heavily grazed middle and upper hill slopes.

Other vegetation types in this picture:

U5 *Nardus-Galium* grassland
U6 *Juncus-Festuca* grassland
U20 *Pteridium-Galium* community
H10 *Calluna-Erica* heath
H12 *Calluna-Vaccinium* heath
H18 *Vaccinium-Deschampsia* heath
M6 *Carex echinata-Sphagnum* mire
M19 *Calluna-Eriophorum* blanket mire
M23 *Juncus-Galium* rush-pasture
M25 *Molinia-Potentilla* mire
MG6 *Lolium-Cynosurus* grassland
U4 is also shown in the pictures for H4, H8, H10, CG10–CG12 and MG5.

## Description

These grasslands are usually short and tightly grazed, and vary in colour from dull, grey-green or pale ochre-green to a brighter rich green that can stand out clearly from the sombre tones of other upland vegetation. The dense turfs of *Festuca ovina, Agrostis capillaris* and *Anthoxanthum odoratum* are trailed through by *Galium saxatile* and *Potentilla erecta*. There is usually a thick mat of mosses around the vascular plants, in which *Rhytidiadelphus squarrosus* is one of the most common species.

333

There are five sub-communities. The Typical sub-community U4a has no distinguishing species of its own. In many places the species-poor sward is thick with mosses: *Hylocomium splendens*, *Rhytidiadelphus loreus*, *Pleurozium schreberi* and *Hypnum jutlandicum* grow in rich golden-green mats and patches. *Potentilla erecta* and *Galium saxatile* are very common here, colouring the turf with their yellow and white flowers in summer. The *Holcus lanatus-Trifolium repens* sub-community U4b has a less mossy sward and can include some species characteristic of more improved grasslands. For example, *Holcus lanatus*, *Trifolium repens*, *Achillea millefolium* and *Cerastium fontanum* are common, and the sward of *Festuca ovina*, *Agrostis capillaris* and *Anthoxanthum* can be augmented by *Cynosurus cristatus*, *Poa pratensis* and even *Lolium perenne* and *Dactylis glomerata*. The *Lathyrus linifolius-Stachys officinalis* sub-community U4c is distinguished by *Stachys officinalis*, *Galium verum*, *Helictotrichon pratense* and *Lathyrus linifolius*. The *Luzula multiflora-Rhytidiadelphus loreus* sub-community U4d is a very mossy grassland in which *Agrostis canina* may be commoner than *A. capillaris*. Quite commonly there are clumps of *Luzula multiflora* and *Deschampsia cespitosa*, and scattered plants of damp-loving species such as *Viola palustris* and *Carex panicea*. The dense, richly coloured mats and turfs of mosses include *Rhytidiadelphus loreus*, *Thuidium tamariscinum* and *Hylocomium splendens*. The *Vaccinium myrtillus-Deschampsia flexuosa* sub-community U4e is also a mossy grassland. Many stands have much *Vaccinium myrtillus* or *Racomitrium lanuginosum* or both, and *Deschampsia flexuosa* and *Nardus stricta* grow here too. These grasslands can have more *R. lanuginosum* than the NVC tables suggest, and in this form they can take the place of *Carex-Racomitrium* moss-heath U10 in Wales, the Lake District and southern Scotland.

There are other forms of *Festuca-Agrostis-Galium* grassland that are not described in the NVC. There is a sparse variant of the *Vaccinium-Deschampsia* sub-community comprising a thin sward of *Festuca vivipara* and *Agrostis canina* growing through a carpet of *Racomitrium lanuginosum* and *Alchemilla alpina* (e.g. Averis and Averis 1999a, 1999b). Other forms of the community resemble the *Vaccinium-Deschampsia* sub-community but have *Diphasiastrum alpinum* (which can be the dominant species), *Huperzia selago* and in some places *Lycopodium clavatum* (e.g. Averis and Averis 2000b). There are various herb-rich forms of *Festuca-Agrostis-Galium* grassland related to the *Lathyrus-Stachys* sub-community. For example, in northern England there are stands transitional between the *Lathyrus-Stachys* sub-community and the *Danthonia* sub-community of *Cynosurus-Centaurea* grassland MG5c, and also stands transitional to *Anthoxanthum-Geranium* grassland MG3 (Richard Jefferson, pers. comm.). Further north there is a type of grassland rather like the *Lathyrus-Stachys* sub-community but with little or no *Stachys officinalis*, and with much *Helianthemum nummularium* and some upland species such as *Persicaria vivipara* (Averis 1999; Smith 2000). In the Highlands there is a herb-rich form of *Festuca-Agrostis-Galium* grassland with tall mesotrophic forbs such as *Filipendula ulmaria*, *Ranunculus acris*, *Geum rivale*, *Alchemilla glabra*, *Parnassia palustris* and *Angelica sylvestris* growing with the grasses, small forbs and mosses typical of the community (e.g. Averis and Averis 1998a, 1999a). In Wales and Scotland there is a form of *Festuca-Agrostis-Galium* grassland with much *Sphagnum capillifolium* and *S. denticulatum* that is apparently derived by grazing from *Calluna-Vaccinium-Sphagnum* damp heath H21 (Alex Turner, pers. comm.; A B G Averis and A M Averis, pers. obs.).

### Differentiation from other communities

*Festuca-Agrostis-Galium* grassland and *Nardus-Galium* grassland U5 share a similar suite of species but can generally be distinguished by the relative amounts of the different grasses: *Festuca* species, *Agrostis* species and *Anthoxanthum* are more common than *Nardus* in *Festuca-Agrostis-Galium* grassland, whereas the converse is true of *Nardus-Galium* grassland. All forms of *Festuca-Agrostis-Galium* grassland can be separated from the similar *Festuca-Agrostis-Thymus* grassland CG10 by the scarcity or absence of *Thymus polytrichus*. In addition, *Galium saxatile* is more common and there are rarely any small calcicolous herbs in *Festuca-Agrostis-Galium* grassland. Some calcicoles do occur in the *Lathyrus-Stachys* sub-community, but this has *Stachys officinalis* and other herbs that are rare in *Festuca-Agrostis-Thymus* grassland.

The Typical sub-community might be confused with the *Potentilla-Galium* sub-community of *Festuca-Agrostis-Rumex* grassland U1e, but has more *Agrostis capillaris* and *Anthoxanthum*, and little or no *Rumex acetosella*.

Semi-improved stands of the *Holcus-Trifolium* sub-community can come close to *Lolium perenne-Cynosurus cristatus* grassland MG6, but have less *Lolium perenne*, *Cynosurus cristatus* and *Dactylis glomerata*, more *Potentilla erecta*, *Galium saxatile* and *Festuca ovina*, and thicker patches of mosses such as *Rhytidiadelphus squarrosus*. Some forms of the *Holcus-Trifolium* sub-community contain *Centaurea nigra*, *Lotus corniculatus*, *Ranunculus acris* and other mesotrophic forbs, and could be confused with *Cynosurus-Centaurea* grassland MG5. *Cynosurus-Centaurea* grassland can usually be distinguished by the scarcity of *Festuca ovina* and *Galium saxatile*, and the greater quantity of *Cynosurus cristatus*.

The mossy *Luzula-Rhytidiadelphus* sub-community can resemble *Deschampsia-Galium* grassland U13. However, *Deschampsia-Galium* vegetation is a snow-bed grassland with less *Festuca ovina*, *Agrostis canina* and *Potentilla erecta*, and more montane species such as *Carex bigelowii*, *Alchemilla alpina*, *Polytrichum alpinum* and *Cetraria islandica*.

Some forms of the *Vaccinium-Deschampsia* sub-community are very similar to the *Galium* sub-community of *Carex-Racomitrium* moss-heath U10a. Montane species such as *Carex bigelowii*, *Diphasiastrum alpinum* and *Salix herbacea* can persist in stands of *Festuca-Agrostis-Galium* grassland that have been derived from summit heath. *Carex-Racomitrium* heath generally has more *Racomitrium lanuginosum* and less *Agrostis capillaris*, *Nardus*, *Anthoxanthum*, *Festuca* species and *Rhytidiadelphus squarrosus*. The form of the *Vaccinium-Deschampsia* sub-community with *Alchemilla alpina* differs from the superficially similar *Festuca-Agrostis-Alchemilla* grassland CG11 in the absence of *Thymus polytrichus* and other small calcicoles.


## Ecology

This is a grassland of acid brown earths and brown podsolic soils that drain freely but can be moist. The Typical sub-community is the most common form. It clothes many hill slopes where the intensity of grazing is moderately high; smaller patches occur high up into the hills in matrices of heaths and bogs, under crags, on alluvial flats and fans, and around rock outcrops. The *Holcus-Trifolium* sub-community, with its more mesotrophic assemblage of species, tends to occur on deeper and richer soils at low altitudes; many stands have probably been limed in the past and some have been treated with fertiliser. The *Lathyrus-Stachys* sub-community occupies soils that appear to be slightly flushed, and are richer than those with the Typical sub-community. The flushed, herb-rich form of *Festuca-Agrostis-Galium* grassland with *Filipendula ulmaria* and other tall mesotrophic herbs occurs on moist soils on cool north-facing slopes in the Highlands. The *Luzula-Rhytidiadelphus* sub-community replaces the Typical sub-community at higher altitudes and on damper soils, and is particularly characteristic of concave slopes. The *Vaccinium-Deschampsia* sub-community covers thin stony soils on convex slopes and summits.

*Festuca-Agrostis-Galium* grasslands have a vast altitudinal range: the various sub-communities cover the whole spread of the uplands from near sea-level to over 1000 m. The community is most common in upland regions where the rocks are acid to at least moderately base-rich and where there has been a long history of grazing. It is rare over highly acid rocks, such as granite and quartzite.

## Conservation interest

Most forms of *Festuca-Agrostis-Galium* grassland have less interest for nature conservation than the heaths and woodlands with which they are typically associated. However, the herb-rich, flushed forms can be very species-rich, including plants such as *Trollius europaeus*, *Coeloglossum viride*, *Gymnadenia conopsea*, *Galium boreale* and *Cirsium heterophyllum*. *Persicaria vivipara* and *Cirsium heterophyllum* grow in some stands of the variant characterised by *Helianthemum nummularium*. These herb-rich grasslands develop only where there is little or no grazing, and can be attractive with their colourful flowers in summer. The herbs are able to flower and set seed, which maintains the genetic diversity of the populations and enables the plants to spread. Other forms of the community generally lack notable plants, although *Gnaphalium supinum* and *Luzula spicata* have been recorded in the *Alchemilla alpina* variant of the *Vaccinium-*

There are *Festuca-Agrostis-Galium* grasslands throughout upland Great Britain from south-west England to Orkney and Shetland; they also occur in some parts of the lowlands. The Typical and *Holcus-Trifolium* sub-communities are common throughout the uplands. The *Lathyrus-Stachys* sub-community occurs mainly in England and Wales. Other herb-rich forms appear to be restricted to Scotland: the variant with *Helianthemum nummularium* is scattered in the Southern Uplands and the southern Highlands, and the form characterised by *Filipendula ulmaria* and other mesotrophic herbs has been found only in the western Highlands and the Hebrides. The *Luzula-Rhytidiadelphus* sub-community has a distinctly northern distribution. The *Vaccinium-Deschampsia* sub-community is also commonest in the north, but stands defined more by *Vaccinium myrtillus* than by *Racomitrium lanuginosum* are widespread in Wales; the form with *Alchemilla alpina* has been found in the Breadalbanes and on Mull and Skye.

There are similar forms of acid grassland in Ireland and also in mainland Europe.

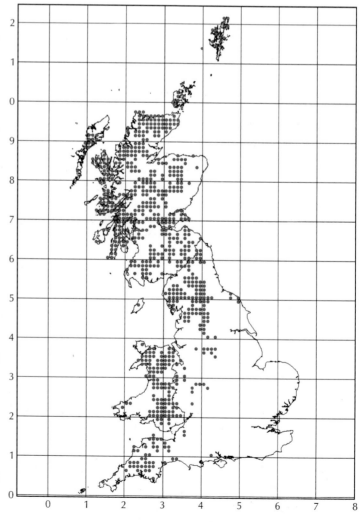

*Deschampsia* sub-community. Stands of *Festuca-Agrostis-Galium* grassland also provide valuable nesting-grounds for Skylark *Alauda arvensis* and Wheatear *Oenanthe oenanthe*. Where they occur over deep soils they can be an important habitat for moles in otherwise rocky or peat-covered uplands where there are few worms.

## Management

These acid grasslands are among the most important types of vegetation for livestock in the rough grazings of the British uplands. They offer a nutritious bite to sheep and cattle as well as to deer, hares, rabbits, and, in some areas, ponies. Most *Festuca-Agrostis-Galium* grassland has evidently been derived, by grazing and other treatments such as burning, from dwarf-shrub heath, herb-rich vegetation, scrub or woodland. The community is maintained by grazing, and in Wales, the Craven Pennines, the Lake District and the Southern Uplands there are vast tracts of *Festuca-Agrostis-Galium* grassland associated with high densities of sheep. When the intensity of sheep grazing is very high the community can be invaded by the unpalatable *Nardus stricta* and the vegetation may develop into *Nardus-Galium* grassland. In some upland areas, the spread of *Nardus* appears to have been encouraged by the loss of mixed grazing systems over the last few centuries – where once there were cattle, sheep, ponies and goats there are now only sheep. In areas where sheep and deer have been fenced out, *Festuca-Agrostis-Galium* grasslands can soon become tall and rank, and the growth of dwarf shrubs and trees is no longer suppressed. If the intensity of grazing is low enough, such grasslands commonly develop into some form of dry heath. In parts of the Highlands, the community has been replaced by *Luzula-Vaccinium* vegetation U16 on open slopes after the number of sheep has been reduced. Most examples of *Festuca-Agrostis-Galium* grassland occur in situations that would once have been covered with native woodland. *Quercus-Betula-Oxalis* woodland W11 and *Quercus-Betula-Dicranum* woodland W17 (both communities of well-drained soils) are the most likely precursors of this type of grassland. Indeed, many hard-grazed woods resemble *Festuca-Agrostis-Galium* grassland beneath a canopy of trees.

The *Vaccinium-Deschampsia* sub-community appears to have taken the place of *Carex-Racomitrium* moss-heath on summits that have been heavily grazed by sheep for many years. There are good examples of this in Wales, the Lake District, the Southern Uplands and the Inner Hebrides. These grassy heaths are believed to be a consequence of dung and urine from the sheep adding so many nutrients to the soil that *Racomitrium lanuginosum* is out-competed by grasses such as *Festuca vivipara*, *F. ovina* and *Agrostis* species. The vegetation can retain a few montane species: a sprinkling of *Carex bigelowii*, *Salix herbacea*, *Diphasiastrum alpinum* and *Polytrichum alpinum*. It is not known whether degraded heaths of this sort would revert to *Carex-Racomitrium* heath if sheep were taken off the hills. In north Wales there are also stands of the *Vaccinium-Deschampsia* sub-community with much *Diphasiastrum alpinum* and *Huperzia selago*. These are thought to be derived from *Vaccinium-Deschampsia* heath H18 or *Vaccinium-Cladonia* heath H19 following the elimination of dwarf shrubs by heavy grazing.

Most stands of *Festuca-Agrostis-Galium* grassland are not fertilised; the natural fertility of the soils, combined with the droppings and urine of grazing animals, is sufficient to maintain them. The *Holcus-Trifolium* sub-community is quite common in enclosed fields where there has been some liming or fertilising, but not enough to produce more improved swards of *Lolium-Cynosurus* grassland.

*Festuca-Agrostis-Galium* grasslands can be invaded by bracken. This is especially likely when stands are grazed solely by sheep rather than being subject to mixed grazing by both sheep and cattle. Cattle trample bracken and keep it in check. Previously, bracken was also regularly cut to provide winter bedding for cattle, but now that there are fewer cattle on the hills this practice has largely died out.

In many cases, management of this community for nature conservation will have the aim of re-establishing dwarf-shrub heath or other types of vegetation by reducing stock numbers. Where stands are deliberately managed as grassland for nature conservation purposes they should be grazed sufficiently hard to prevent the invasion of dwarf shrubs, *Luzula sylvatica*, trees and shrubs, but not so hard that the herbaceous element of the grassland is unable to flower and set seed. Low-intensity management of this general kind appears to be required for the maintenance of the more herb-rich semi-natural forms of *Festuca-Agrostis-Galium* grassland, such as the *Lathyrus-Stachys* sub-community and related vegetation.

# U5 *Nardus stricta-Galium saxatile* grassland

## Synonyms
*Nardetum* sub-alpinum (McVean and Ratcliffe 1962); *Nardo-Juncetum squarrosi* (Birks 1973); Sub-montane *Nardus* grassland C2a and Species-rich *Nardus* grassland C2c (Birks and Ratcliffe 1980).

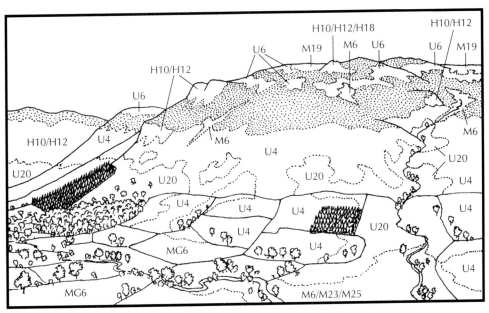

Typical habitat of U5 *Nardus-Galium* grassland (stippled): on moist acidic soils on gentle to steep, heavily grazed upper and middle slopes.

Other vegetation types in this picture:

U4 *Festuca-Agrostis-Galium* grassland
U6 *Juncus-Festuca* grassland
U20 *Pteridium-Galium* community
H10 *Calluna-Erica* heath
H12 *Calluna-Vaccinium* heath
H18 *Vaccinium-Deschampsia* heath
M6 *Carex echinata-Sphagnum* mire
M19 *Calluna-Eriophorum* blanket mire
M23 *Juncus-Galium* rush-pasture
M25 *Molinia-Potentilla* mire
MG6 *Lolium-Cynosurus* grassland
U5 is also shown in the pictures for CG10 and M20.

## Description
Pale swards of the dense, wiry clumps of *Nardus stricta* clothe many upland slopes, and can form vast stands. Several other grasses can grow with the *Nardus*, the most common of which are *Agrostis capillaris* and *Festuca ovina*. The short turf is starred with *Potentilla erecta* and *Galium saxatile*, their flowers bright in summer. There is usually some *Vaccinium myrtillus*, its short, bright-green shoots pushing up through the mat of grasses. At the feet of the vascular plants there are the usual mosses of acid grassland: *Hypnum jutlandicum*, *Pleurozium schreberi*, *Rhytidiadelphus loreus*, *R. squarrosus* and *Hylocomium splendens*.

There are five sub-communities. The Species-poor sub-community U5a takes in the most impoverished *Nardus* grasslands and has no distinguishing species of its own. The *Agrostis canina-Polytrichum commune* sub-community U5b occurs on damper soils, and has *Agrostis canina, Luzula multiflora, Juncus squarrosus* and *Polytrichum commune;* various Sphagna are common in some stands. The *Carex panicea-Viola riviniana* sub-community U5c extends the range of the community onto flushed, mildly base-rich soils where there can be an array of mesotrophic species such as *Ranunculus acris, Alchemilla glabra, Thalictrum alpinum, Geum rivale, Filipendula ulmaria, Trollius europaeus* and *Persicaria vivipara.* The *Calluna vulgaris-Danthonia decumbens* sub-community U5d has a mixed sward of *Nardus, Festuca ovina, Agrostis capillaris, Anthoxanthum odoratum, Deschampsia flexuosa* and *Danthonia decumbens,* together with patches of short *Calluna vulgaris, Erica cinerea* and *E. tetralix.* The *Racomitrium lanuginosum* sub-community U5e has an extensive silvery carpet of *Racomitrium lanuginosum;* the sward of *Nardus* can be sparse, and may be interspersed with *Trichophorum cespitosum, Calluna* and, more locally, a few montane or northern species such as *Diphasiastrum alpinum, Huperzia selago, Carex bigelowii* and *Vaccinium vitis-idaea.*

## Differentiation from other communities

*Nardus-Carex* grass-heath U7 can be confused with sub-montane *Nardus-Galium* grassland, but it occurs in places where snow lies late, and has montane species such as *Carex bigelowii, Alchemilla alpina, Diphasiastrum alpinum, Vaccinium uliginosum* and *Cetraria islandica* that are uncommon in *Nardus-Galium* grassland. It can be particularly hard to separate the *Racomitrium* sub-community of *Nardus-Galium* grassland from the *Empetrum-Cetraria* sub-community of *Nardus-Carex* grass-heath, especially in the western Highlands where both are common and can occur in close association. The main difference is that the snow-bed grassland generally has a thicker sward of *Nardus* containing some *C. bigelowii* and other montane species.

*Nardus-Galium* grassland can be very similar to *Festuca-Agrostis-Galium* grassland U4 and *Juncus-Festuca* grassland U6. All three communities share a similar range of species, although the relative proportions differ and a few plants are more common in one type than in the others. Generally, *Nardus-Galium* grassland is the only one that is clearly dominated by *Nardus.*

The *Racomitrium* sub-community can be very heathy, with some resemblance to *Calluna-Racomitrium* heath H14 and to low and open forms of the *Cladonia* sub-community of *Trichophorum-Erica* wet heath M15c. *Nardus-Galium* grassland has more *Nardus* than either of the heath communities, less *Trichophorum cespitosum, Molinia caerulea* and *Erica tetralix* than the wet heath, and the heather is generally not as severely wind-pruned as it is in the montane *Calluna* heath.

## Ecology

*Nardus-Galium* grassland clothes damp mineral soils which have peaty upper horizons, and can occur in very large stands on upland hillsides. It typically occupies slopes where the depth and wetness of the soil are intermediate between the drier podsols under *Festuca-Agrostis-Galium* grasslands and the wet shallow peats under *Juncus-Festuca* grassland. The underlying rock can be anything from acid to basic, but the soils are generally acid with some podsolisation. The community can also occur on deep peat from which the original mire vegetation has been lost (e.g. through burning, heavy grazing or drainage). Another typical habitat is well-drained but moist alluvial soil along the margins of streams, and it is quite common to see the pale sinuous lines of such grasslands cutting across areas of heath or bog.

This is a community of mid-altitude slopes, ranging from about 300 m to above 700 m, with the *Racomitrium* sub-community extending to over 850 m. Most stands are anthropogenic, but many examples of the *Racomitrium* sub-community at high altitudes are likely to be near-natural.

## Conservation interest

*Nardus-Galium* grasslands are used by upland birds including Skylarks *Alauda arvensis*, Meadow Pipits *Anthus pratensis* and, in rocky places, Wheatear *Oenanthe oenanthe*. Along with *Festuca-Agrostis-Alchemilla* grasslands CG11, sub-montane *Nardus* grasslands are probably the most important habitat in Great Britain for the scarce Mountain Ringlet butterfly *Erebia epiphron*. Its few British populations are almost all associated with stands of these grassland communities at moderate to high altitudes in the Lake District and the central Highlands.

Many stands of the *Racomitrium* sub-community are at high altitudes and occur in mosaics with near-natural montane heaths and grasslands; some of these stands are themselves probably near-natural. In the far west and on the Hebrides, this sub-community occurs in hollows where snow might be expected to linger. The *Racomitrium* sub-community can include a number of uncommon montane or upland plants, such as

This community is ubiquitous throughout the British uplands, but is most common where there has been a long history of sheep grazing. Some of the largest stands are in the Breadalbane region of Scotland, the Southern Uplands, the Lake District and Wales. It is scarce in south-west England. Most of the sub-communities are widely distributed, but the *Racomitrium* sub-community occurs mainly in the cool oceanic north-west Highlands and Hebrides, and the *Carex-Viola* sub-community mainly in the Breadalbanes.

Vegetation similar to *Nardus-Galium* grassland occurs rarely in the west of mainland Europe.

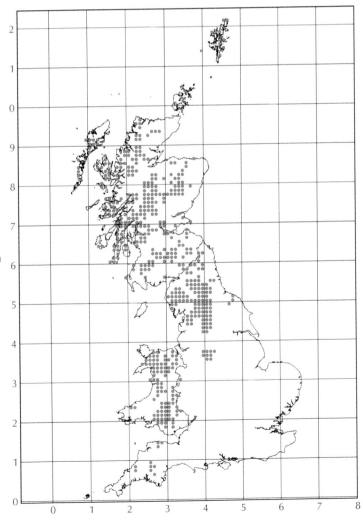

*Loiseleuria procumbens, Juniperus communis* ssp. *nana* and the liverwort *Anastrophyllum donnianum*. Species-rich examples of the *Carex-Viola* sub-community typically occur in mosaics with other species-rich grasslands and mires, and can be home to the rarities *Carex vaginata* and *Salix arbuscula*, and other scarce plants such as *Galium boreale, Persicaria vivipara, Trollius europaeus* and *Coeloglossum viride*.

Although some forms of *Nardus-Galium* grassland are valuable for nature conservation, over much of its range this community is regarded as a conservation problem rather than an asset. Stands have usually been derived from dwarf-shrub heaths as a result of burning and grazing or directly from woodlands (Rodwell 1992). In both cases the former vegetation would have been more diverse and usually of greater value for nature conservation.

## Management

This community is maintained by grazing. The leaves of *Nardus* are coarse, hard and fibrous, and are neither inviting nor palatable to livestock. *Nardus* is hardly grazed by ewes and lambs, although cattle, wethers and ponies will eat the green centres of the tussocks, especially in spring when the young growth is most tender. On uplands that are grazed primarily by breeding ewes, for example in southern Scotland, northern England and Wales, *Nardus-Galium* grasslands have become widespread. In upland areas with fewer sheep and more cattle or ponies, *Nardus-Galium* grassland is scarcer, and stands tend to be smaller and restricted to flushed slopes. The scarcity of *Nardus-Galium* grassland in south-west England may be attributable in part to the large numbers of ponies there. Where the intensity of grazing is reduced, *Nardus-Galium* grasslands can develop into dry *Calluna vulgaris* or *Vaccinium myrtillus* heaths. This has been clearly demonstrated in Snowdonia, where grazing animals have been excluded from plots of ground since the 1950s or 1960s (Hill *et al.* 1992). However, once the thick grassy swards have become well-established it is difficult for trees and shrubs to recolonise without planting.

The spread of unpalatable *Nardus* into more mixed grasslands is a problem for farmers who see their pastures becoming less and less useful for grazing stock. On unenclosed land this concentrates the grazing animals onto the remaining more palatable herbage, which ultimately causes further losses of heaths and richer grasslands.

# U6 *Juncus squarrosus-Festuca ovina* grassland

## Synonyms
*Juncetum squarrosi* sub-alpinum and *Juncus squarrosus* bog *p.p.* (McVean and Ratcliffe 1962); *Nardo-Juncetum squarrosi* (Birks 1973); *Junco squarrosi-Festucetum tenuifoliae* (Birse and Robertson 1976); *Juncus squarrosus* grasslands C3 (Birks and Ratcliffe 1980).

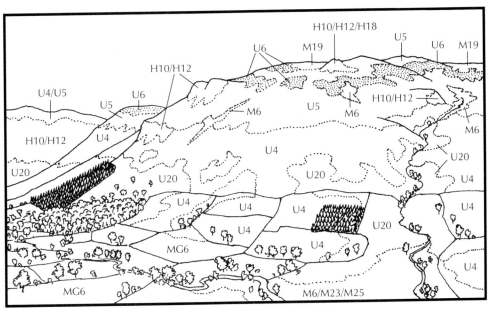

Typical habitats of U6 *Juncus-Festuca* rush-heath (stippled): among grasslands on moist to wet, peaty soils on gentle, heavily grazed middle to upper slopes. Also replacing M19 *Calluna-Eriophorum* blanket mire on level, deep peat (upper-middle to upper-right of picture) as a result of heavy grazing.

Other vegetation types in this picture:

U4 *Festuca-Agrostis-Galium* grassland
U5 *Nardus-Galium* grassland
U20 *Pteridium-Galium* community
H10 *Calluna-Erica* heath
H12 *Calluna-Vaccinium* heath
H18 *Vaccinium-Deschampsia* heath
M6 *Carex echinata-Sphagnum* mire
M19 *Calluna-Eriophorum* blanket mire
M23 *Juncus-Galium* rush-pasture
M25 *Molinia-Potentilla* mire
MG6 *Lolium-Cynosurus* grassland
U6 is also shown in the pictures for CG10 and M20.

## Description
These deep-green, tussocky swards stand out clearly from the surrounding vegetation on hillsides and plateaux. The dense rosettes of *Juncus squarrosus* are entwined with trailing shoots of *Potentilla erecta* and *Galium saxatile*, and enmeshed with the mosses *Pleurozium schreberi*, *Polytrichum commune*, *Hylocomium splendens*, *Rhytidiadelphus squarrosus* and *R. loreus*.

Four sub-communities are described in the NVC, showing a general trend from wet to drier substrates. The *Sphagnum* species sub-community U6a occurs on deep wet peat, and the stiff leaves of the *Juncus* push through cushions and carpets of *Sphagnum fallax*, *S. capillifolium* and, in some places, *S. papillosum*. The *Carex nigra-Calypogeia azurea* sub-community U6b is defined by the mosses *Plagiothecium undulatum* and *Aulacomnium palustre*, and the uncommon liverwort *Calypogeia azurea*. The *Vaccinium myrtillus* sub-community U6c has a mixed sward of *J. squarrosus* and *Deschampsia flexuosa* with scattered sprigs of *Vaccinium myrtillus*; large pleurocarpous mosses such as *Rhytidiadelphus loreus*, *Pleurozium schreberi* and *Hypnum jutlandicum* are more common in this sub-community than in other forms of the community. The *Agrostis capillaris-Luzula multiflora* sub-community U6d is the grassiest form, and comprises variegated swards in which *Agrostis canina*, *Anthoxanthum odoratum*, *Deschampsia flexuosa*, *Nardus stricta* and *Festuca vivipara* either form discrete patches or grow in mixtures with *J. squarrosus*.

Across much of the geographical range of *Juncus-Festuca* grassland, variation within the community does not correspond closely to the four sub-communities described in the NVC. The *Sphagnum* sub-community is generally distinct, and the *Agrostis-Luzula* sub-community can be recognised without difficulty in many places. However, one of the most common types of *Juncus-Festuca* grassland on damp peaty soils is species-poor vegetation conforming to the description given above for the community as a whole. The nearest fit is with the *Carex-Calypogeia* or *Vaccinium* sub-communities but the vegetation cannot be assigned unequivocally to either of these. In addition, there are two distinctive forms of *Juncus-Festuca* grassland that are very different from any of the NVC sub-communities. One is a heathy form, usually resembling the *Sphagnum* sub-community but with much *Calluna vulgaris* or *Vaccinium myrtillus* (e.g. Averis and Averis 1999a, 2000b). The other is a flushed, species-rich form with mesotrophic herbs such as *Ranunculus acris*, *Thalictrum alpinum*, *Trollius europaeus*, *Geum rivale*, *Alchemilla glabra*, *Crepis paludosa* and *Parnassia palustris* (McVean and Ratcliffe 1962; Birks and Ratcliffe 1980; Averis and Averis 1999a, 1999b).

## Differentiation from other communities

*Juncus squarrosus* is common in some stands of *Nardus-Galium* grassland U5 and *Festuca-Agrostis-Galium* grassland U4, and both of these communities can blend gradually into the *Agrostis-Luzula* sub-community of *Juncus-Festuca* grassland. There is a thicker, more continuous sward of *J. squarrosus* in *Juncus-Festuca* grassland; the other communities are clearly dominated by their characteristic grasses. One could argue that the *Agrostis-Luzula* sub-community is simply an intermediate stage between *Juncus-Festuca* grassland and *Festuca-Agrostis-Galium* grassland, and that if it is worthy of recognition as a distinct vegetation type, then so are many other 'intermediates'.

The *Sphagnum* sub-community can resemble various types of bog vegetation, particularly the *Juncus-Rhytidiadelphus* sub-community of *Trichophorum-Eriophorum* blanket bog M17c, but bog species such as *Eriophorum vaginatum*, *E. angustifolium*, *Trichophorum cespitosum*, *Erica tetralix* and *Narthecium ossifragum* are sparse or absent.

## Ecology

This is a vegetation type of damp peaty soils or gleyed podsols on flat or gently sloping ground. The soils are moist and can be waterlogged. The community is generally indifferent to underlying geology, but the scarce species-rich stands are associated with base-rich rocks, particularly the Dalradian schist and limestone of the central Highlands. Most stands occur between 400 m and 800 m (McVean and Ratcliffe 1962; Rodwell 1992).

There is an ecological gradient from *Festuca-Agrostis-Galium* grassland on mineral soils on steep free-draining hillsides, through *Nardus-Galium* grassland on moister, peat-topped soils on somewhat gentler slopes, to *Juncus-Festuca* grassland on wet, peaty plateaux and shallow slopes. Commonly, there is a parallel increase in altitude, so that *Juncus-Festuca* grasslands experience the coolest and wettest climate and have the shortest growing-season. This sequence can be seen very well on many hill slopes in the Lake District.

## Conservation interest

Most stands of *Juncus-Festuca* grassland, like *Festuca-Agrostis-Galium* and *Nardus-Galium* acid grasslands, are anthropogenic in origin (Rodwell 1992). Like *Nardus-Galium* grassland, this type of vegetation is often seen as a problem by both nature conservationists and farmers. However, these species-poor swards are apparently unique to Great Britain and Ireland, where they owe their existence to the combination of an oceanic climate and a long history of burning and grazing.

Some forms of *Juncus-Festuca* grassland at higher altitudes may be near-natural (McVean and Ratcliffe 1962; Derek Ratcliffe pers. comm.), for example, stands within *Racomitrium* heaths on high plateaux, where there are poorly drained hollows with seepage from melting snow-beds (McVean and Ratcliffe 1962). In the western Highlands it is quite common for the *Sphagnum* sub-community to form small patches of blanket

*Juncus-Festuca* grassland occurs throughout the north and west of Great Britain, but is scarce in south-west England, where it is confined to the higher parts of Dartmoor. It is especially common in upland areas with a long history of heavy grazing and frequent burning: Wales, northern England, the Southern Uplands and the southern Highlands. Stands tend to be smaller in the north-west Highlands.

Species-poor *Juncus squarrosus* grasslands appear to be confined to the uplands of Great Britain and Ireland. Similar grasslands occur very locally in south-west Norway and in some more southern parts of western mainland Europe, but these European grasslands tend to be floristically richer than our *Juncus-Festuca* vegetation (Rodwell 1992). On the Faroes there are *Juncus squarrosus* blanket bogs with some resemblance to *Juncus-Festuca* grassland (Hobbs and Averis 1991a).

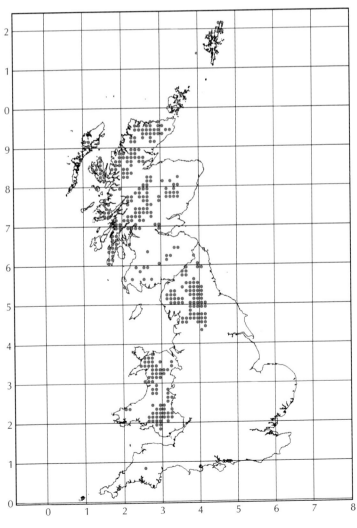

bog at high altitudes. A few rare species occur in these bogs, notably the bryophytes *Barbilophozia lycopodioides*, *Aulacomnium turgidum* and *Anastrophyllum joergensenii*. On Ben Lui the herb-rich form of *Juncus-Festuca* grassland is an important habitat for the rare sedge *Carex vaginata* (Averis and Averis 1999a).

## Management

Where *Juncus-Festuca* grassland covers peaty soils on upper slopes and plateaux it has usually replaced some sort of bog or wet heath from which the dwarf shrubs and other characteristic species have been lost by repeated burning and grazing. *Juncus squarrosus* is fairly unpalatable to sheep, although it is grazed by cattle and ponies. As appears to be the case with *Nardus-Galium* grassland, the abandonment of mixed grazing in the uplands may have encouraged the spread of *Juncus-Festuca* vegetation at the expense of *Festuca-Agrostis-Galium* grassland. However, stands of *Juncus-Galium* vegetation are neither as pernicious nor as intractable as *Nardus-Galium* grassland. In the absence of grazing, *J. squarrosus* is soon out-competed by other graminoids, and after only a year or two of reduced grazing the rosettes of *J. squarrosus* begin to be overgrown by other species (MacDonald *et al.* 1998*)*. If this treatment was continued these swards would probably develop into more mixed grasslands and eventually into damp heath, wet heath or blanket bog. At lower altitudes it might be possible for woodland to become established (Rodwell 1992).

# U7 *Nardus stricta-Carex bigelowii* grass-heath

## Synonyms
Low-alpine *Nardus* noda (McVean and Ratcliffe 1962); *Nardus stricta-Vaccinium myrtillus* Association (Birks 1973); Snow-bed *Nardus* grassland C2b (Birks and Ratcliffe 1980).

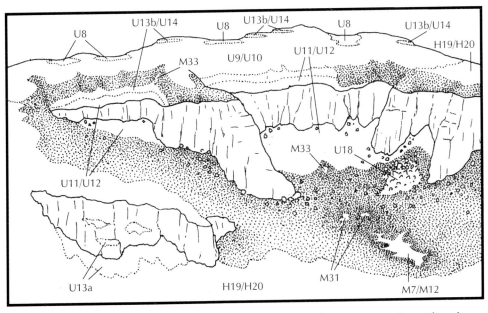

Typical habitat of U7 *Nardus-Carex* grass-heath (stippled): in high corries and on other slopes where snow lies late into the year.

Other vegetation types in this picture:

U8 *Carex-Polytrichum* heath
U9 *Juncus-Racomitrium* rush-heath
U10 *Carex-Racomitrium* moss-heath
U11 *Polytrichum-Kiaeria* snow-bed
U12 *Salix-Racomitrium* snow-bed
U13 *Deschampsia-Galium* grassland
U14 *Alchemilla-Sibbaldia* dwarf-herb community
U18 *Cryptogramma-Athyrium* snow-bed
M7 *Carex-Sphagnum russowii* mire
M12 *Carex saxatilis* mire
M31 *Anthelia-Sphagnum* spring
M33 *Pohlia wahlenbergii* spring
H19 *Vaccinium-Cladonia* heath
H20 *Vaccinium-Racomitrium* heath

U7 is also shown in the pictures for H19–H22.

## Description
These pale patches of *Nardus stricta* grassland form distinct and usually clean-edged stands on steep high slopes, in gullies and hollows, and in shaded corries. The dense, short, tufted sward of *Nardus* is set with small montane species such as *Carex bigelowii*, *Diphasiastrum alpinum*, *Polytrichum alpinum* and *Cetraria islandica*. The turf is pressed flat by the weight of winter snow and is saturated by melt-water in spring and early

347

summer; it looks dull and bedraggled, and in spring is usually a tawny-brown colour until the new leaves of *Nardus* start to grow.

There are three sub-communities. The *Empetrum nigrum* ssp. *hermaphroditum-Cetraria islandica* sub-community U7a characteristically has *Trichophorum cespitosum*, *Juncus squarrosus*, *Vaccinium uliginosum*, and, in some places, a few oceanic liverworts such as *Anastrophyllum donnianum* and *Pleurozia purpurea*. *Trichophorum* can be co-dominant with *Nardus* here, and *Racomitrium lanuginosum* may also be very common. The Typical sub-community U7b has no particular distinguishing species and includes stands that do not fit either of the other two sub-communities. The *Alchemilla alpina-Festuca ovina* sub-community U7c has a more mixed sward including *Festuca ovina*, *Agrostis canina*, *Carex pilulifera* and *Alchemilla alpina*.

### Differentiation from other communities

*Nardus-Carex* grass-heath is most readily confused with sub-montane *Nardus-Galium* grassland U5, but has more montane plants such as *Carex bigelowii*, *Polytrichum alpinum* and *Cetraria islandica*. Forms of the *Empetrum-Cetraria* sub-community co-dominated by *Nardus* and *Trichophorum* can have some resemblance to *Trichophorum-*

*Nardus-Carex* grass-heath occurs in the uplands from north Wales and the Lake District northwards to Orkney and Shetland; it is most common and extensive in the Scottish Highlands. In the north-west, there are particularly big stands in corries that hold moderately late snow, and there are good examples on Ben More Assynt, the Fannich Hills, Beinn Eighe and in the Inverlael Forest.

Similar vegetation occurs in western Scandinavia and the Faroes. The more oceanic form of the *Empetrum-Cetraria* sub-community, with much *Racomitrium lanuginosum* and oceanic liverworts, is apparently restricted to Scotland and the Faroe Islands.

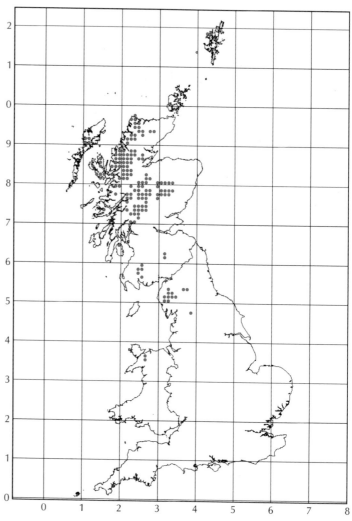

*Erica* wet heath M15, but have more *Nardus* and montane species, and lack *Erica tetralix, Calluna vulgaris* and *Molinia caerulea*.

## Ecology

This is a montane grassland of medium to high altitudes; most stands lie above 500 m. It marks out places where snow lies moderately late in spring. Winter snow is redistributed by the wind, swirling into hollows on the high plateaux, and being blown off the edges of plateaux and high ridges to build up into deep drifts on the slopes below, where it fills hollows, chokes gullies, and lies thickly on the floors of shaded corries. Because the prevailing strong winds are from the west and south-west, snow accumulates on the northern and eastern sides of the hills. These slopes are also the most shaded; where they are very steep they may not receive any direct sunlight during the winter. The growing season is short and cool. Most commonly the soils are peats or humic gravels, and even where the underlying rocks are basic there is so much leaching that the substrate is generally acid and wet.

## Conservation interest

This community is of interest as a near-natural form of grassland that is typically associated with montane heaths, snow-beds and other notable vegetation types. Rare species recorded in *Nardus-Carex* grass-heath include the moss *Kiaeria starkei* and, particularly in the damper *Empetrum-Cetraria* sub-community, the oceanic liverworts *Anastrophyllum donnianum, A. joergensenii, Scapania nimbosa, S. ornithopodioides, Bazzania pearsonii* and *Plagiochila carringtonii*. The montane lichen *Cladonia maxima* is confined to this type of vegetation in Great Britain (Fryday 1997). The rare lichen *Cetraria delisei* has also been recorded here, but interestingly this species tends to grow around the margins of snow-beds, along the boundary between *Nardus-Carex* grass-heath and the surrounding vegetation (Fryday 1997).

## Management

*Nardus-Carex* grass-heath is a climax community associated with a harsh climate and poor soils. However, it is usually grazed lightly by sheep and deer in summer. It could potentially be threatened by climate change.

# U8 *Carex bigelowii-Polytrichum alpinum* sedge-heath

## Synonyms

*Dicraneto-Caricetum bigelowii* and *Polytricheto-Caricetum bigelowii p.p.* (McVean and Ratcliffe 1962); *Carex bigelowii* snow-beds/heaths C6 (Birks and Ratcliffe 1980).

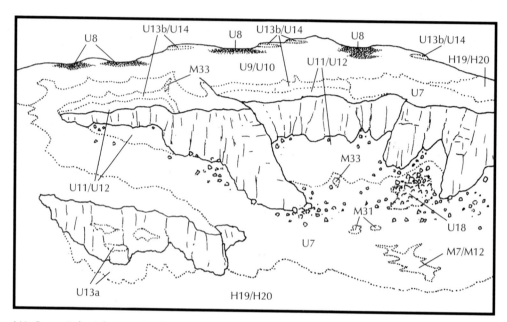

U8 *Carex-Polytrichum* sedge-heath (stippled): in shallow depressions with prolonged snow-lie, on high-plateaux.

Other vegetation types in this picture:

U7 *Nardus-Carex* grass-heath
U9 *Juncus-Racomitrium* rush-heath
U10 *Carex-Racomitrium* moss-heath
U11 *Polytrichum-Kiaeria* snow-bed
U12 *Salix-Racomitrium* snow-bed
U13 *Deschampsia-Galium* grassland
U14 *Alchemilla-Sibbaldia* dwarf-herb community
U18 *Cryptogramma-Athyrium* snow-bed
M7 *Carex-Sphagnum russowii* mire
M12 *Carex saxatilis* mire
M31 *Anthelia-Sphagnum* spring
M33 *Pohlia wahlenbergii* spring
H19 *Vaccinium-Cladonia* heath
H20 *Vaccinium-Racomitrium* heath

## Description

In late-lying snow-beds at high altitudes, the montane sedge *Carex bigelowii* grows in a spiky, grey-green sward over a velvety dark green turf of the mosses *Polytrichum alpinum* or *Dicranum fuscescens* or both. The ground is scattered with lichens; the most common species are *Cladonia arbuscula* and the montane *Cetraria islandica*. This vegetation type is a species-poor assemblage in which few other plants play a part.

There are two sub-communities. The *Polytrichum alpinum-Ptilidium ciliare* sub-community U8a has a thick turf of *C. bigelowii* growing through a layer of *Polytrichum alpinum* that is variegated with other species such as *Dicranum scoparium*, the liverwort *Ptilidium ciliare*, and the lichens *Cladonia pyxidata* and *Cetraria delisei*. The *Dicranum fuscescens-Racomitrium lanuginosum* sub-community U8b has a sparser turf of *C. bigelowii*, and *P. alpinum* tends to occur as scattered tufts in a carpet of *Dicranum fuscescens*. The sward is more varied than in the *Polytrichum-Ptilidium* sub-community, with much *Racomitrium lanuginosum* and species such as *Festuca vivipara*, *Agrostis capillaris* and *Vaccinium myrtillus*. Many stands of *Carex-Polytrichum* sedge-heath, especially in the west, have few of the species that distinguish either form of the community, and cannot therefore be assigned clearly to a sub-community.

### Differentiation from other communities
These sedge-heaths comprise dense swards of *Carex bigelowii* and mosses, and are usually sharply defined from other communities on the ground. They can be confused only with other montane vegetation types in which *C. bigelowii* is common. *Nardus-Carex* grass-heath U7 has more *Nardus stricta*, and less *Polytrichum alpinum* and *Dicranum fuscescens*. *Carex-Racomitrium* moss-heath U10 has more *Racomitrium lanuginosum*, and less *P. alpinum*. The *Rhytidiadelphus* sub-community of *Deschampsia-Galium*

*Carex-Polytrichum* sedge-heath is one of the more northern, continental types of upland vegetation in Great Britain and is more common in the east than in the west. It occurs on the higher hills in the central and eastern Highlands of Scotland, from Ben Lui and the Cowal Peninsula in the south to Ben Wyvis and the Fannich Hills in the north. There are particularly impressive examples on the highest plateaux of the Cairngorms and Lochnagar, on the hills around the head of Caenlochan Glen, on Creag Meagaidh, and on the summit of Ben Wyvis. Very similar vegetation, but lacking the more exacting montane lichens, occurs on high slopes just below ridges and plateaux on Carnedd Llewelyn, Yr Wyddfa (Snowdon) and Crib y Ddysgl in north Wales (A M Averis and A Turner pers. obs.).

Similar snow-bed heaths occur locally in Norway and Sweden.

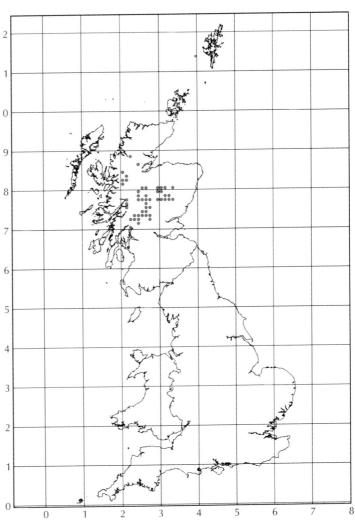

grassland U13b can be distinguished by the deep mats of the mosses *Rhytidiadelphus loreus* and *Hylocomium splendens*.

## Ecology

*Carex-Polytrichum* sedge-heath is a community of late-lying snow-beds at high altitudes, usually above 850 m. It is most typical of those parts of the Highlands where there is a cold montane climate with much of the winter precipitation falling as snow. It occurs on the acid Cairngorm granite, and on the Moine and Dalradian rocks of the central Highlands. The soils are usually podsolised shallow peats with a pH between 3.6 and 4.4 (McVean and Ratcliffe 1962). Drainage is impeded and the substrate is wet, often being irrigated by melting snow. The community occupies a rather similar ecological niche to the *Rhytidiadelphus* sub-community of *Deschampsia-Galium* grassland, but it is more dependent on snow-cover. It occurs most commonly on ground that is level or at most only slightly sloping, and stands can be large where there is deep snow throughout the winter. It is also common around the margins of bryophyte-dominated snow-beds belonging to the *Polytrichum-Kiaeria* U11 and *Salix-Racomitrium* U12 communities (McVean and Ratcliffe 1962; Rodwell 1992).

## Conservation interest

Like other forms of snow-bed, *Carex-Polytrichum* sedge-heath is of high value for nature conservation. It is a rare community in Great Britain, and forms an important link with the vegetation of mountains in mainland Europe. The uncommon *Luzula spicata* and *Cetraria delisei* occur in some stands.

## Management

This is a near-natural plant community that is not usually affected by management. It may be threatened by a warmer climate with less snow. The lichen flora of stands on popular hills could be damaged by human trampling and by the use of all-terrain vehicles on the high ground.

# U9 *Juncus trifidus-Racomitrium lanuginosum* rush-heath

## Synonyms
*Cladineto-Juncetum trifidi* (McVean and Ratcliffe 1962); *Juncus trifidus* heaths C7 (Birks and Ratcliffe 1980).

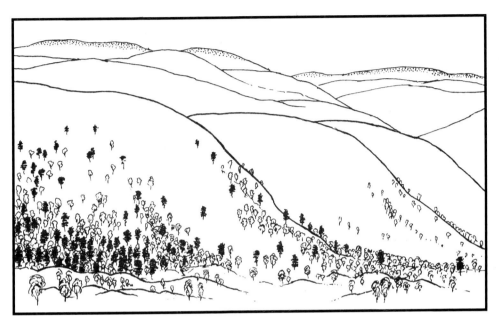

U9 *Juncus-Racomitrium* rush-heath (stippled): on high, stony, exposed summits and plateaux.

U9 is also shown in the pictures for U7–U8, U11–U14 and U18.

## Description
The most striking plant in these montane heaths is the small, slender rush *Juncus trifidus*, which grows in sparse swards, circular clumps or scattered thin turfs on bare stony plateaux. The clumps of rush are typically interleaved with a little *Carex bigelowii* and can have an understorey of *Racomitrium lanuginosum*. The *J. trifidus* leaves are bright shining green in their vigorous phase of growth in early spring, but by mid-summer the shoot tips begin to turn orange, and in the autumn the vegetation is tinged a conspicuous rich russet.

There are two sub-communities: one lichen-rich and one more mossy. The *Cladonia arbuscula-Cetraria islandica* sub-community U9a has lichens such as *Cladonia arbuscula*, *C. uncialis* ssp. *biuncialis*, *C. bellidiflora*, *C. gracilis*, *Ochrolechia frigida*, *Stereocaulon saxatile* and *Cetraria islandica*. These lichens can be dominant in small patches. Very open stands of *J. trifidus* belong in this sub-community. The *Salix herbacea* sub-community U9b takes in thicker turfs in which the sedges and rushes are joined by *Vaccinium myrtillus*, *Salix herbacea* and *Galium saxatile*; this sub-community is not as lichen-rich as the *Cladonia-Cetraria* sub-community.

## Differentiation from other communities
The *Salix* sub-community of *Juncus-Racomitrium* rush-heath can be confused with various forms of *Carex-Racomitrium* moss-heath U10. It can usually be easily distinguished

from the *Galium* and Typical sub-communities U10a and U10b, as *Juncus trifidus* and *Salix herbacea* are much more common, and *Festuca vivipara*, *Deschampsia flexuosa* and *Galium saxatile* are scarce. However, there are much greater similarities to the *Silene* sub-community U10c. This sub-community encompasses a wide range of vegetation, including herb-rich swards with many small calcicoles, slightly less rich examples with a scattering of herbs within a mossy carpet, and sparse open stands of mosses and herbs on wind-scoured ground. These variants all have much *J. trifidus* and *S. herbacea*; they can be distinguished from the *Salix* sub-community of *Juncus-Racomitrium* rush-heath by the greater quantities of *Silene acaulis*, *Thymus polytrichus*, *Alchemilla alpina*, *Ranunculus acris*, *Persicaria vivipara*, *Plantago maritima* and *Polytrichum alpinum*. Stands with few calcicolous species and those with a very sparse sward have more *S. herbacea* than *Juncus-Racomitrium* rush-heath.

Nardus-*Carex* grass-heaths U7 can resemble *Juncus-Racomitrium* rush-heath but have much less *J. trifidus* and much more *Nardus stricta*.

## Ecology

*Juncus-Racomitrium* rush-heath is a montane community that occupies some of the most inhospitable and extreme environments in Great Britain. It occurs on barren

The community is most common on the highest granite plateaux of the Cairngorm hills, but also occurs further west in the western Grampians, the Breadalbanes, the Torridon hills and on Arran.

Similar vegetation occurs in Norway, although these stands generally hold more snow during the winter, and occur at much higher altitudes.

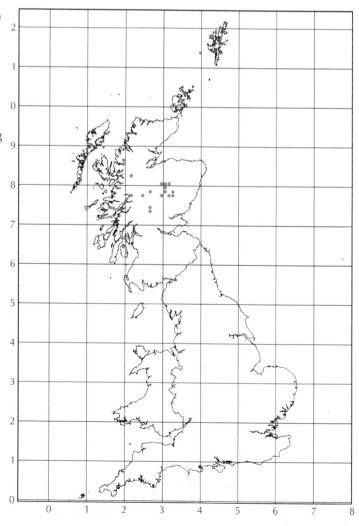

wildernesses of stone on exposed plateaux and mountain summits at 1000 m or more above sea-level. The community forms part of the uppermost altitudinal zone of vegetation in Great Britain – beyond this the climate is so severe that only scattered plants are able to grow amid a bare waste of stone. The substrate is usually unstable loose gravel, most commonly granite but in some places quartzite or Moine or Dalradian schist. The soils are mixtures of gravel and humus (McVean and Ratcliffe 1962) and are pervious and free-draining. *Juncus trifidus* can tolerate a moderate amount of snow-cover, and although many examples of *Juncus-Racomitrium* rush-heath are blown clear of snow by strong winds, some stands have an intermittent cover of up to 50 cm of snow throughout the winter and into early spring (McVean and Ratcliffe 1962).

### Conservation interest

*Juncus-Racomitrium* rush-heath is a climax vegetation type, and is very scarce in Great Britain. It is the main habitat of *Juncus trifidus* and *Luzula arcuata* in Scotland. Other rare plants include *L. spicata* and the lichen *Cetraria delisei*. The *Cladonia-Cetraria* sub-community has a rich and characteristic flora of crustose microlichens, of which the most common species are *Lecidea limosa*, *Lecidoma demissum* and *Catillaria contristans* (Fryday 1997). Purple Sandpiper *Calidris maritima* and Dotterel *Charadrius morinellus* sometimes nest in this habitat.

### Management

This community is evidently the natural vegetation of poor soils in a harsh climate at high altitude. It is not maintained by any kind of management. Since it seems to depend on exposure to strong winds rather than snow-lie, it is perhaps less threatened by climatic warming than the communities of late snow-beds. In the Cairngorms, some *Juncus-Racomitrium* heaths close to well-used paths have been damaged by human trampling (Legg 2000). They have a more sporadic carpet of *Racomitrium* than untrampled stands – the ground can be almost bare – and *Juncus trifidus* grows in circular patches like fairy rings. The humus layer can be less continuous under some of these trampled stands than in untrampled locations (e.g. Gordon *et al.* 1998, 2000).

# U10 *Carex bigelowii-Racomitrium lanuginosum* moss-heath

## Synonyms

*Cariceto-Rhacomitretum lanuginosi* (McVean and Ratcliffe 1962; Birks 1973); *Polytricheto-Rhacomitretum lanuginosi* and *Juncus trifidus-Festuca ovina* nodum (McVean and Ratcliffe 1962); *Agrostis montana-Rhacomitrium lanuginosum* community (Birse and Robertson 1976); Species-poor *Racomitrium* heath E1a, Species-rich *Racomitrium* heath E1b, *Festuca ovina-Deschampsia flexuosa-Racomitrium* heath E1c *p.p.* and *Juncus trifidus-Racomitrium* heath E1e (Birks and Ratcliffe 1980).

Typical habitats of U10 *Carex-Racomitrium* moss-heath (stippled):

1  Exposed mountain summits
2  Exposed high cols
3  Exposed shoulders on upper slopes

U10 is also shown in the pictures for H12, H18–H22, CG10–CG12, U7–U8, U11–U14 and U18.

## Description

Silvery-grey-green, tightly woven carpets of the woolly shoots of *Racomitrium lanuginosum* cover vast areas of the tops of high hills in the British uplands. These carpets are spiked through by slender grey-green shoots of the montane sedge *Carex bigelowii*, sprigs of *Vaccinium myrtillus*, fine leaves of *Deschampsia flexuosa* and *Festuca vivipara*, and scrambling stems of *Galium saxatile* and *Salix herbacea*.

There are three sub-communities. The grassy *Galium saxatile* sub-community U10a and the Typical sub-community U10b can look very similar. Both can have thick mats of *R. lanuginosum*, but there is more *Deschampsia flexuosa*, *Festuca ovina* and *Galium saxatile* in the *Galium* sub-community. In the south of Scotland, northern England and Wales there is usually very little *Racomitrium* in these heaths. The *Silene acaulis* sub-community U10c encompasses three quite different types of moss-heath. One type is species-rich; the sombre grey-green turf of *R. lanuginosum* and *C. bigelowii* is variegated

356

by emerald-green, pink-flowered clumps of *Silene acaulis* and small trailing shoots of *Salix herbacea*, and there is a great array of herbs such as *Alchemilla alpina*, *Persicaria vivipara*, *Minuartia sedoides*, *Armeria maritima*, *Sedum rosea*, *Luzula spicata*, *Saussurea alpina* and *Ranunculus acris*. The second type is species-poor, resembling the Typical sub-community, but with a sprinkling of *Alchemilla alpina*, *Silene acaulis* and *Armeria maritima* indicating a slightly more base-rich substrate. The third type consists of a very open scatter of *Juncus trifidus*, *R. lanuginosum*, *Carex bigelowii*, *Festuca vivipara* and *Salix herbacea*, together with a few other species including *A. maritima* and *Ochrolechia frigida*.

On Aonach Mòr next to Ben Nevis there are moss-heaths of *Racomitrium ericoides* rather than *R. lanuginosum*, covering sandy soil at about 1220 m (McVean and Ratcliffe 1962). There are also *R. ericoides* heaths on gravel and rock debris on Stob Coire nan Lochan, one of the subsidiary peaks of Bidean nam Bian in Glen Coe.

### Differentiation from other communities

*Carex-Racomitrium* moss-heath can be mistaken for other montane heaths containing *Racomitrium lanuginosum*. *Carex-Polytrichum* sedge-heath U8 has much less *R. lanuginosum*, fewer grasses, and more lichens, *Dicranum fuscescens* and *Polytrichum alpinum*. *Juncus-Racomitrium* rush-heath U9 has much more *Juncus trifidus* and lichens, although the flora of the *Salix* sub-community U9b is very similar to that of species-poor forms of the *Silene* sub-community of *Carex-Racomitrium* heath (see comments on p. 354). *Vaccinium-Racomitrium* heath H20 can resemble *Carex-Racomitrium* heath but has more dwarf shrubs and pleurocarpous mosses, and *Calluna-Racomitrium* heath H14 has more *Calluna vulgaris*.

Some forms of the *Racomitrium* sub-community of *Nardus-Galium* grassland U5e come close to *Carex-Racomitrium* moss-heath but they have more *Nardus stricta*, and can have *Calluna* too. Stands of the *Vaccinium-Deschampsia* sub-community of *Festuca-Agrostis-Galium* grassland U4e in exposed hilltop habitats can have a similar flora to the *Galium* sub-community of *Carex-Racomitrium* moss-heath, including montane plants such as *Carex bigelowii*, *Diphasiastrum alpinum* and *Salix herbacea*. Generally, the moss-heath community has more *R. lanuginosum* and less *Agrostis capillaris*, *Nardus*, *Anthoxanthum odoratum* and *Rhytidiadelphus squarrosus* than any forms of *Festuca-Agrostis-Galium* grassland.

*R. lanuginosum* can grow in thick pure mats on dry scree and even on walls in upland districts. In contrast to stands of *Carex-Racomitrium* heath, these moss mats are deep and not wind-pruned, and montane species such as *Carex bigelowii* are absent; this vegetation is not included within the NVC.

### Ecology

*Carex-Racomitrium* moss-heath clothes the rocky inhospitable terrain of most of the higher summits and ridges in Great Britain and Ireland, extending from below 300 m in the far north-west to over 1000 m in the eastern Highlands. The community occupies ground that is blown free of snow by winter winds, leaving it exposed to freezing, desiccating winds. These heaths experience huge ranges of temperature over the year, from the bitterest winter winds to the most scorching summer sun. They endure weeks of blowing mist and rain, but can be dried out in clear, warm weather. In many places the ground is patterned by solifluction and frost-heave, so that plants have to contend with shifting unstable soils as well as a severe climate. Soils are generally shallow rankers or podsols, and because of strong leaching are mostly acidic. In the south-west Highlands the *Silene* sub-community occurs primarily on hills with base-rich rocks, but further north it also occurs on more acid substrates where mineral availability is increased by frost-heaving. The species-poor form of the *Silene* sub-community containing *Juncus trifidus* and *Salix herbacea* occurs on rocky, soliflucted or wind-blasted ground with a firm surface of gravel and humus.

### Conservation interest

Great Britain's *Carex-Racomitrium* heaths are of international significance. The community is also of interest as one of the most extensive forms of climax vegetation in the British uplands. Several rare plants occur in these summit heaths. The only known British locality for *Diapensia lapponica* is in a stand of the Typical sub-community, and the rare arctic species *Artemisia norvegica* and *Luzula arcuata* also grow in some very stony, open stands of this sub-community. The *Silene acaulis* sub-community is the habitat of the lichen *Nephroma arcticum*, and some examples include other rarities such as *Minuartia sedoides*, *Cerastium arcticum*, *Aulacomnium turgidum*, *Antitrichia curtipendula*, *Herzogiella striatella* and *Hypnum hamulosum*. *Carex-*

This summit heath is widespread throughout the uplands northwards from north Wales to Harris and north-west Sutherland. It is common on most of the higher upland massifs in the Scottish Highlands and it is here that the largest stands occur. The largest known single stand of *Carex-Racomitrium* moss-heath is on the summit of Ben Wyvis, where it stretches almost unbroken for more than 8 km (Ratcliffe 1977); there are also substantial tracts on the Drumochter hills, the Affric-Cannich hills and the Cairngorms. South of the Highlands, the community is almost entirely represented by the grassy *Galium* sub-community. The *Silene* sub-community is known only in Scotland.

Racomitrium heath is confined to arctic environments in some of the cooler oceanic parts of the world: Great Britain, Ireland, the Faroes, Norway, Iceland, Greenland and some of the sub-Antarctic islands (Ratcliffe and Thompson 1988). There is probably more of it in Great Britain than anywhere else in the world. The variant of *Carex-Racomitrium* heath dominated by *Racomitrium ericoides* instead of *R. lanuginosum* forms a floristic link with moss-heaths in Iceland and Jan Mayen (McVean and Ratcliffe 1962). *R. ericoides* is also locally dominant within stands of *R. lanuginosum* heath on the Faroes (Hobbs and Averis 1991a).

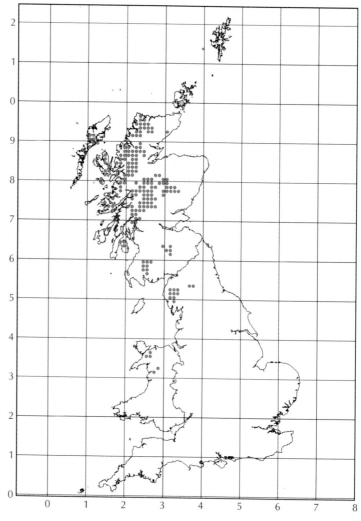

*Racomitrium* heaths are also an important nesting habitat for Dotterel *Charadrius morinellus* and Ptarmigan *Lagopus mutus*.

## Management

*Carex-Racomitrium* moss-heath is an example of near-natural vegetation that occurs only in arctic-like conditions in oceanic climates where natural herbivores are rare. The greatest threat to this type of vegetation appears to be heavy grazing, especially south of the Scottish Highlands (Pearsall and Pennington 1973; Thompson *et al.* 1987; Ratcliffe and Thompson 1988; Thompson and Brown 1992; Thompson *et al.* 2001). In the Lake District and north Wales, there is now very little *Racomitrium lanuginosum* in many sheep-grazed *Carex-Racomitrium* heaths, but it remains common among coarse, stabilised screes that are much less accessible to grazing animals (e.g. Averis and Averis 2000b). This suggests that *R. lanuginosum* has become rare in the summit heaths because of grazing and associated trampling and manuring, which tend to favour grasses at the expense of bryophytes. Heavy grazing may lead to the development of grasslands such as the *Vaccinium-Deschampsia* sub-community of *Festuca-Agrostis-Galium* grassland U4e, dominated by *Festuca* species and *Deschampsia flexuosa* with some *Agrostis* species and small grazing-resistant herbs such as *Galium saxatile* (A M Averis and D B A Thompson, unpublished data). There is usually very little *R. lanuginosum* in modified vegetation of this sort, but *Carex bigelowii* can be locally abundant along paths and in places where the vegetation is trampled by sheep.

*Carex-Racomitrium* heaths may also be damaged by acid deposition, particularly as this is more severe at high altitudes because of occult deposition by mist (Thompson and Baddeley 1991). In some upland areas, acid deposition may have exacerbated the effects of heavy grazing, and it has probably contributed to the replacement of *Carex-Racomitrium* heath by more grassy vegetation in the south of Scotland, northern England and Wales.

# U11 *Polytrichum sexangulare-Kiaeria starkei* snow-bed

## Synonyms

*Polytricheto-Dicranetum starkei* (McVean and Ratcliffe 1962); *Dicranum starkei* snow-bed heaths E3 *p.p.* (Birks and Ratcliffe 1980).

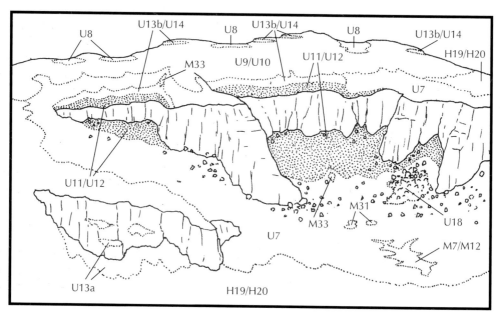

Typical habitats of U11 *Polytrichum-Kiaeria* snow-bed (stippled): on sheltered high slopes with the most prolonged snow-lie.

Other vegetation types in this picture:

U7 *Nardus-Carex* grass-heath
U8 *Carex-Polytrichum* heath
U9 *Juncus-Racomitrium* rush-heath
U10 *Carex-Racomitrium* moss-heath
U12 *Salix-Racomitrium* snow-bed
U13 *Deschampsia-Galium* grassland
U14 *Alchemilla-Sibbaldia* dwarf-herb community
U18 *Cryptogramma-Athyrium* snow-bed
M7 *Carex-Sphagnum russowii* mire
M12 *Carex saxatilis* mire
M31 *Anthelia-Sphagnum* spring
M33 *Pohlia wahlenbergii* spring
H19 *Vaccinium-Cladonia* heath
H20 *Vaccinium-Racomitrium* heath

## Description

In this snow-bed community the dense bright green shoots of the moss *Kiaeria starkei*, in many stands accompanied by *K. falcata* and *K. blyttii*, grow in carpets patterned with other montane bryophytes such as the mosses *Polytrichum sexangulare*, *Andreaea alpina* and *Oligotrichum hercynicum*, and the liverworts *Lophozia opacifolia*, *Anthelia juratzkana* and *Pleurocladula albescens*. This low variegated velvety turf is studded with small montane vascular plants including *Deschampsia cespitosa* ssp. *alpina*, *Saxifraga stellaris*, *Salix herbacea* and *Gnaphalium supinum*.

Two sub-communities are described in the NVC. The moss carpets of the Typical sub-community U11a are interspersed with *Carex bigelowii*, *Huperzia selago*, *Nardus stricta*, *Alchemilla alpina*, *Polytrichum alpinum* and *Racomitrium lanuginosum*, together with a few scarce montane bryophytes such as *Conostomum tetragonum* and *Moerckia blyttii*. *Racomitrium fasciculare* and *R. ericoides* are common in some stands. The Species-poor sub-community U11b does not have this rich sward, but is described in the NVC as having more of the liverwort *Lophozia sudetica* than the Typical sub-community.

The NVC account of the community (Rodwell 1992) has been recognised to be inadequate because of the scarcity of samples (Rodwell *et al.* 2000). Many stands of *Polytrichum-Kiaeria* snow-bed outside the east-central Highlands do not fit well into either sub-community. Rothero (1991) has recorded many more samples and distinguished three main types: a Typical form; a hepatic-rich type with *Barbilophozia floerkei*, *Nardia scalaris*, *Cephalozia bicuspidata* and *Pleurocladula albescens*; and a *Racomitrium* type with *R. heterostichum*.

### Differentiation from other communities

*Polytrichum-Kiaeria* snow-bed has many species in common with *Salix-Racomitrium* snow-bed U12, but *Kiaeria* species and *Polytrichum sexangulare* are less common in the *Salix-Racomitrium* community, which is characterised by *Racomitrium heterostichum* and crusts of tiny, dark-coloured liverworts. The Typical sub-community can be confused with the *Alchemilla-Sibbaldia* community U14: *Alchemilla alpina*, *Sibbaldia procumbens*, *Silene acaulis*, *Juncus trifidus*, *Deschampsia cespitosa* and *Nardus stricta* occur in both vegetation types but are commoner in the *Alchemilla-Sibbaldia* community, where the place of the *Kiaeria* species is taken by *Racomitrium fasciculare* or other mosses.

### Ecology

These ragged patches of mossy vegetation cling to the steep headwalls of shaded corries where snow lies late and deep far into spring. The vegetation may be uncovered for only a few weeks each year: for the rest of the time it is covered by snow in cold, dark and saturated conditions. The snow insulates the vegetation from really low temperatures, and the temperature experienced by the vegetation is rarely more than a few degrees below zero; as little as 5 cm of new snow is enough to provide effective insulation (Fryday 1997). The soils are moist but free-draining, leached and acid. Almost all stands are above 900 m and the community extends above 1200 m on the highest Scottish hills. At over 1330 m on Ben Nevis, patches of *Polytrichum-Kiaeria* snow-bed form the highest-altitude vegetation in Great Britain.

### Conservation interest

Snow-beds such as these are characteristic of some of the most extreme habitats in the world: continuously cold and moist and with a short growing-season. They are an important link with continental mountain vegetation, and have been described as 'the last relics of genuine late-glacial vegetation remaining in Great Britain' (Gilbert and Fox 1985). The community is rare in Great Britain, and is valuable as a habitat for the rare *Luzula arcuata*, *Carex lachenalii*, *Kiaeria glacialis*, *K. falcata*, *Pleurocladula albescens* and *Moerckia blyttii*. It is also one of the main habitats for rare montane species of *Andreaea*, in particular *A. nivalis*, *A. blyttii* and *A. sinuosa* (Murray 1988). Insects often become trapped on the surface of the patches of snow that linger here, and are caught by Snow Buntings *Plectrophenax nivalis* and other upland birds (Nethersole-Thompson and Watson 1981). Birds may also feed on late-emerging craneflies.

## U11 *Polytrichum sexangulare-Kiaeria starkei* snow-bed

The *Polytrichum-Kiaeria* snow-bed community is confined to the Scottish Highlands. There are some especially fine stands in the eastern and central Highlands, particularly on the Cairngorms, Ben Alder, Creag Meagaidh and the Affric-Cannich hills, and in the west on the Glen Coe hills and the Ben Nevis range. Elsewhere most stands are small.

Very similar snow-bed vegetation occurs in the mountains of mainland Europe, from Norway and Sweden to the Alps.

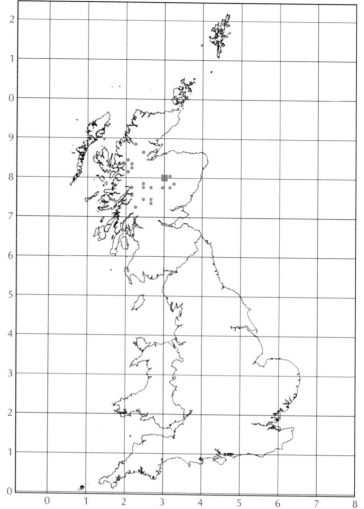

## Management

*Polytrichum-Kiaeria* snow-bed is climax vegetation that is scarcely affected by management. It is one of the plant communities most likely to be adversely affected by climatic warming, and it would be useful to undertake studies to investigate how the vegetation responds to changing amounts and duration of snow-lie in different years. Atmospheric pollutants are concentrated in snow, and melt-water can be highly acid: this has been shown to damage some of the characteristic bryophytes (Woolgrove and Woodin 1995). There may also be threats to the community from skiing developments.

# U12 *Salix herbacea-Racomitrium heterostichum* snow-bed

## Synonyms

*Rhacomitreto-Dicranetum starkei* and *Gymnomitreto-Salicetum herbaceae* (McVean and
Ratcliffe 1962); *Dicranum starkei* snow-bed heaths E3 *p.p.* (Birks and Ratcliffe 1980).

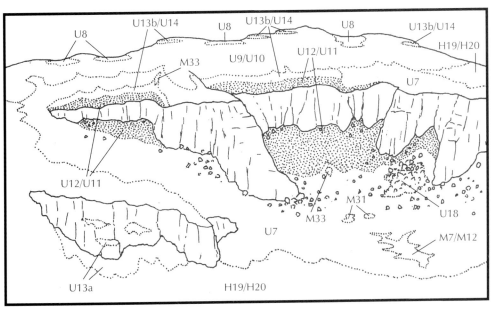

Typical habitats of U12 *Salix-Racomitrium* snow-bed (stippled): on sheltered high slopes with
the most prolonged snow-lie.

Other vegetation types shown in this picture:

U7 *Nardus-Carex* grass-heath
U8 *Carex-Polytrichum* sedge-heath
U9 *Juncus-Racomitrium* rush-heath
U10 *Carex-Racomitrium* moss-heath
U11 *Polytrichum-Kiaeria* snow-bed
U13 *Deschampsia-Galium* grassland
U14 *Alchemilla-Sibbaldia* dwarf-herb community
U18 *Cryptogramma-Athyrium* snow-bed
M7 *Carex-Sphagnum russowii* mire
M12 *Carex saxatilis* mire
M31 *Anthelia-Sphagnum* spring
M33 *Pohlia wahlenbergii* spring
H19 *Vaccinium-Cladonia* heath
H20 *Vaccinium-Racomitrium* heath

## Description

This is snow-bed vegetation in which the dark-green moss *Racomitrium heterostichum*
and the diminutive willow *Salix herbacea* grow in mixed mats and crusts of bryophytes.
The bryophyte mats are dotted with tiny vascular plants: silver-grey rosettes of
*Gnaphalium supinum* and *Saxifraga stellaris*, yellow-green shoots of *Diphasiastrum
alpinum*, and silvery-green leaves of *Alchemilla alpina*. The colourful lichen *Solorina
crocea* occurs in some stands.

There are three sub-communities. The *Silene acaulis-Luzula spicata* sub-community
U12a has a rather open sward of *Silene acaulis*, *Luzula spicata*, *Persicaria vivipara*,

363

*Carex bigelowii* and *Saxifraga stellaris* speckling a discontinuous layer of either *Racomitrium lanuginosum*, *R. heterostichum* and *R. fasciculare* or tiny liverworts such as *Gymnomitrion concinnatum*, *Nardia scalaris* and *Anthelia juratzkana*. The *Gymnomitrion concinnatum* sub-community U12b consists of distinctive dark grey-black patches of diminutive liverworts in intricately mixed mats. There may be as many as eight species of liverwort within a square centimetre (Rothero 1991); the most common species are *Gymnomitrion concinnatum*, *Nardia scalaris*, *Anthelia juratzkana* and *Diplophyllum albicans*. The *Marsupella brevissima* sub-community U12c is characterised by the liverworts *Marsupella brevissima* and *Lophozia sudetica*.

Since the analysis undertaken for the NVC, further sampling has revealed that this type of vegetation is more variable than was originally realised (Rodwell *et al.* 2000). Rothero (1991) has suggested that the *Marsupella* sub-community, represented by only three quadrat samples in the NVC, should be recognised as a separate community comprising three sub-communities: a Typical sub-community with *Kiaeria falcata*; a *Salix herbacea* sub-community with much *Salix herbacea* and *Ditrichum zonatum*; and a *Cephalozia bicuspidata* ssp. *bicuspidata* sub-community with *Cephalozia bicuspidata*, *Nardia scalaris*, *Pleurocladula albescens* and *Kiaeria starkei*. Some Cairngorm snow-beds with much *Deschampsia flexuosa* in a mixed sward with *Juncus trifidus*, *Huperzia selago*, *Galium saxatile*, *Polytrichum alpinum*, *Dicranum fuscescens*, *Rhytidiadelphus loreus* and *Barbilophozia floerkei* are now considered to warrant a new sub-community of *Salix-Racomitrium* snow-bed (Rodwell *et al.* 2000). Snow-beds comprising an almost pure sward of *Racomitrium heterostichum*, with a speckling of *Deschampsia cespitosa* ssp. *alpina* but little or no *Salix herbacea*, may also merit recognition. These correspond to the *Rhacomitreto-Dicranetum starkei* of McVean and Ratcliffe (1962), but their distinctiveness is not apparent in the NVC where they have been subsumed within the broader *Silene-Luzula* sub-community.

Rothero (1991) has sampled a *Pohlia ludwigii* snow-bed that is related to the *Salix-Racomitrium* community but which is not included in the NVC. This vegetation is dominated by *P. ludwigii*, and contains other species such as *Polytrichum sexangulare*, *Nardia scalaris*, *Deschampsia cespitosa* and *Marsupella sphacelata*, and less commonly *Andreaea nivalis*, *Racomitrium heterostichum*, *Lophozia sudetica*, *Marsupella brevissima*, *Saxifraga rivularis* and *Salix herbacea*.

## Differentiation from other communities

The *Salix-Racomitrium* community can be confused with other snow-beds and moss-heaths. The *Gymnomitrion* sub-community, with its dark crusts of liverworts dotted with *Salix herbacea*, is unlike any other type of British vegetation. Stands with a continuous layer of *Anthelia juratzkana* may superficially resemble *Anthelia-Sphagnum* springs M31, but this community is characterised by *A. julacea* rather than *A. juratzkana*, has much *Sphagnum denticulatum*, *Scapania undulata* and *Marsupella emarginata*, and lacks *Salix herbacea* and *Gymnomitrion concinnatum*.

The herb-rich *Silene-Luzula* sub-community might be confused with the *Silene* sub-community of *Carex-Racomitrium* moss-heath U10c, with the *Salix* sub-community of *Juncus-Racomitrium* rush-heath U9b, or with *Alchemilla-Sibbaldia* dwarf-herb vegetation U14. *Carex bigelowii*, *Salix herbacea*, *Juncus trifidus*, *Luzula spicata* and *Silene acaulis* can grow in all of these vegetation types. The two montane heath types can usually be distinguished by having more *Racomitrium lanuginosum* and *Vaccinium myrtillus* than the snow-bed vegetation, little *R. heterostichum*, and none of the characteristic crust-forming liverworts. *R. heterostichum* can be common in *Alchemilla-Sibbaldia* vegetation, and *R. fasciculare* usually grows here too, but *Alchemilla alpina* and *Sibbaldia procumbens* are far more common than they are in *Salix-Racomitrium* snow-bed.

Species-poor examples of the *Salix-Racomitrium* community dominated by *R. heterostichum* are distinctive, but they can be small and hard to pick out where they occur within larger stands of *Carex-Racomitrium* heath. Compared to the surrounding heath vegetation, *Salix-Racomitrium* snow-beds have more *R. heterostichum* and less *R. lanuginosum* and *Vaccinium* species. *Carex-Racomitrium* heath occurs in exposed places that are blown clear of snow, whereas the *Salix-Racomitrium* community occupies hollows or slopes where snow accumulates.

## Ecology

Like *Polytrichum-Kiaeria* snow-beds, the *Salix-Racomitrium* community occurs in places where growing conditions are inimical to most plants. These places are cold and wet, have a short growing-season, and the vegetation is often disturbed by solifluction, flooded with icy water and buried beneath snow and wind-blown soil and debris. The community is recorded only above 600 m and extends to over 1200 m. Both types of snow-bed occur on the coldest, highest parts of the hills where snow persists into the summer, but the *Salix-Racomitrium* community occupies wetter, less stable ground than the *Polytrichum-Kiaeria* community. It is most common on the steep upper slopes of corries where the ground is disturbed by amorphous solifluction as well as by the

This community is the more widespread of the two moss-dominated snow-beds. It is distributed from Ben More Assynt in the north-west Highlands to the Breadalbanes in the southern Highlands, with outliers on Ben More in Mull and in the Southern Uplands. It is particularly common in the Cairngorms and there are also good stands on other hills with high plateaux, such as Ben Alder and Beinn Dearg at the head of Loch Broom.

There is similar vegetation in Scandinavia, where it is much more extensive because snow lies for longer over larger areas of ground.

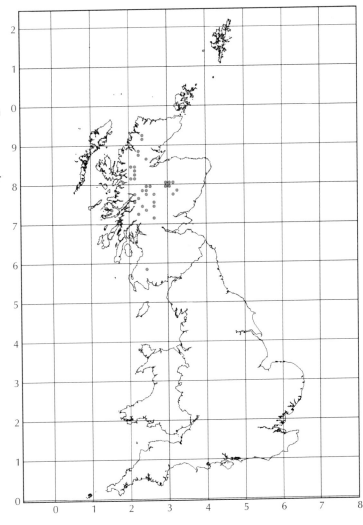

tearing effects of the snow-field itself moving down the slope as its weight increases. The thin skin of vegetation is also often disrupted by water running from the melting snow. It is quite common for the liverwort crust to roll over on itself, burying some of the plants but exposing new soil for recolonisation (McVean and Ratcliffe 1962). The community can occur on slopes of any aspect, but is virtually restricted to sheltered places where snow can accumulate. It occurs rarely on exposed summits such as Aonach Beag in the Nevis range.

## Conservation interest

This is a rare vegetation type and is of considerable value for nature conservation. Like other forms of snow-bed, the *Salix-Racomitrium* community is the Scottish equivalent of the type of late-glacial vegetation that is more common at high altitudes in mainland Europe. It demonstrates that although our hills are relatively low and cover a relatively small area, conditions can be as severe as on the higher, more northern Scandinavian ranges. There are hills on which one can stand in a patch of this type of vegetation and look down on woodland in the sheltered glens far below. This shows just how steep the ecological gradients of temperature, precipitation and wind speed are in the Scottish hills.

Although they consist of little more than a thin skin of vegetation over stony ground, *Salix-Racomitrium* snow-beds are home to many rare montane species. Vascular plant rarities include *Luzula arcuata, Saxifraga rivularis, S. cernua* and *Carex lachenalii*. Some of the bryophytes that grow in this community are extreme specialists confined to late snow-beds; examples are the mosses *Kiaeria glacialis, K. falcata, Andreaea nivalis, A. blyttii,* and *A. sinuosa,* and the liverworts *Marsupella brevissima, M. condensata, M. boeckii, Cephalozia ambigua, Gymnomitrion corallioides, G. apiculatum, Diplophyllum taxifolium* and *Moerckia blyttii*.

## Management

The *Salix-Racomitrium* community is a natural type of vegetation and is not maintained by management. The main threats, shared with *Polytrichum-Kiaeria* snow-beds, are climatic warming and possibly in some places, the development of sites for downhill skiing. There is also the potential threat of damage from acid melt-water resulting from the concentration of atmospheric pollutants in snow (Woolgrove and Woodin 1995).

# U13 *Deschampsia cespitosa-Galium saxatile* grassland

## Synonyms

*Deschampsietum caespitosae alpinum* and *Deschampsieto-Rhytidiadelphetum* p.p. (McVean and Ratcliffe 1962); Species-poor *Deschampsia cespitosa* grassland C5a and *Rhytidiadelphus loreus-Deschampsia cespitosa* heaths E2 (Birks and Ratcliffe 1980).

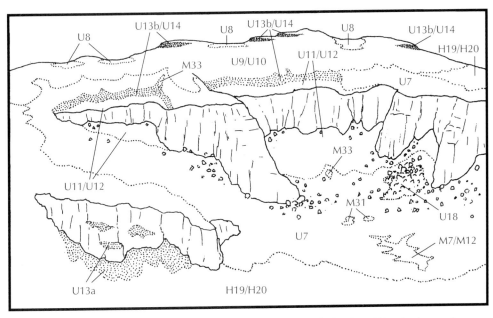

Typical habitats of U13 *Deschampsia-Galium* grassland (stippled): the U13a *Anthoxanthum-Alchemilla* sub-community on cliff ledges and below cliffs, where snow lies moderately late in the year and ground is kept moist by flushing; the U13b *Rhytidiadelphus* sub-community at higher altitudes on snow-bound corrie rims and terraces on high plateaux.

Other vegetation types shown in this picture:

U7 *Nardus-Carex* grass-heath
U8 *Carex-Polytrichum* sedge-heath
U9 *Juncus-Racomitrium* rush-heath
U10 *Carex-Racomitrium* moss-heath
U11 *Polytrichum-Kiaeria* snow-bed
U12 *Salix-Racomitrium* snow-bed
U14 *Alchemilla-Sibbaldia* dwarf-herb community
U18 *Cryptogramma-Athyrium* snow-bed
M7 *Carex-Sphagnum russowii* mire
M12 *Carex saxatilis* mire
M31 *Anthelia-Sphagnum* spring
M33 *Pohlia wahlenbergii* spring
H19 *Vaccinium-Cladonia* heath
H20 *Vaccinium-Racomitrium* heath

U13 is also shown in the pictures for CG10–CG12.

## Description

This NVC community brings together an upland grassland and a snow-bed: two types of vegetation that differ ecologically but which share the species *Deschampsia cespitosa*,

*Galium saxatile* and *Agrostis capillaris*, and the mosses *Rhytidiadelphus loreus*, *Hylocomium splendens* and *Polytrichum alpinum*.

The two sub-communities of *Deschampsia-Galium* grassland look very different even though their flora has so much in common. The *Anthoxanthum odoratum-Alchemilla alpina* sub-community U13a clothes steep, irrigated, shaded slopes. It is a short, dense, tufted, dark-green grassland composed of the tough, sharp-ridged leaves of *D. cespitosa*. *Agrostis canina*, *Anthoxanthum odoratum* and *Festuca vivipara* grow with *D. cespitosa*, and the turf is flecked with a few small species such as *Viola palustris*, *Potentilla erecta*, *Alchemilla alpina*, *Galium saxatile* and *Blechnum spicant*. At the foot of the vascular plants there is a thin layer of bryophytes, of which the most common are *Hylocomium splendens*, *Thuidium tamariscinum*, *Rhytidiadelphus loreus*, *R. squarrosus*, *Diplophyllum albicans* and *Marsupella emarginata*. In contrast, the *Rhytidiadelphus loreus* sub-community U13b is a golden-yellow, mossy snow-bed that occupies steep shaded slopes at high altitudes, where it commonly forms crescent-shaped patches. The deep, lustrous blanket of *R. loreus* is variegated with a little *Hylocomium splendens*, *Pleurozium schreberi* and *Polytrichum alpinum*, and is dotted with small tufts of *D. cespitosa* (many plants belonging to ssp. *alpina*), *Carex bigelowii*, *Saxifraga stellaris*, *Nardus stricta*, *Galium saxatile* and *Potentilla erecta*.

There is also a herb-rich form of the *Anthoxanthum-Alchemilla* sub-community in which the sward is enriched with small base-tolerant species such as *Thalictrum alpinum*, *Persicaria vivipara*, *Carex pulicaris*, *C. viridula* ssp. *oedocarpa* and even *Thymus polytrichus*. This is not described in the NVC.

### Differentiation from other communities

The *Anthoxanthum-Alchemilla* sub-community is distinct from other upland acid grasslands in being dominated by *Deschampsia cespitosa*. Additionally, *Festuca ovina*, *Nardus stricta* and *Juncus squarrosus*, which play such an important part in most submontane acid grasslands, are scarce in this vegetation type. It is possible to confuse the herb-rich form of the *Anthoxanthum-Alchemilla* sub-community with stands of the *Agrostis-Rhytidiadelphus* sub-community of *Luzula-Geum* tall-herb vegetation U17c, but this occurs on more calcareous soils and includes tall mesotrophic forbs such as *Alchemilla glabra*, *Angelica sylvestris*, *Trollius europaeus*, *Crepis paludosa*, *Filipendula ulmaria*, *Geranium sylvaticum* and *Cirsium heterophyllum*. These tall forbs do not usually grow in *Deschampsia-Galium* grassland, and if they do occur it is as a few small individuals that do not make a great contribution to the sward.

The *Rhytidiadelphus* sub-community is usually distinctive in having so much *Rhytidiadelphus loreus*; this species is more common here than in any other recognised British vegetation type. Some stands of the sub-community can look rather like *Carex-Racomitrium* moss-heath U10, but contain more *R. loreus* and *Hylocomium splendens*, and less *Racomitrium lanuginosum*.

### Ecology

The *Anthoxanthum-Alchemilla* sub-community is a montane grassland of high, shaded slopes, mostly above 500 m. It typically occurs on the steep headwalls of corries or under cliffs, where acid soils are constantly flushed with cold water, often running from melting snow. *Deschampsia cespitosa* can out-compete other grasses on these saturated soils in a climate where the summers are short and cool and much of the winter precipitation falls as snow. The herb-rich form occurs where there is irrigation by base-rich water.

The *Rhytidiadelphus* sub-community generally occurs at altitudes over 600 m, on shaded slopes around the upper rims of slopes and corries where snow lies deep far into the spring. It can also form fine-scale mosaics with *Carex-Racomitrium* heath; in such situations, *Deschampsia-Galium* grassland occupies shallow sheltered hollows, with *Carex-Racomitrium* heath on the surrounding ground where conditions are more exposed. This sub-community seems to need a cool oceanic climate with a high precipitation; many of its sites receive over 220 wet days a year, and in winter much of the precipitation falls as snow.

## Conservation interest

This community is of interest as a near-natural example of acid grassland vegetation. Few rare plant species find a home in *Deschampsia-Galium* grassland, but there can be populations of *Sibbaldia procumbens*, *Diphasiastrum alpinum*, *Saussurea alpina*, *Minuartia sedoides*, *Persicaria vivipara*, *Barbilophozia lycopodioides*, *Anastrepta orcadensis* and *Kiaeria starkei*. The *Rhytidiadelphus* sub-community is one of the few types of upland vegetation known from Great Britain and Ireland but not mainland Europe.

The *Anthoxanthum-Alchemilla* sub-community is fairly common throughout the Scottish Highlands, and there are outliers in the Southern Uplands, the Lake District and north Wales. It is most extensive in the western and central Highlands, on high hills such as Ben Nevis, the Affric-Cannich hills, Foinaven and Ben Alder. The herb-rich form is rare in the west Highlands. The *Rhytidiadelphus* sub-community is restricted to the Highlands, extending south to Ben Lomond. It is most common in the north-west Highlands, especially in the high north-facing corries of the Fannich Hills, the Monar Forest hills and on Beinn Dearg at the head of Loch Broom; there are also large expanses on Ben Wyvis.

Deschampsia cespitosa grass-lands similar to the *Anthoxanthum-Alchemilla* sub-community occur in Scandinavia, but the *Rhytidiadelphus* sub-community is apparently confined to Great Britain and Ireland.

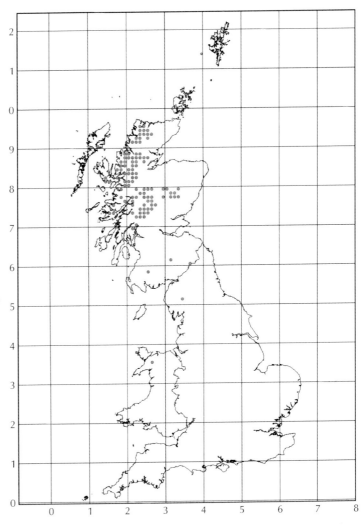

**Management**

Both the grassland and snow-bed forms of this community are primarily climax vegetation types: stable assemblages of plants owing more to climate, soils and topography than to the actions of people and grazing animals. They provide a striking contrast to the sub-montane *Agrostis-Festuca*, *Nardus* and *Juncus squarrosus* grasslands, which if left ungrazed would almost all revert to dwarf-shrub heath, scrub or woodland given a local source of seed. However, it is possible that there would be more tall herbs and ferns in many stands of the *Anthoxanthum-Alchemilla* sub-community if the vegetation were grazed less heavily; some stands might even develop into communities dominated by tall herbs (*Luzula-Vaccinium* U16 or *Luzula-Geum* U17).

# U14 *Alchemilla alpina-Sibbaldia procumbens* dwarf-herb community

## Synonyms
*Alchemilla-Sibbaldia* nodum (McVean and Ratcliffe 1962); *Dicranum starkei* snow-bed heaths E3 *p.p.* (Birks and Ratcliffe 1980).

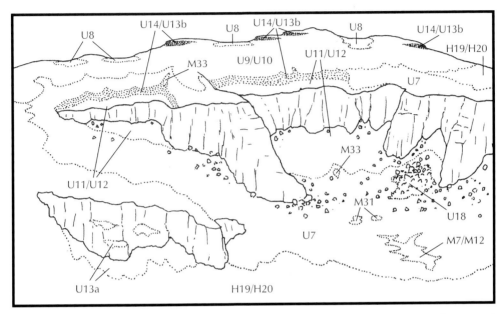

Typical habitats of U14 *Alchemilla-Sibbaldia* dwarf-herb community (stippled), on the rims of corries where snow lies late, and on terraces on high plateaux.

Other vegetation types shown in this picture:

U7 *Nardus-Carex* grass-heath
U8 *Carex-Polytrichum* sedge-heath
U9 *Juncus-Racomitrium* rush-heath
U10 *Carex-Racomitrium* moss-heath
U11 *Polytrichum-Kiaeria* snow-bed
U12 *Salix-Racomitrium* snow-bed
U13 *Deschampsia-Galium* grassland
U18 *Cryptogramma-Athyrium* snow-bed
M7 *Carex-Sphagnum russowii* mire
M12 *Carex saxatilis* mire
M31 *Anthelia-Sphagnum* spring
M33 *Pohlia wahlenbergii* spring
H19 *Vaccinium-Cladonia* heath
H20 *Vaccinium-Racomitrium* heath

## Description
This community consists of a loosely woven turf of *Sibbaldia procumbens* and *Alchemilla alpina*, both of which have delicate, silvery-green leaves and drifts of yellow-green flowers in summer. Other characteristic vascular plants include *Carex bigelowii*, *C. pilulifera*, *Galium saxatile*, *Deschampsia cespitosa*, *D. flexuosa*, *Gnaphalium supinum* and *Thymus polytrichus*. These plants are commonly set in a matrix of golden-brown *Racomitrium fasciculare* or greyish *R. ericoides*, which is tufted with small shoots of a

few other snow-tolerant bryophytes such as *Conostomum tetragonum* and *Oligotrichum hercynicum*.

## Differentiation from other communities

The *Alchemilla-Sibbaldia* dwarf-herb community may be confused with the similar *Festuca-Alchemilla-Silene* dwarf-herb community CG12, but it has fewer grasses, fewer calcicoles such as *Silene acaulis*, *Saxifraga hypnoides* and *S. oppositifolia* and more *Galium saxatile*. Some stands resemble forms of the *Silene* sub-community of *Carex-Racomitrium* moss-heath U10c, but the stricter calcicoles and *Racomitrium lanuginosum* that usually characterise the moss-heath vegetation are scarcer in the *Alchemilla-Sibbaldia* community. The habitats of these two vegetation types also differ: *Carex-Racomitrium* moss-heath occurs on very exposed ground, whereas the *Alchemilla-Sibbaldia* community is a snow-bed and occupies more sheltered slopes and hollows. *Sibbaldia* and *Alchemilla alpina* can grow in the Typical sub-community of *Polytrichum-Kiaeria* snow-bed U11a, but this has a thicker and more variegated carpet of mosses including *Kiaeria* species, as well as many other snow-tolerant species.

The *Alchemilla-Sibbaldia* dwarf-herb community is widespread but rather scarce in the Scottish Highlands. Most individual stands are small. Similar vegetation lacking *Sibbaldia* but with *Alchemilla alpina* and *Potentilla erecta* occurs at high altitude in the Lake District and in the western Highlands (McVean and Ratcliffe 1962), where it may be as dependent on intermittent irrigation as it is on late-lying snow.

There is similar vegetation in Norway and Sweden.

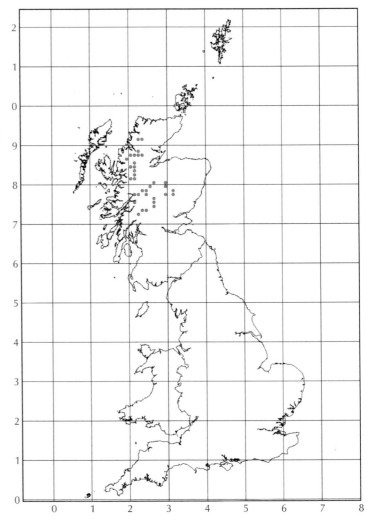

## Ecology

This snow-bed community occurs at high altitudes (generally over 600 m) where snow accumulates in gullies or hollows on high ridges and plateaux, in corries, and on sheltered slopes that face north or east. It is most common on base-rich rocks such as the Dalradian schist of the Breadalbanes, but also occurs over acid rocks such as the Cairngorm granite, where the vegetation is irrigated by water running from melting snow. Although the community is herb-rich and can contain calcicoles such as *Thalictrum alpinum* and *Persicaria vivipara*, the soils are usually incipient podsols with a pH between 4.8 and 5.4 (McVean and Ratcliffe 1962). A common habitat is the unstable upper rim of high corries where the soil is disturbed by solifluction during the winter. This disturbance redistributes nutrients in the soil and may enable calcicoles to grow on substrates that would otherwise be too acid.

## Conservation interest

This is a rare plant community that typically occurs in close association with other forms of near-natural vegetation influenced by late snow-lie. These vegetation mosaics are of great interest for nature conservation. The rare *Cerastium alpinum*, *C. arcticum*, *Sagina saginoides*, *Aulacomnium turgidum*, *Hypnum hamulosum* and *Moerckia blyttii* have been recorded in the *Alchemilla-Sibbaldia* dwarf-herb community.

## Management

The *Alchemilla-Sibbaldia* dwarf-herb community is a near-natural type of vegetation that is not maintained by management, although stands may be lightly grazed in summer (Rodwell 1992). As it depends on snow-lie, it could be adversely affected by a warmer climate, and acidic deposition concentrated in melting snow may damage bryophytes and lichens.

# U15 *Saxifraga aizoides-Alchemilla glabra* banks

## Synonyms

*Saxifragetum aizoidis* (McVean and Ratcliffe 1962; Birks 1973); Mixed Saxifrage facies (McVean and Ratcliffe 1962); *Saxifraga aizoides* banks D5 (Birks and Ratcliffe 1980).

U15 *Saxifraga-Alchemilla* banks (stippled): on steep, dripping banks and basic cliff ledges, inaccessible to grazing animals and typically associated with the U17 *Luzula-Geum* tall-herb community.

## Description

This community consists of dripping mats of diverse, species-rich vegetation sprawling over wet rocks and shelving ledges. The most common species is *Saxifraga aizoides* with its grey-green leaves and the dazzling spectacle of its cascading yellow flowers in summer. The stands are tangled with trailing stems of *S. oppositifolia*, lush green leaves of *Alchemilla glabra*, small plants of *Persicaria vivipara* and *Thalictrum alpinum*, yellow-green spikes of *Selaginella selaginoides*, and the sedges *Carex viridula* ssp. *oedocarpa*, *C. dioica* and *C. pulicaris*. Forming layers on the rock surfaces is a rich array of bryophytes including *Ctenidium molluscum*, *Orthothecium rufescens*, *Palustriella commutata* and *Molendoa warburgii*.

## Differentiation from other communities

This vegetation type is distinctive but could be confused with *Luzula-Geum* tall-herb vegetation U17. *Saxifraga-Alchemilla* banks contain more *Saxifraga aizoides*, *S. oppositifolia*, *Pinguicula vulgaris*, *Alchemilla alpina*, *Selaginella selaginoides*, *Thalictrum alpinum* and *Carex pulicaris* than the *Luzula-Geum* community, and less *Luzula sylvatica*, *Geum rivale* and *Angelica sylvestris*. *Saxifraga-Alchemilla* banks may also resemble densely vegetated examples of *Carex-Saxifraga* mire M11, but have more forbs such as *Thalictrum alpinum*, *Alchemilla glabra*, *A. alpina*, *Persicaria vivipara* and *Saxifraga oppositifolia*, and less *Carex viridula* ssp. *oedocarpa*, *C. panicea*, *Blindia acuta*, *Drepanocladus revolvens*, *Campylium stellatum* and *Aneura pinguis*.

## Ecology

The *Saxifraga-Alchemilla* community is almost confined to steep faces of basic rock with a copious supply of base-rich water dripping through the foliage. Stands can also occur on steep, flushed gravelly ground below cliffs and on wet grassy banks. The essential condition is constant irrigation by base-rich water. The community occurs on slopes of all aspects, but most sites face north or east (since these tend to be the wettest slopes). Many of the constituent species are montane plants, and this is generally a high-level community growing at altitudes between 200 m and almost 900 m, although it does occur locally near sea-level on the Isle of Skye. The soil is usually a wet, silty humus that can accumulate to a depth of 30 cm (McVean and Ratcliffe 1962), but some stands form hanging curtains over steep, wet rocks with almost no soil at all.

## Conservation interest

This scarce community is an important locus for some of our rarest montane species, such as *Cystopteris montana*, *Juncus biglumis*, *Poa alpina*, the mosses *Bryoerythrophyllum caledonicum*, *Hylocomium pyrenaicum*, *Mnium thomsonii* and

This is a community of the Scottish Highlands, with a few outlying stands in the Lake District. Similar vegetation but without *Saxifraga aizoides* occurs in Snowdonia. Some of the best and most extensive stands of *Saxifraga-Alchemilla* vegetation are in the Breadalbanes, on the calcareous Dalradian schist that extends from the Cowal peninsula to Caenlochan. There are also some fine examples on base-rich exposures of the Moine schist, such as in the Monar Forest and on Beinn Dearg at the head of Loch Broom in Ross, and on the base-rich Fucoid beds and Serpulite grits that outcrop in north-west Ross and Sutherland, notably on Meall Horn in the Reay Forest. The stands in the Lake District are on the base-rich Borrowdale Volcanic rocks, and illustrate the importance of the Lake District hills as a refugium for arctic-alpine plants.

There is vegetation resembling the *Saxifraga-Alchemilla* community in Norway and Sweden, although it is ecologically rather different, most examples being associated with deep snow rather than with steep, irrigated slopes. A community more like the British vegetation occurs on flushed basalt cliffs and slopes in the Faroes (Hobbs and Averis 1991b).

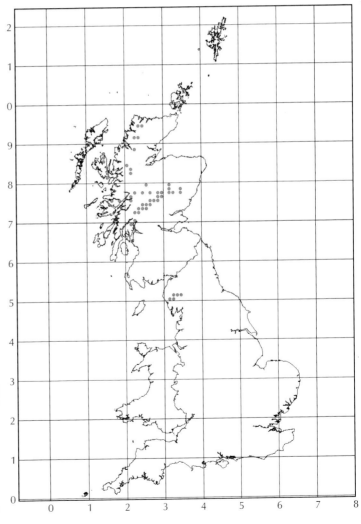

*Hypnum bambergeri*, and the liverworts *Barbilophozia quadriloba, Jungermannia polaris, J. borealis* and *Scapania calcicola*.

## Management

The *Saxifraga-Alchemilla* community is a natural vegetation type that is scarcely affected by management. Most stands are in situations that cannot be reached by grazing animals, and because of the lack of grazing, sedges, rushes and small forbs are able to flower and set seed, maintaining the genetic diversity of their populations. It is possible that disturbance by climbers, particularly in the Lake District, may damage the vegetation.

# U16 *Luzula sylvatica-Vaccinium myrtillus* tall-herb community

## Synonyms
*Luzula sylvatica* Grassland nodum and *Betuletum Oxaleto-Vaccinietum, Vaccinium-Luzula* treeless facies (McVean and Ratcliffe 1962); *Luzula sylvatica-Vaccinium myrtillus* Association (Birks 1973); *Luzula sylvatica-Dryopteris* communities D6 *p.p.* (Birks and Ratcliffe 1980).

Typical habitats of U16 *Luzula-Vaccinium* tall-herb community (stippled):

1 Ungrazed cliff ledges
2 Lightly grazed or ungrazed open slopes
3 Ungrazed island in loch
4 Ungrazed upper part of large boulder

## Description
This community consists of swards of tall herbs and ferns with much *Luzula sylvatica* and usually *Vaccinium myrtillus*. Beneath the vascular plants there are big sprawling wefts of mosses such as *Rhytidiadelphus loreus*, *R. squarrosus*, *Hypnum jutlandicum*, *Hylocomium splendens*, *Pleurozium schreberi* and *Scleropodium purum*.

There are three sub-communities. The *Dryopteris dilatata-Dicranum majus* sub-community U16a is the most lush and diverse, and is largely confined to inaccessible ledges. Here the luxuriant rich green herbage of *L. sylvatica* and *V. myrtillus* is enriched with large ferns; *Dryopteris dilatata*, *D. borreri*, *Athyrium filix-femina* and *Oreopteris limbosperma* are among the most common species. At a lower level in the sward there is a layer of smaller plants including *Oxalis acetosella* with its delicate folded leaves, tough clumps of *Blechnum spicant*, and delicate soft-green fronds of *Phegopteris connectilis* and *Gymnocarpium dryopteris*; there can also be a little *Alchemilla alpina*. The *Anthoxanthum odoratum-Festuca ovina* sub-community U16b is a more grassy assemblage of lightly grazed or ungrazed hillsides, where it occurs in mosaics with heaths and grasslands. In these stands there are fewer ferns, and the clumps of *L. sylvatica* and *V. myrtillus* are entwined with grasses such as *Festuca ovina*, *Anthoxanthum odoratum* and *Agrostis canina*, as well as the common grassland forbs *Potentilla erecta* and *Galium saxatile*. The Species-poor sub-community U16c, as its name implies, takes in the impoverished, almost pure swards of *L. sylvatica* that grow on cliffs, scree and open hillsides.

## Differentiation from other communities

The *Luzula-Vaccinium* tall-herb community can be quite similar to the *Luzula-Geum* tall-herb community U17, but it is less species-rich and generally lacks *Geum rivale*, *Angelica sylvestris*, *Sedum rosea*, *Filipendula ulmaria* and calcicole species. Although *Vaccinium myrtillus* can be common in *Luzula-Vaccinium* vegetation, there should not be any confusion with the various forms of sub-montane and montane *Vaccinium* heath, which have much less *Luzula sylvatica* and fewer ferns. The *Oreopteris-Blechnum* community U19 can look rather like the more fern-rich stands of the Species-poor sub-community of *Luzula-Vaccinium* vegetation, but is dominated by *Oreopteris limbosperma*. *Dryopteris affinis* vegetation (Rodwell *et al.* 2000) can also resemble the Species-poor sub-community, but is distinguished by the distinctive golden-green tussocks of *D. affinis* growing among mixtures of grasses, forbs and bryophytes. *L. sylvatica* can form pure swards in lightly grazed or ungrazed birch and oak woodlands on acid ground, but such vegetation is immediately distinguished from the *Luzula-Vaccinium* community by the presence of a tree canopy.

## Ecology

The succulent shoots of *Luzula sylvatica* are relished by grazing animals and *Luzula-Vaccinium* vegetation is rare except in places where grazing is light or absent. The community is the classic tall-herb vegetation of acid rock ledges that are inaccessible to grazing animals. More impoverished stands occur on accessible ground, and even on heavily grazed hills the community can persist on very steep, rocky slopes, on the tops of large boulders, or on islands in lochs. On Orkney, *Luzula-Vaccinium* vegetation occurs in the unusual habitat of open moorland as long as the intensity of grazing is low, but rising numbers of sheep soon eliminate it.

   The community is most common around or above the altitude of the former tree-line between 400 m and 600 m, where it probably represents a near-natural form of vegetation, but it descends almost to sea-level in the cool oceanic climate of north-west Scotland. Stands at higher altitudes can be covered by snow for part of the winter. Soils are usually base-poor raw humus and can be up to 60 cm deep (McVean and Ratcliffe 1962). *Luzula-Vaccinium* vegetation usually occurs on shaded slopes facing between north and east and can be mildly flushed. Gradations into the richer *Luzula-Geum* community are quite common, reflecting changes in the underlying rock or flushing with base-rich water.

## Conservation interest

The *Luzula-Vaccinium* community is perhaps best known as the habitat of the rare *Cicerbita alpina*, which is confined to vegetation of this type on the cliffs of Lochnagar, Caenlochan Glen and Glen Clova. Other montane species that can grow in this community include the rare willows *Salix lapponum*, *S. myrsinites*, *S. lanata* and *S. reticulata* (Averis and Averis 1999a), and *Gnaphalium norvegicum*. The willows would perhaps be more common in this type of vegetation if they were more widespread generally, less confined to small, isolated populations, and better able to reproduce by cross-pollination. However, the community is still an important refuge for these plants, helping to maintain the genetic diversity of the tiny and dwindling populations. Cliff-nesting Golden Eagles *Aquila chrysaetos* and White-tailed Eagles *Haliaeetus albicilla* often have

The *Luzula-Vaccinium* community occurs on acid cliff ledges
and open slopes throughout the
Highlands, Orkney, Shetland,
the Inner and Outer Hebrides,
the Southern Uplands, the Lake
District and north Wales. It is
uncommon in the Pennines and
south Wales, and is rare in
south-west England. Some of the
largest stands are in the far north
and west, for example on the
Cowal Hills in Argyll, Ben Hope
and Ben Loyal in Sutherland,
and St Kilda.

There is similar vegetation on
cliffs in Ireland, for example on
the Mourne Mountains (Hobbs
and Averis 1991b) and in parts
of Kerry and Cork. Related vegetation also occurs in ungrazed
sub-alpine birch woodlands and
on steep treeless slopes in
western Norway and on the
Faroes.

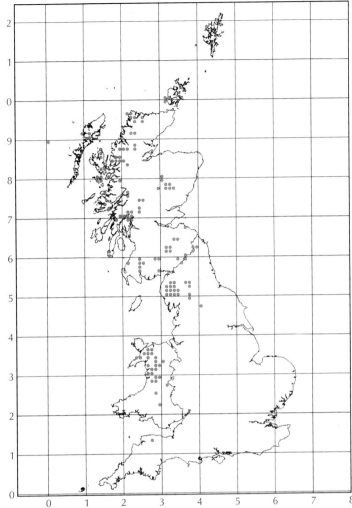

their eyries in stands of *Luzula-Vaccinium* vegetation, and *Luzula sylvatica* is used as
material to construct nests.

## Management

This community is highly susceptible to grazing and burning. Isolated patches of
*Luzula-Vaccinium* vegetation, clinging to ungrazed rock ledges or grazed into fragments
on more open slopes, help to give a picture of what our hillsides might have looked like
in the days when there were no sheep in the uplands and there were probably not so
many deer. McVean and Ratcliffe (1962) considered this community to be a *Vaccinium*-
rich birch wood (where the ground flora can be dominated by *Luzula sylvatica*) without
trees. Some stands are above what appears to be the natural altitudinal limit of woodland, but many are below the tree-line and have a speckling of birch and rowan. It is
likely that some examples on open slopes mark sites of former woodland. In support of
this hypothesis, *Cicerbita alpina* is primarily a woodland plant in Scandinavia and the
Alps. It would be interesting to fence some of the larger and more accessible stands
against grazing livestock and see whether woodland developed. Even where the current
intensity of grazing is light, further reducing the numbers of sheep and deer may well
allow more ferns and herbs to flourish. Stands on cliff ledges are unmanaged and only

suffer when grazed by animals falling or climbing to the cliff ledges. Goats are able to reach ledges that are inaccessible to other animals, and an increase in the numbers of goats in the uplands could have a serious effect on the *Luzula-Vaccinium* community.

# U17 *Luzula sylvatica-Geum rivale* tall-herb community

## Synonyms
Tall-herb nodum and *Deschampsietum caespitosae alpinum* (McVean and Ratcliffe 1962); *Luzula sylvatica-Silene dioica* Association and *Sedum rosea-Alchemilla glabra* Association (Birks 1973); *Sedum rosea-Alchemilla glabra* communities D1 and Species-rich *Deschampsia cespitosa* grassland C5b (Birks and Ratcliffe 1980).

Typical habitats of U17 *Luzula-Geum* tall-herb community (stippled):

1   Ungrazed cliff ledges
2   Lightly grazed or ungrazed open slopes

## Description
The *Luzula-Geum* tall-herb community comprises luxuriant herb-rich vegetation on ungrazed or lightly grazed basic cliffs and hillsides. As with the *Luzula-Vaccinium* community U16, the most species-rich and lush stands are on inaccessible ledges where the plants are safe from the depredations of sheep and deer. In such situations there can be stunningly rich mixtures of tall herbs and ferns, growing in deep meadows on larger ledges and hanging over sheer rock faces in shaggy mats and curtains dripping with water. The most common species in the community are *Luzula sylvatica*, *Geum rivale*, *Deschampsia cespitosa*, *Alchemilla glabra*, *A. alpina*, *Trollius europaeus*, *Oxyria digyna*, *Crepis paludosa*, *Filipendula ulmaria*, *Sedum rosea*, *Angelica sylvestris*, *Ranunculus acris* and *Thymus polytrichus*, all with a glowing profusion of bright flowers in summer. They are accompanied by a great array of smaller plants, including *Selaginella selaginoides*, *Saxifraga oppositifolia*, *S. aizoides*, *Carex pulicaris* and *Thalictrum alpinum*. There are also mats and cushions of bryophytes among the larger plants and hanging over the ledges, with *Tortella tortuosa*, *Ctenidium molluscum*, *Neckera crispa*, *Anoectangium aestivum* and other calcicoles.

The community varies in floristic composition over its geographical range and the four sub-communities reflect this. The *Alchemilla glabra-Bryum pseudotriquetrum* sub-community U17a clothes ledges where the seepage of water is more or less continuous, and includes plants of wet ground such as *Chrysosplenium oppositifolium*, *Saxifraga aizoides* and the moss *Philonotis fontana*. The richest swards belong to the *Geranium sylvaticum* sub-community U17b; most of the magnificent tall-herb swards of the mica-schist ledges of the Breadalbane hills are of this type. *Rubus saxatilis*, *Geranium sylvaticum*, *Trollius europaeus*, *Thalictrum alpinum*, *Heracleum sphondylium*, *Polystichum lonchitis*, *Carex pulicaris*, *Ctenidium molluscum*, *Dicranum scoparium* and *Plagiochila asplenioides* are all more common here than in the other forms of the community. The *Agrostis capillaris-Rhytidiadelphus loreus* sub-community U17c is grassier. The tall herbs are not so tightly packed, and the vegetation includes some typical grassland species such as *Rumex acetosa*, *Ranunculus acris* and *Plantago lanceolata*. This sub-community can occur on open slopes where grazing is not heavy, and is the herb-rich counterpart of *Deschampsia-Galium* grassland U13. The *Primula vulgaris-Hypericum pulchrum* sub-community U17d has a lowland and woodland element in the flora, including *Calluna vulgaris*, *Primula vulgaris*, *Teucrium scorodonia*, *Hyacinthoides non-scripta* and *Allium ursinum*.

## Differentiation from other communities

The most rich and luxuriant stands of the *Geranium* sub-community are quite unmistakable: no other type of vegetation in the British uplands has so great a profusion of tall flowering herbs. Some examples of *Luzula-Geum* vegetation resemble the *Luzula-Vaccinium* tall-herb community (its counterpart on acid soils), but have more *Geum rivale*, *Sedum rosea*, *Angelica sylvestris*, *Alchemilla glabra*, *Filipendula ulmaria* and other mesotrophic forbs. The *Saxifraga-Alchemilla* community U15 shares many species with the *Alchemilla-Bryum* sub-community of *Luzula-Geum* vegetation, but is usually distinct in having more *Saxifraga aizoides*, *S. oppositifolia*, *Pinguicula vulgaris*, *Alchemilla alpina*, *Selaginella selaginoides*, *Thalictrum alpinum* and *Carex pulicaris*, and less *Luzula sylvatica*, *Angelica sylvestris* and *Geum rivale*. Lightly grazed stands of the *Agrostis-Rhytidiadelphus* sub-community can resemble the *Anthoxanthum-Alchemilla* sub-community of *Deschampsia-Galium* grassland U13a, but *Deschampsia* grassland is less species-rich and has a sparser scatter of tall herbs. Heathy stands of the *Primula-Hypericum* sub-community have some similarities to the *Thymus-Carex* sub-community of *Calluna-Erica* heath H10d, but tall forbs such as *Angelica sylvestris*, *Geum rivale*, *Sedum rosea* and *Filipendula ulmaria* are more common. *Fraxinus-Sorbus-Mercurialis* woodland W9 and *Salix-Luzula* scrub W20 can have a similar flora to the *Luzula-Geum* community but with the addition of a canopy of trees or shrubs.

## Ecology

The *Luzula-Geum* community is the mesotrophic equivalent of *Luzula-Vaccinium* tall-herb vegetation and is likewise confined to places where grazing is light or absent. It is most widespread on cliff ledges and in ravines, and is only locally extensive on open slopes. Most stands are at altitudes above 300 m, but the *Primula-Hypericum* sub-community can occur at lower altitudes in the Hebrides. The soils are fertile brown loams with a pH between 4.8 and 5.4 (McVean and Ratcliffe 1962) and are usually flushed and moist. The community can occur on slopes of any aspect but most examples are on north-facing or east-facing cliff ledges and steep hillsides that are shaded and often dripping with water. There is always some base-enrichment, either directly from the rock or from irrigating water. Some stands have a considerable cover of snow in winter, although this does not usually lie very late in the spring (McVean and Ratcliffe 1962).

*Luzula-Geum* vegetation is widely distributed in upland areas from south Wales to Shetland. Some of the sub-communities have distinctive distributions. For example, the *Alchemilla-Bryum* sub-community is the most common form south of the Highlands, and the *Primula-Hypericum* sub-community is the most common type in the far west Highlands and the Hebrides. Stands are rarely large, but they can form a high proportion of the total cliff vegetation on the Dalradian schist and limestone in the Breadalbanes, as seen on Ben Lui, Beinn Heasgarnich and Ben Lawers.

Similar forms of tall-herb vegetation occur in Norway, Sweden, Ireland, the Faroe Islands, Greenland and Iceland.

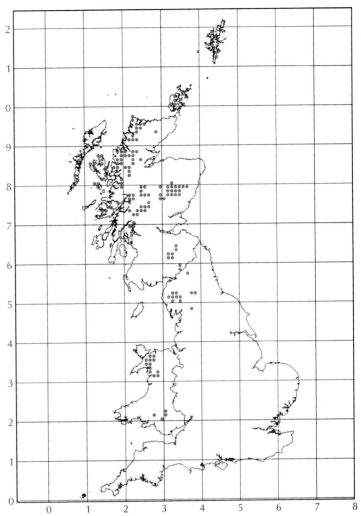

## Conservation interest

Many of our rarest plants survive in this type of vegetation, for example *Carex atrata, C. vaginata, Poa alpina, P. glauca, Polystichum lonchitis, Salix lanata, S. lapponum, S. reticulata, S. arbuscula, S. myrsinites, Potentilla crantzii, Woodsia alpina, Cystopteris montana, Oxytropis campestris* and *Bartsia alpina.* Many rare bryophytes grow here too, including *Bryoerythrophyllum caledonicum, Mnium thomsonii, Molendoa warburgii, Hymenostylium insigne, Paraleptodontium recurvifolium, Herbertus stramineus, Scapania ornithopodioides, Bazzania pearsonii* and *Mastigophora woodsii*. The community is also valuable because the vascular plants can flower and reproduce sexually, so that genetic diversity is maintained or increased.

## Management

Like the *Luzula-Vaccinium* community, the *Luzula-Geum* community persists only where grazing is absent or light. If stands are grazed too heavily, the characteristic tall herbs are soon lost, usually leading to the development of some form of calcareous grassland. As with the *Luzula-Vaccinium* community, goats are a threat to this type of vegetation as they can climb onto ledges that are inaccessible to sheep and deer.

The most species-rich and lush stands of *Luzula-Geum* vegetation on ledges and in ravines are evidence of what the vegetation of larger areas of the uplands on base-rich ground must have looked like before the depredations of sheep and deer. They are a reminder of how much has been lost but also show the value and the rewards of managing upland areas in such a way that the vegetation becomes more natural. Tall-herb vegetation similar to this must once have formed an understorey to sub-montane woodlands (such as the *Fraxinus-Sorbus-Mercurialis* community) and montane willow scrub (*Salix-Luzula* vegetation) on base-rich soils, as well as forming herb-rich meadows on high slopes. This state of affairs can be recreated only if grazing is eliminated or drastically reduced in the uplands. Excluding stock from selected areas would allow tall-herb vegetation to spread. This is already being done with some success on Ben Lawers and Caenlochan. On Ben Lui sheep have been taken off the north-facing slopes, and on Beinn an Lochain in the Cowal hills sheep numbers on the hill have decreased because the lower ground is afforested. In both places tall-herb vegetation is becoming re-established on the open hillsides. This is also expected to happen in Cwm Idwal in north Wales, where sheep grazing ceased in 1998. Potentially, such changes could occur on many upland sites where there are suitable base-rich soils at high altitude. This would result in larger populations of many uncommon plants and would also benefit invertebrates, birds and small mammals. In some places it would be possible to re-establish continuous zonations of vegetation ranging from woodland up to montane cliff ledges. There are already hillsides, such as the northern slopes of Ben Lui and adjacent hills (Averis and Averis 1999a), where herb-rich woodland, willow scrub and tall-herb vegetation form mosaics on ungrazed cliffs.

# U18 *Cryptogramma crispa-Athyrium distentifolium* snow-bed

## Synonyms
*Cryptogrammeto-Athyrietum chionophilum* (McVean and Ratcliffe 1962); Northern, snow-bed *Cryptogramma-Athyrium alpestre* community D2b (Birks and Ratcliffe 1980).

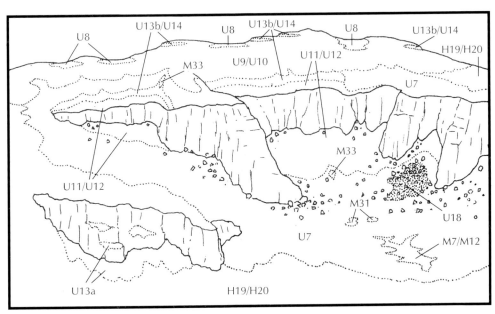

Typical habitat of U18 *Cryptogramma-Athyrium* snow-bed (stippled) among boulders on high slope with prolonged snow-lie.

Other vegetation types shown in this picture:

U7 *Nardus-Carex* grass-heath
U8 *Carex-Polytrichum* sedge-heath
U9 *Juncus-Racomitrium* rush-heath
U10 *Carex-Racomitrium* moss-heath
U11 *Polytrichum-Kiaeria* snow-bed
U12 *Salix-Racomitrium* snow-bed
U13 *Deschampsia-Galium* grassland
U14 *Alchemilla-Sibbaldia* dwarf-herb community
M7 *Carex-Sphagnum russowii* mire
M12 *Carex saxatilis* mire
M31 *Anthelia-Sphagnum* spring
M33 *Pohlia wahlenbergii* spring
H19 *Vaccinium-Cladonia* heath
H20 *Vaccinium-Racomitrium* heath

## Description
This vegetation type comprises dense, lush patches of ferns tucked between boulders and stones; the vivid green fronds of summer and the dead rusty-red collapsed remains in winter stand out equally strongly against the stony habitat. The two most common ferns are *Athyrium distentifolium* and *Cryptogramma crispa*, and there can also be a few plants of *Oreopteris limbosperma*, *Dryopteris dilatata* or the rare *D. oreades*. Bryophytes are very common, with species such as *Rhytidiadelphus loreus*, *Hylocomium splendens* and *Barbilophozia floerkei* growing in mixed mats. There are usually tufts of

385

*Deschampsia cespitosa*, together with other plants that can tolerate late-lying snow including *Gnaphalium supinum*, *Sibbaldia procumbens*, *Kiaeria starkei*, *Polytrichum sexangulare*, *Oligotrichum hercynicum* and *Lophozia sudetica*.

### Differentiation from other communities

This form of snow-bed vegetation is likely to be confused only with the less montane *Cryptogramma-Deschampsia* community U21. *Cryptogramma-Athyrium* snow-bed has montane species such as *Athyrium distentifolium*, *Dryopteris oreades*, *Alchemilla alpina* and *Carex bigelowii* (although *A. alpina* also grows in *Cryptogramma-Deschampsia* vegetation in the Lake District and on Skye), and more extensive and diverse bryophyte mats.

### Ecology

The *Cryptogramma-Athyrium* community occurs on stable scree, among boulders, on ledges and on steep rocky ground (McVean and Ratcliffe 1962). It is restricted to high altitudes, generally above 800 m, in places where snow lies moderately late and protects the frost-sensitive ferns from freezing. Most stands are on cold sheltered slopes facing between north and east. The community usually occurs over acid rocks, but there are many stands on moderately basic outcrops of Moine Schist. The soils are free-draining

This community is widespread but scarce in the Scottish Highlands, occurring from Ben More Assynt in the north to Ben Lui in the south. It is most common on the high Moine and Dalradian schist hills, notably Ben Nevis, Bidean nam Bian, Creag Meagaidh and Ben Alder, and on the hills that lie along the main watershed of western Scotland: the Fannich Hills, the Monar Forest and the Affric-Cannich hills. There are outliers in the Cairngorms and on Lochnagar.

There is similar vegetation in Scandinavia, although *Cryptogramma crispa* is largely confined there to the west of Norway (McVean and Ratcliffe 1962).

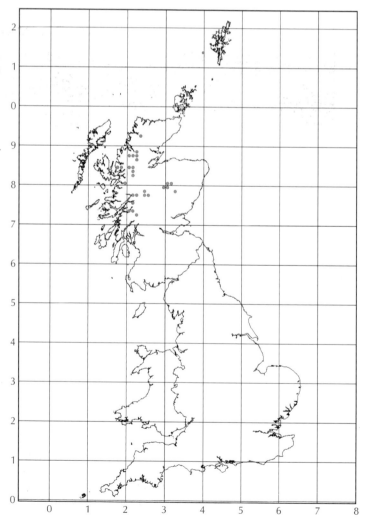

acid mixtures of humus, sand and silt with a pH around 4.1 (McVean and Ratcliffe 1962). Melting snow helps to keep the soils moist throughout the growing season.

## Conservation interest

Like other forms of snow-bed in Great Britain, the *Cryptogramma-Athyrium* community is a rare, near-natural vegetation type, and is of considerable value for nature conservation. The vegetation has a distinctly oceanic feel. The rare oceanic liverworts *Anastrophyllum donnianum*, *A. joergensenii*, *Bazzania pearsonii*, *Plagiochila carringtonii*, *Scapania nimbosa* and *S. ornithopodioides* occur in some stands, and records of these species in the eastern Highlands are mostly from this type of vegetation. The rare mosses *Brachythecium reflexum* and *B. glaciale* are particularly strongly associated with the *Cryptogramma-Athyrium* community (Rothero 1991). The rare *Athyrium flexile* replaces *A. distentifolium* in a few stands in the central Highlands (Page 1982).

## Management

Like all forms of snow-bed, this is a near-natural type of vegetation that is hardly affected by management, although sheep and deer may graze the fern fronds.

# U19 *Oreopteris limbosperma-Blechnum spicant* community

## Synonyms
*Thelypteris oreopteris* community (McVean and Ratcliffe 1962); *Luzula sylvatica-Dryopteris* communities D6 *p.p.* (Birks and Ratcliffe 1980); *Thelypteris limbosperma-Blechnum spicant* community (Rodwell 1992).

Typical habitat of U19 *Oreopteris-Blechnum* community: on well-drained slopes and stream-banks facing between north-west and east.

## Description
This vegetation type comprises dense, lush, yellow-green stands of the tufted fern *Oreopteris limbosperma*, growing in patches on steep sub-montane slopes or running in linear stands up shallow grassy gullies. The lemon scent of the *Oreopteris* is most noticeable when the fronds are brushed in passing, but it can be detected from a distance on still warm days. Under the fragrant canopy of fern fronds there is a sprinkling of *Blechnum spicant*, and in some places *Oreopteris* is accompanied by some *Dryopteris dilatata* or *D. filix-mas*. On the ground, a thick golden-green carpet of the pleurocarpous mosses *Hylocomium splendens*, *Pleurozium schreberi* and *Rhytidiadelphus loreus* is pricked through by small vascular plants such as *Oxalis acetosella*, *Galium saxatile*, *Festuca ovina*, *Agrostis capillaris* and *Anthoxanthum odoratum*.

## Differentiation from other communities
This community is very distinctive as it is the only type of British non-woodland vegetation that is dominated by *O. limbosperma*.

## Ecology
The *Oreopteris-Blechnum* community occurs on steep slopes with shallow, acid, well-drained peaty soil. It commonly forms dense stands in sheltered gullies and ravines but can also cover large areas on open slopes, especially those facing north or east. Steep

stream banks are another favoured habitat. The ferns are sensitive to frost and desiccation, so the community is most common in sheltered situations, and rarely extends above 600 m or the altitudinal limit of woodland. *Oreopteris-Blechnum* vegetation usually occurs in a matrix of sub-montane heaths and grasslands.

Similar vegetation can make up the ground-layer of birch, oak, Scots pine and juniper woodlands, and can form pure stands around their upper margins and in clearings, for example on Creag Fhiaclach in the Cairngorms and in the Morrone Birkwood near Braemar.

## Conservation interest

Few rare species grow in *Oreopteris-Blechnum* vegetation, although *Athyrium distentifolium* occurs in some stands at higher altitudes (Rodwell 1992). *Draba incana*, *Hymenophyllum wilsonii*, and the liverworts *Plagiochila spinulosa*, *Drepanolejeunea hamatifolia* and *Bazzania tricrenata* have been recorded in this community on Skye (Averis and Averis 1997b). Although it is not uncommon, the community makes a valuable contribution to the mosaic of upland habitats, and provides shade and shelter for woodland plants, small birds and invertebrates.

The community is widespread throughout the British uplands from north-west Sutherland to Wales and south-west England. It is very common in the Breadalbane region and in Argyll.

There is similar vegetation in western Norway, where it may form a fringe at the upper edges of birch woodland under an open canopy of *Juniperus communis* and *Salix lapponum*. It also occurs as dense stands on open north-facing slopes in the extreme west of Norway.

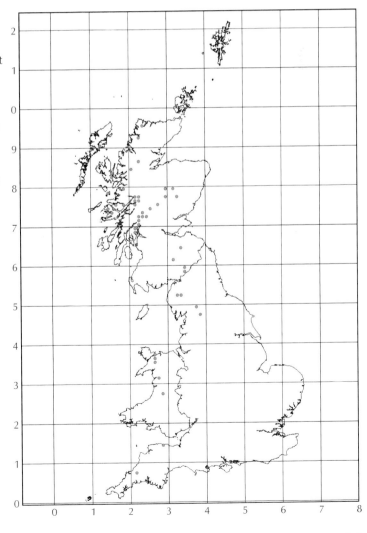

**Management**

The *Oreopteris-Blechnum* community is maintained by light grazing. Heavy grazing will tend to eliminate *Oreopteris*, and the vegetation will eventually be replaced by sub-montane grassland, such as *Festuca-Agrostis-Galium* grassland U4. Most stands have probably been derived from woodland or scrub, and regeneration of trees and shrubs would be likely to occur if the intensity of grazing was reduced. The most likely form of woodland would probably be oak-birch vegetation such as *Quercus-Betula-Oxalis* W11 or *Quercus-Betula-Dicranum* W17, or, more locally, pine or juniper communities.

# U20 *Pteridium aquilinum-Galium saxatile* community

## Synonyms
*Pteridium aquilinum* communities D7 (Birks and Ratcliffe 1980).

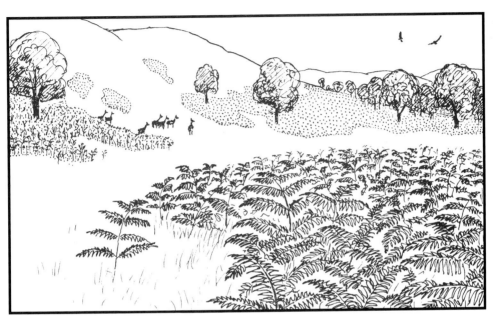

Typical habitat of the U20 *Pteridium-Galium* community (in foreground and stippled):
patches of bracken-dominated vegetation scattered among grassland, heath, woodland and
mires on lower hill slopes.

U20 is also shown in the pictures for U1, U3–U6, CG10–CG12, MG5, H4, H8–H10, H12
and H18.

## Description
The *Pteridium-Galium* community is vegetation dominated by *Pteridium aquilinum*, and
is one of the easiest upland plant communities to recognise. The patches of bracken,
which may be very extensive, are a rich dark green in summer, adding texture rather
than colour to the landscape. They turn bright red-gold in winter, when they stand out
against the sombre bleached grasses and dark heather on lower hill slopes.

There are three sub-communities. In the *Anthoxanthum odoratum* sub-community
U20a there is a grassy sward beneath the bracken fronds consisting of *Agrostis capillaris*,
*Festuca ovina*, *Anthoxanthum odoratum*, *Holcus lanatus*, *Galium saxatile*, *Potentilla
erecta*, *Rumex acetosa* and *Viola riviniana*, and the mosses *Rhytidiadelphus squarrosus*,
*Hypnum jutlandicum* and *Scleropodium purum*. There can be a few woodland herbs
such as *Hyacinthoides non-scripta* and *Oxalis acetosella*. The heathy *Vaccinium
myrtillus-Dicranum scoparium* sub-community U20b includes stands in which an
uneven layer of *Vaccinium myrtillus* and, in some places, *Calluna vulgaris* grows
beneath the bracken in a short mossy turf with much *Dicranum scoparium*,
*Rhytidiadelphus loreus* and *Pleurozium schreberi*. In the Species-poor sub-community
U20c the tall, dense fronds of bracken rise from a thick carpet of their own leaf litter, and
there is little room for anything else to grow.

In some places in the Highlands there are patches of bracken-dominated vegetation related
to the *Pteridium-Galium* community but with much *Molinia caerulea*. This vegetation is

usually closely associated with *Molinia-Potentilla* grassland M25 and *Trichophorum-Erica* wet heath M15.

## Differentiation from other communities
This community can be confused with *Pteridium-Rubus* scrub W25, in which bracken is also dominant. Compared with the *Pteridium-Galium* community, *Pteridium-Rubus* scrub usually has more *Rubus fruticosus* and *R. idaeus*, and a more herb-rich flora, including woodland plants such as *Hyacinthoides non-scripta*, *Teucrium scorodonia*, *Urtica dioica* and *Stachys sylvatica*. Conversely, *Galium saxatile* and *Potentilla erecta* are very common in *Pteridium-Galium* vegetation, especially in the *Anthoxanthum* sub-community, but are scarce in *Pteridium-Rubus* scrub.

## Ecology
The *Pteridium-Galium* community is typical of the zone where the farmed lowlands adjoin the unenclosed uplands. It is most common on lower hill slopes and on marginal ground, including abandoned fields, where it forms mosaics with heaths, grasslands and woodlands. The community covers fairly deep, well-drained but moist, base-poor and infertile soils. It is absent from wet ground and strongly flushed slopes. Bracken is intolerant of frost and its altitudinal range is therefore limited by exposure. Soils at higher altitudes also tend to be too shallow, rocky or peaty, although the community can develop on dry peat where bogs have been cut-over or drained. The upper altitudinal limit of the *Pteridium-Galium* community appears broadly to correspond with that of native woodland at around 600 m; it is most extensive below 450 m. Stands of *Pteridium-Galium* vegetation can cover huge areas of hillside, but it is also common to see small, discrete patches that are perhaps the beginning of new colonies.

## Conservation interest
Because of its poor flora and the difficulty of controlling its spread, *Pteridium-Galium* vegetation is often regarded with despair by both nature conservationists and farmers. However, stands of bracken contribute to the diversity of vegetation on lower hill slopes, and are not completely devoid of wildlife interest. Whinchat *Saxicola rubetra* is one of the few birds strongly associated with bracken (Allen 1995), but other birds also use the habitat for breeding (MacDonald *et al.* 1998; Ratcliffe and Thompson 1988). About 40 species of invertebrate feed on bracken, 11 of which are specialists (Lawton 1976). Although some notable invertebrates (e.g. High Brown Fritillary *Argynnis adippe*) are associated with *Pteridium-Galium* vegetation, grasslands, heaths and woodlands usually have a more varied fauna (Pakeman and Marrs 1992).

## Management
Up until the middle of the 20th century it was standard practice to cut and bale bracken to use as winter bedding for cattle. Indeed in some places, such as parts of Mull, patches of bracken were carefully maintained for this purpose. In recent years, with the decline of cattle farming in the uplands and reduced numbers of farm labourers, this practice is no longer economic and has largely ceased. This, coupled with poor burning and grazing practices, has led to a massive increase in the area covered by bracken within living memory. According to Mackey *et al.* (1998), there was a net increase of 79% in the area of Scotland covered by bracken between the 1940s and the 1980s: an annual increase of around 2%. Figures for England and Wales are believed to be similar.

Chemical or mechanical control of bracken is difficult, expensive and labour-intensive (see Rodwell (1992) and Paterson *et al.* (1997) for a review). Once *Pteridium-Galium* vegetation has become established, restoration of grassland or heathland is problematic.

This community is common throughout the British uplands, and is thinly scattered in the lowlands. It is one of the more extensive types of vegetation in many parts of England, Wales and the south and west of Scotland, including the Inner Hebrides. It thins out somewhat in the north, possibly because the climate is too harsh, and is rare on the very acid, wet and peaty islands of Lewis and Harris in the Outer Hebrides.

Essentially similar vegetation occurs in the uplands of Ireland. Although bracken is cosmopolitan it rarely forms the dense stands characteristic of the *Pteridium-Galium* community in the more continental climate of mainland Europe (Rodwell 1992), where it is a plant of woodland clearings and margins. It can, however, form large stands in the Pyrenees (Angus MacDonald, pers. comm.).

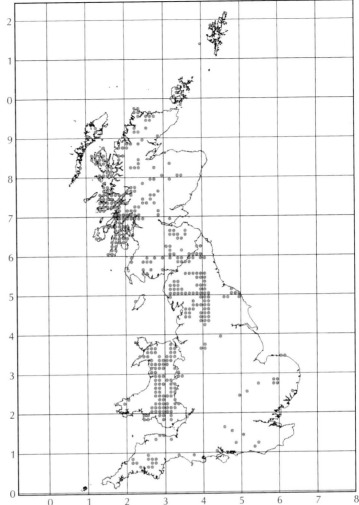

The Species-poor sub-community is especially intractable. If very dense bracken is eradicated by spraying or cutting, the leaf litter decays after a few years to leave a bare, fibrous soil, which is easily eroded if the slope is steep or if it is trampled by sheep. This bare soil can be colonised by weedy species such as *Urtica dioica*, *Digitalis purpurea*, *Deschampsia flexuosa* and the moss *Campylopus introflexus*.

Most stands of bracken are in places where the soils and climate appear to be suitable for the establishment of native trees. Particularly in the uplands of Wales and the Lake District, restoring native woodland to areas currently covered by bracken would simultaneously increase the diversity of upland vegetation and provide shelter for domestic livestock. Rodwell (1992) notes that bracken is much reduced where ground with the community has been planted with conifers; perhaps the spread of the fern could eventually be controlled by restoring the native woodlands whose removal may have helped to encourage its spread in the first place.

# U21 *Cryptogramma crispa-Deschampsia flexuosa* community

## Synonyms
Southern, scree *Cryptogramma* community D2a (Birks and Ratcliffe 1980).

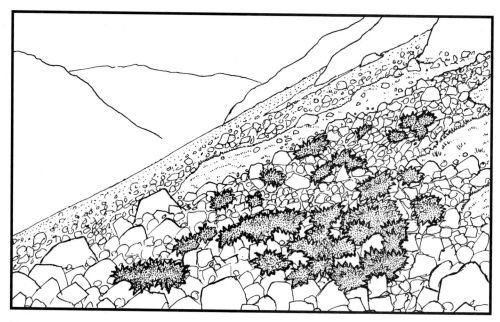

U21 *Cryptogramma-Deschampsia* community forming small patches of vegetation among screes on steep slopes.

## Description
This is typically rather open vegetation on scree or among boulders. In summer the crisp, ruffled, bright-green fronds of *Cryptogramma crispa* are conspicuous. The fern dies back in the winter, but the dead, red-brown tufts of old fronds are still distinctive. These tufts become bigger each year, eventually forming humus in which plants such as *Deschampsia flexuosa*, *Festuca ovina* and *Galium saxatile* become established.

## Differentiation from other communities
The *Cryptogramma-Deschampsia* community can be confused only with *Cryptogramma-Athyrium* snow-bed U18. The snow-bed has more montane plants such as *Athyrium distentifolium*, *Dryopteris oreades*, *Carex bigelowii*, *Alchemilla alpina* and *Saxifraga stellaris*, although *A. alpina* also grows in *Cryptogramma-Deschampsia* vegetation in the Lake District and on Skye. In addition, the bryophyte layer of *Cryptogramma-Athyrium* snow-bed is generally more species-rich than that in the *Cryptogramma-Deschampsia* community.

## Ecology
This is pioneer vegetation of open, rocky ground, usually on steep slopes at moderate altitudes. Like many ferns, *Cryptogramma crispa* cannot tolerate much frost or drought, so it grows in places where there is an equable, humid climate. However, the *Cryptogramma-Deschampsia* community can occur at altitudes of 900 m or more where there are suitable scree slopes, for example on Yr Wyddfa (Snowdon) in north Wales. The

rocks are typically hard and acid, and the soil is rarely more than a discontinuous raw humus derived from the decaying fern fronds. *Cryptogramma* seems unable to colonise very fine and loose scree, but can be very common on more stable slopes. A fragmentary form of the community can develop on artificial habitats, such as dry-stone walls, bridges and old buildings (Rodwell 1992).

Grass-covered screes occur on which *Cryptogramma* grows as sparse tufts within a matrix of *Festuca-Agrostis-Galium* grassland U4. This suggests that *Cryptogramma-Deschampsia* vegetation may eventually be replaced by grassland as screes become stabilised and soil accumulates. However, in many places the community appears to be held as an arrested succession by sheep grazing, especially in north Wales and the Lake District.

## Conservation interest

The *Cryptogramma-Deschampsia* community is not known as a habitat for rare plants or animals, but is of interest as a vegetation type that can develop under more or less natural conditions. It is one of the few upland plant communities that have a British distribution centred on England and Wales, rather than Scotland.

The *Cryptogramma-Deschampsia* community is most common in north Wales and the Lake District, where it is especially associated with hard volcanic rocks. It also occurs on sedimentary rocks in the north Pennines, the Cheviot Hills and the Southern Uplands, and on volcanic and metamorphic rocks in the western Highlands and on Skye.

Related vegetation occurs in the mountains of central Europe and in parts of western Norway.

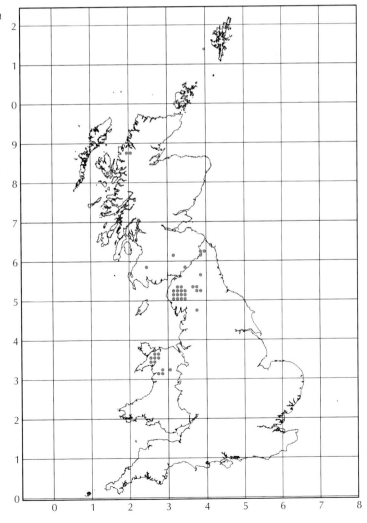

## Management

The habitat of the *Cryptogramma-Deschampsia* community usually protects it from burning. Many stands are grazed but the extent to which this damages the vegetation is unclear. On the Lake District fells, sheep have access to most screes, and the characteristic ferns persist in the presence of large numbers of grazing animals. Sheep have been excluded from some areas of *Cryptogramma-Deschampsia* vegetation in north Wales, and the plots are being monitored to see whether any changes take place. In the absence of grazing there might eventually be a succession to woodland. Indeed, grazing may help to keep the screes unstable and prevent succession to grassland, heath or woodland. The *Cryptogramma-Deschampsia* community would probably always have formed pioneer vegetation on unstable scree, but may not have persisted so long in some places in the absence of heavy grazing. In some southern parts of Great Britain, the distribution of the community may be restricted by high levels of atmospheric pollution (Rodwell 1992).

# S9 *Carex rostrata* swamp

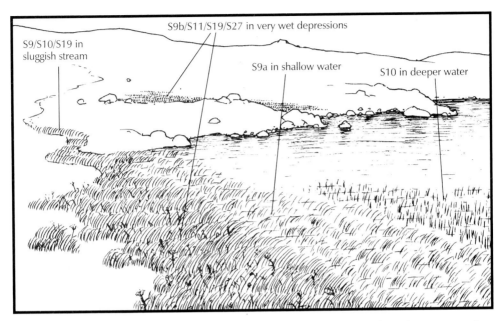

S9 *Carex rostrata* swamp in very wet depressions, in shallow water in sluggish stream, and around edge of loch. The *Carex rostrata* sub-community (S9a) typically occurs in deeper water than the *Menyanthes-Equisetum* sub-community (S9b).

Other vegetation types shown in this picture:

S10 *Equisetum fluviatile* swamp
S11 *Carex vesicaria* swamp
S19 *Eleocharis palustris* swamp
S27 *Carex-Potentilla* tall-herb fen

S9 is also shown among fen, mire and woodland in the picture for M26 *Molinia-Crepis* mire.

## Description

This vegetation type comprises tall, grey-green swards of *Carex rostrata* emerging from shallow water.

There are two sub-communities: one species-poor, the other more species-rich. The *Carex rostrata* sub-community S9a takes in stands where there is little or nothing more than *C. rostrata*. The *Menyanthes trifoliata-Equisetum fluviatile* sub-community S9b is usually not as tall and dense, and has a more varied array of associated species, the most common of which are *Equisetum fluviatile*, *Menyanthes trifoliata* and *Potentilla palustris*. *Potamogeton polygonifolius*, *Caltha palustris*, *Pedicularis palustris*, *Carex nigra* and *Ranunculus flammula* are also fairly common in this sub-community. On sunny days the reflection of the light from the water and the shining leaves of *P. polygonifolius* and *Caltha* can be almost dazzling.

## Differentiation from other communities

The *Carex rostrata* sub-community can hardly be mistaken for any other type of vegetation. The more herb-rich *Menyanthes-Equisetum* sub-community can be confused with several other vegetation types: the *Carex-Calliergon* sub-community of *Carex-*

*Calliergonella* mire M9b, the *Carex rostrata* sub-community of *Equisetum fluviatile* swamp S10b, the *Carex rostrata* sub-community of *Carex vesicaria* swamp S11c, and the *Carex-Equisetum* sub-community of *Carex-Potentilla* tall-herb fen S27a. These five vegetation types are very close to each other floristically, and can have similar amounts of *Carex rostrata*, *Menyanthes trifoliata* and *Potentilla palustris*. The *Carex-Calliergon* sub-community of *Carex-Calliergonella* mire is a sedge-mire or fen and typically has a carpet of bryophytes such as *Calliergonella cuspidata* and *Calliergon giganteum*, whereas the other vegetation types are swamps with tall emergent plants amid floating mats of vegetation in which bryophytes are generally scarce. Although *C. rostrata* can be common in both *Equisetum fluviatile* swamp and *Carex vesicaria* swamp, it is subordinate to either *E. fluviatile* or *C. vesicaria*, whereas it is the most common species in *Carex rostrata* swamp. Some stands of the *Carex-Equisetum* sub-community of *Carex-Potentilla* tall-herb fen have a richer and more varied flora than even the most species-rich forms of *Carex rostrata* swamp, with *Carex vesicaria*, *C. aquatilis*, *Juncus acutiflorus* and bryophytes such as *Brachythecium rutabulum*, *B. rivulare* and *Rhizomnium punctatum*, but where these species are absent it can be impossible to assign stands to one vegetation type or the other.

*Carex rostrata* swamp is widespread in upland regions and some adjacent lowland areas. It is absent from most of southern and eastern England.

There is similar vegetation in western Ireland (Horsfield *et al.* 1991).

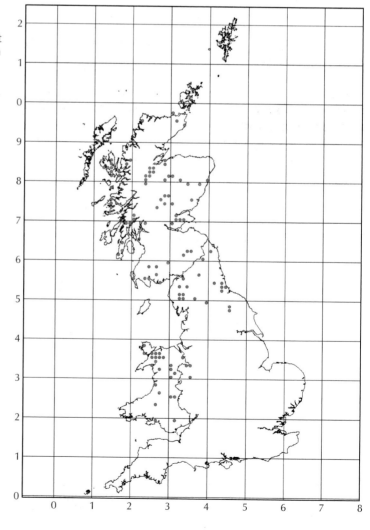

## Ecology

*Carex rostrata* swamp is a community of shallow acid water with a bottom of mud or peat rather than stones. The *Carex rostrata* sub-community can grow in water as deep as a metre or more, but the shorter herbs of the *Menyanthes-Equisetum* sub-community are restricted to water less than about 30 cm deep. Most stands are below 600 m. They are usually small, covering only a few square metres at the edges of sheltered lochans, in deeper bog pools, in seepage channels in blanket bogs, and in the slow-moving deeper reaches of streams.

## Conservation interest

This community forms a characteristic component of zonations of aquatic, swamp and mire vegetation in little-disturbed situations around oligotrophic upland lakes. Stands of *Carex rostrata* swamp can provide shelter and feeding grounds for ducks, such as Mallard *Anas platyrhynchos* and Teal *Anas crecca*, and also Moorhens *Gallinula chloropus* and Coots *Fulica atra*. On Skye, some swamps of this type are home to the rare *Eriocaulon aquaticum*. Stands of the *Menyanthes-Equisetum* sub-community are valuable as places where tall forbs and sedges are able to flower and set seed every year, and where genetic diversity can therefore be maintained.

## Management

These swamps are generally too wet to be grazed or burned, and are threatened only by drainage or by changes to the water chemistry caused by, for example, acidification or agricultural run-off.

# S10 *Equisetum fluviatile* swamp

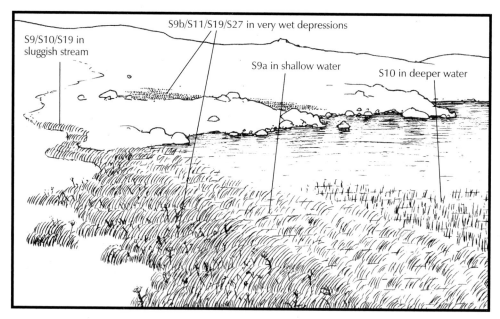

S10 *Equisetum fluviatile* swamp in water around edge of loch (forming outermost swamp zone) and in sluggish stream.

Other vegetation types shown in this picture:

S9 *Carex rostrata* swamp
S11 *Carex vesicaria* swamp
S19 *Eleocharis palustris* swamp
S27 *Carex-Potentilla* tall-herb fen

## Description

These are swamps in which the thin grey stems of *Equisetum fluviatile*, rustled by the wind and reflected in the water, stand over an open, species-poor assemblage of plants.

There are two sub-communities. The species-poor *Equisetum fluviatile* sub-community S10a includes pure stands of *E. fluviatile* in Scottish lochs, as well as examples from periodically inundated places in the lowlands with species such as *Persicaria hydropiper*, *P. amphibia*, *Rorippa islandica* and *Ranunculus flammula*. The richer *Carex rostrata* sub-community S10b is a more mixed form of *Equisetum* swamp with *Carex rostrata*, *Menyanthes trifoliata*, *Potentilla palustris*, *Galium palustre* and other plants.

## Differentiation from other communities

The *Equisetum* sub-community is distinctive, but the more species-rich *Carex rostrata* sub-community can resemble various other forms of swamp. It can be separated from the *Menyanthes-Equisetum* sub-community of *Carex rostrata* swamp S9b and the *Carex rostrata* sub-community of *Carex vesicaria* swamp S11c because there is more *Equisetum fluviatile* than either *Carex rostrata* or *C. vesicaria*. The flora of the *Carex rostrata* sub-community overlaps with that of the *Carex-Equisetum* sub-community of *Carex-Potentilla* tall-herb fen S27a, and the two vegetation types can share *E. fluviatile*, *C. rostrata*, *Menyanthes trifoliata* and *Potentilla palustris*. The richer forms of *Carex-Potentilla* tall-herb fen are distinctive, but it is not always possible to assign stands of vegetation to a particular NVC type.

There are *Equisetum fluviatile* swamps throughout the uplands from south-west England to northern Scotland. They also occur in lowland Great Britain.

Swamps dominated by *E. fluviatile* occur widely but locally in mainland Europe.

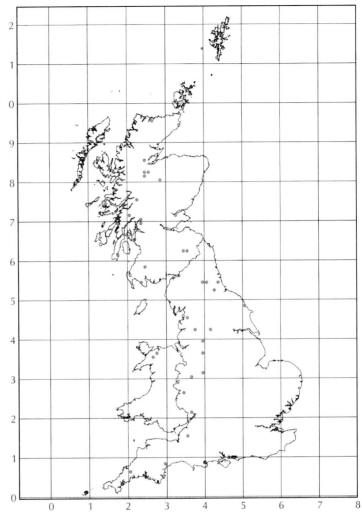

## Ecology
This is a community of open water in lochs, peaty pools, water tracks through bogs, and slow-moving streams; stands of the *Equisetum* sub-community can grow in water more than a metre deep, and can extend into deeper water than *Carex rostrata* swamp. The bottom is usually peat, mud or silt rather than stones, and the water chemistry varies from acid to basic. Most stands occur at altitudes below 600 m.

## Conservation interest
*Equisetum fluviatile* swamps provide shelter and food for ducks and waders. This is one of a small number of near-natural swamp communities in the British uplands.

## Management
The aquatic habitat of the community provides protection from grazing and burning. It is susceptible to drainage and to changes in water chemistry.

# S11 *Carex vesicaria* swamp

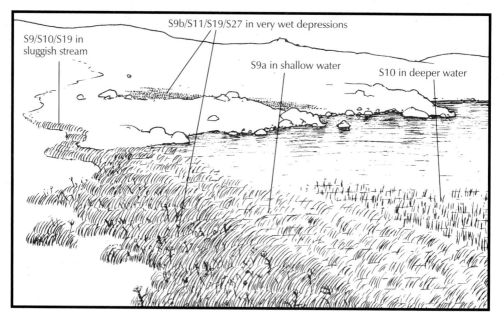

S11 *Carex vesicaria* swamp in mixed swamp mosaics in very wet depressions.

Other vegetation types shown in this picture

S9 *Carex rostrata* swamp
S10 *Equisetum fluviatile* swamp
S19 *Eleocharis palustris* swamp
S27 *Carex-Potentilla* tall-herb fen

## Description

This form of swamp is dominated by tall tussocks of *Carex vesicaria*. The stout, yellow-green shoots of the sedge grow in conspicuous swards which stand out prominently from the surrounding vegetation.

There are three sub-communities. The *Carex vesicaria* sub-community S11a includes pure stands of *C. vesicaria*, and species-poor swards with a scattering of other plants such as *Myosotis scorpioides* and *Persicaria amphibia*. The *Mentha aquatica* sub-community S11b is a more lowland form of swamp with *Galium palustre*, *Equisetum fluviatile*, *Mentha aquatica*, *Juncus effusus*, and in some places tall mesotrophic forbs such as *Filipendula ulmaria*, *Veronica scutellata*, *Lythrum salicaria* and *Angelica sylvestris*. The *Carex rostrata* sub-community S11c is a more upland type with *Carex rostrata*, *Menyanthes trifoliata* and *Potentilla palustris*.

## Differentiation from other communities

The *Carex vesicaria* sub-community is distinctive, as there are no other swamps in which *Carex vesicaria* forms almost pure stands. The other two sub-communities can resemble the *Carex-Equisetum* sub-community of *Carex-Potentilla* tall-herb fen S27a, which can be dominated by *C. vesicaria*. *Mentha aquatica*, *Juncus effusus*, *Myosotis scorpioides*, *Filipendula ulmaria*, *Phalaris arundinacea* and *Veronica scutellata* are more common in the *Mentha* sub-community of *Carex vesicaria* swamp. However, the *Carex rostrata* sub-community of *Carex vesicaria* swamp and the *Carex-Equisetum* sub-community of *Carex-*

The community is widespread but rare, occurring from south Wales northwards. It is most common in Scotland.

   *Carex vesicaria* swamps occur locally in parts of Scandinavia.

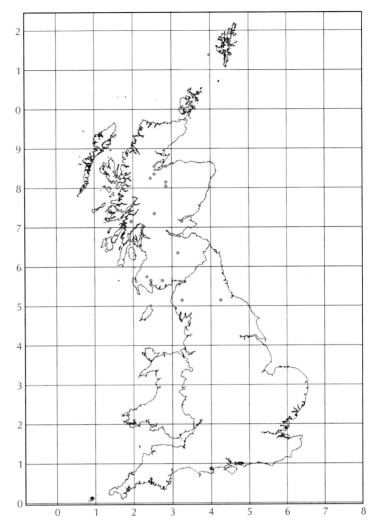

*Potentilla* tall-herb fen share so many species that some stands could be equally well assigned to either vegetation type.

## Ecology
Most *Carex vesicaria* swamps are in shallow standing water, but they can also occur on wet ground that is only periodically flooded.

## Conservation interest
The uncommon *Carex aquatilis* can occur in *Carex vesicaria* swamp. Some stands provide shelter for a range of upland birds, including ducks and waders. As is the case with other swamp communities in the uplands, it is an example of vegetation that occurs in more or less natural situations.

## Management
Aquatic stands of *Carex vesicaria* swamp are shielded from grazing and burning, but terrestrial examples are often grazed, especially by cattle (Rodwell 1995). The community is threatened by drainage and by agricultural run-off.

# S19 *Eleocharis palustris* swamp

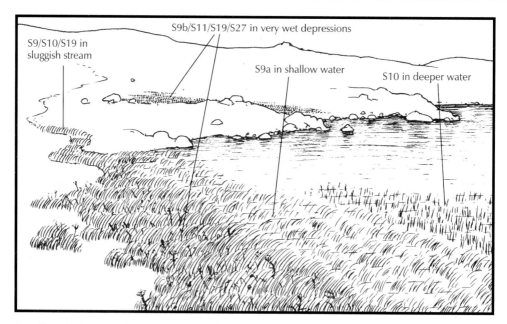

S19 *Eleocharis palustris* swamp in mixed swamp mosaics in very wet depressions and in shallow water in sluggish stream.

Other vegetation types shown in this picture:

S9 *Carex rostrata* swamp
S10 *Equisetum fluviatile* swamp
S11 *Carex vesicaria* swamp
S27 *Carex-Potentilla* tall-herb fen

## Description

*Eleocharis palustris* grows here in an even sward up to about 60 cm high. The dense, vivid green patches of *E. palustris* can be a conspicuous component of mixtures of mires and swamps in wet upland habitats.

There are three sub-communities, each defined by an almost mutually exclusive assemblage of species. The *Eleocharis palustris* sub-community S19a is the most species-poor form, but there can be a little *Alisma plantago-aquatica*, *Mentha aquatica* and *Myosotis laxa* ssp. *cespitosa*. The *Littorella uniflora* sub-community S19b takes in more upland stands of vegetation, containing species such as *Littorella uniflora*, *Equisetum fluviatile*, *Juncus bulbosus*, and, more locally, *Lobelia dortmanna* and *Potamogeton natans*. The *Agrostis stolonifera* sub-community S19c is a maritime form, and is characterised by *Agrostis stolonifera*, *Potentilla anserina*, *Glaux maritima*, *Triglochin maritimum* and *Juncus gerardii*.

## Differentiation from other communities

This is a distinct community. The dense swards of *Eleocharis palustris* are not likely to be confused with any other NVC type.

## Ecology

*Eleocharis palustris* swamp occurs in shallow water up to about 50 cm deep, in lochans, small pools, flushed channels and shallow streams. The *Agrostis* sub-community is known only from salt-marshes. Most stands of the community are below 600 m.

This community is widespread in Great Britain, occurring in both upland and lowland situations. In the uplands, the *Littorella* sub-community is the most usual form. The other two sub-communities are predominantly lowland, and the *Agrostis* sub-community is strictly coastal.

Similar types of swamp vegetation occur widely in western parts of mainland Europe.

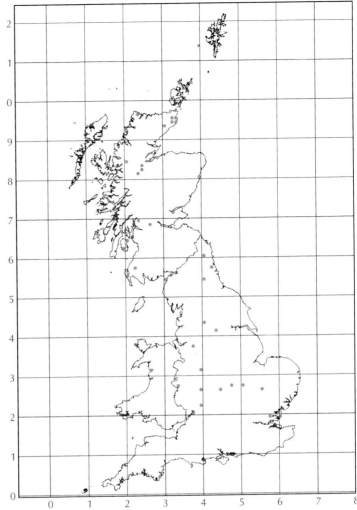

## Conservation interest

*Eleocharis palustris* swamp is a near-natural vegetation type, which occurs in close association with other swamp communities and aquatic vegetation in distinctive zonations.

## Management

This swamp community is protected by its wet habitat from burning and most grazing. However, it is susceptible to drainage and to changes in water chemistry.

# S27 *Carex rostrata-Potentilla palustris* tall-herb fen

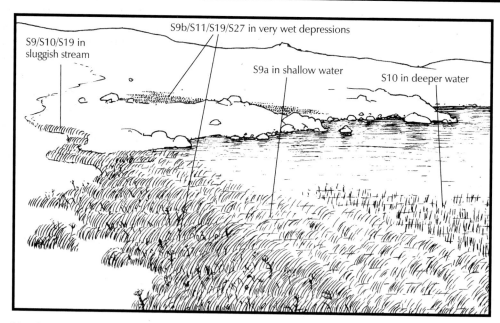

S27 *Carex-Potentilla* tall-herb fen in mosaics with various swamps in very wet low-lying depressions.

Other vegetation types shown in this picture:

S9 *Carex rostrata* swamp
S10 *Equisetum fluviatile* swamp
S11 *Carex vesicaria* swamp
S19 *Eleocharis palustris* swamp

S27 is also shown among fen, mire and woodland in the picture for M26 *Molinia-Crepis* mire.

## Description

These are herb-rich fens in which the thin, grey-green sward of *Carex rostrata* is mixed with tall or mat-forming mesotrophic herbs in lush stands which can contrast markedly with the surrounding heaths, bogs, rush-mires or sedge-mires. *Galium palustre*, *Menyanthes trifoliata* and *Potentilla palustris* are the most common associated herbs.

The two sub-communities distinguish vegetation dominated by *C. rostrata* from tall-herb fens in which dominance is shared by other species. The *Carex rostrata-Equisetum fluviatile* sub-community S27a is usually species-poor. Here, *C. rostrata* is accompanied by *E. fluviatile*, *Caltha palustris*, *Potamogeton polygonifolius*, *Juncus effusus* and *Ranunculus flammula*; *Carex vesicaria*, *C. aquatilis* and *Mentha aquatica* occur in some stands. The *Lysimachia vulgaris* sub-community S27b is more species-rich, including, in addition to the characteristic species of the *Carex-Equisetum* sub-community, *Angelica sylvestris*, *Phragmites australis*, *Carex nigra*, *Lychnis flos-cuculi*, *Iris pseudacorus* and *Filipendula ulmaria*. *Phragmites*, *C. nigra*, *J. effusus* or *Eriophorum angustifolium* usually dominate the sward in place of *C. rostrata*. Stands intermediate between the two sub-communities can be found in the uplands.

## Differentiation from other communities

The herb-rich swards of the *Lysimachia* sub-community are quite distinctive, especially where *Carex rostrata* is not solely dominant. The *Carex-Equisetum* sub-community and stands intermediate between the two sub-communities include vegetation characterised by *C. rostrata*, *Menyanthes trifoliata* and *Potentilla palustris* growing with a wide range of other herbs, and with a few bryophytes such as *Calliergonella cuspidata*, *Calliergon giganteum* and *Brachythecium rutabulum*. As such, the vegetation encompasses the floristic range of several other upland mires, fens and swamps: the *Menyanthes-Equisetum* sub-community of *Carex rostrata* swamp S9b, the *Carex rostrata* sub-community of *Equisetum fluviatile* swamp S10b, the *Carex rostrata* sub-community of *Carex vesicaria* swamp S11c, and the *Carex-Calliergon* sub-community of *Carex-Calliergonella* mire M9b. These vegetation types have so much in common that some stands could equally well be assigned to more than one type.

## Ecology

In the uplands, *Carex-Potentilla* tall-herb fen occurs around the margins of lochs and in pools and seepage lines within stands of blanket bog, rush-mire and sedge-mire. Most stands are at altitudes below about 300 m.

*Carex-Potentilla* tall-herb fen is widespread in the uplands and lowlands of Wales, northern England and Scotland. It is also common in East Anglia.

Similar vegetation occurs widely in mainland Europe, especially in parts of Norway.

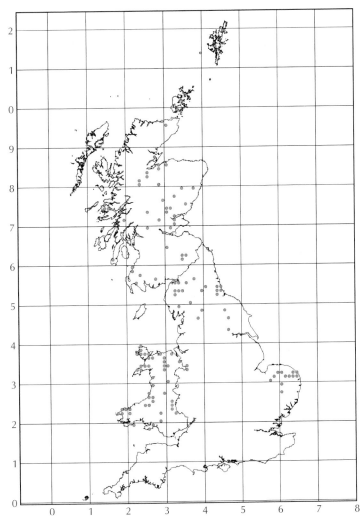

**Conservation interest**

Around upland lakes, *Carex-Potentilla* tall-herb fen commonly occurs with other forms of wetland vegetation in mosaics that are scarcely affected by management. The community is fairly species-rich, and there are occasional records for the uncommon plants *Carex limosa*, *C. aquatilis*, *Dactylorhiza incarnata* and *D. purpurella*. Some stands provide feeding habitat for ducks and waders. Most examples are so wet that they are hardly grazed; consequently, the herbs can flower and set seed, so maintaining genetic diversity.

**Management**

This community is damaged by drainage and probably also by changes to water chemistry resulting from agricultural run-off. The wetness of the substrate prevents burning and usually affords protection against grazing. Some stands are accessible to livestock and this may be help to stop the vegetation developing into wet scrub or woodland.

# OV37 *Festuca ovina-Minuartia verna* community

Typical habitats of OV37 *Festuca-Minuartia* community (stippled):

1   On spoil of old lead mines
2   On thin soil derived from veins of metalliferous rock within Carboniferous limestone

## Description

This is open, grassy vegetation of metalliferous soils. It usually looks distinctively sparse from a distance, standing out from the denser surrounding swards. The short, patchy turf of *Festuca ovina* and *Agrostis capillaris* is intermixed with low mats of *Thymus polytrichus*, and is dotted with the small, white, starry flowers of the metallophyte *Minuartia verna* which enliven the drab greyish vegetation in early summer. *Thlaspi caerulescens*, another characteristic plant of metalliferous soils, can also be locally common. There are usually a few other small herbs such as *Campanula rotundifolia*, *Rumex acetosa*, *R. acetosella* and *Linum catharticum*.

There are three sub-communities. The Typical sub-community OV37a has no special distinguishing species except that *Festuca rubra* is common in some stands. The *Achillea millefolium-Euphrasia officinalis* agg. sub-community OV37b has a thicker sward including *Achillea millefolium*, *Euphrasia officinalis* agg., *Anthoxanthum odoratum*, *Plantago lanceolata*, *Trifolium repens*, *Lotus corniculatus* and *Rhytidiadelphus squarrosus*. The *Cladonia* species sub-community OV37c resembles the Typical sub-community but has more lichens; the most common species are *Cladonia rangiformis*, *C. chlorophaea*, *C. pyxidata*, *C. portentosa* and *Coelocaulon aculeatum*. There can be a little *Calluna vulgaris* here too.

On gravelly serpentine debris in Shetland and the eastern Highlands there are related assemblages of very open vegetation with scattered *M. verna* or variable mixtures of *Arenaria norvegica*, *Cerastium alpinum*, *C. nigrescens*, *C. fontanum* ssp. *scotica*, *Lychnis alpina*, *Cochlearia micacea*, *Silene uniflora*, *Anthyllis vulneraria* and *Arabis petraea*. These are not included in the NVC (see p. 422).

## Differentiation from other communities

The most distinctive plants of the *Festuca-Minuartia* community are *Minuartia verna* and *Thlaspi caerulescens*, both of which can tolerate high concentrations of heavy metals. The presence of these species, and the scarcity of *Potentilla erecta*, is generally sufficient to distinguish the community from swards of *Festuca-Agrostis-Thymus* grassland CG10. Other metal-tolerant species, such as *Armeria maritima*, *Cochlearia pyrenaica*, *Botrychium lunaria* and *Viola lutea*, are also characteristic of the *Festuca-Minuartia* community.

## Ecology

*Festuca-Minuartia* vegetation occurs on fragmentary mineral soils and gravelly debris with a high concentration of heavy metals. It is the characteristic community of waste from lead mines and outcrops of metalliferous rock. It also occurs on metal-enriched river shingles and alluvium in the northern and central Pennines (Richard Jefferson, pers. comm.). Lead and more particularly zinc occur here in amounts that are toxic to most plants, which is why the vegetation is sparse and open.

The *Festuca-Minuartia* community has been recorded on the spoil of lead mines and on veins of metalliferous rock within the Carboniferous limestone in the Pennines, Derbyshire, north-east Wales and the Mendip Hills. It also occurs on the waste of copper mines in Snowdonia, and has been recorded on the Lendalfoot Hills in Ayrshire. Related vegetation occurs on serpentine in Scotland (see p. 422).

Similar forms of vegetation are known from Ireland and mainland Europe.

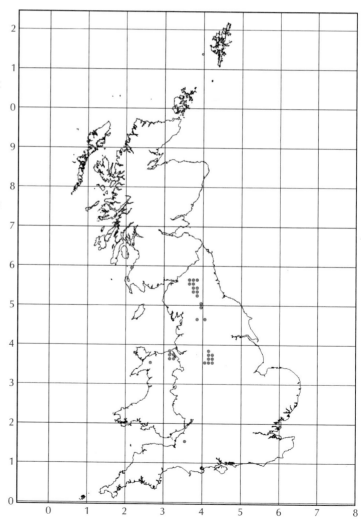

## Conservation interest

Metalliferous rocks are rare in Great Britain, and so *Festuca-Minuartia* vegetation and the related serpentine communities are also rare; near-natural examples of the community are especially scarce. Several uncommon vascular species occur in stands of this community and related Scottish vegetation: *Minuartia verna*, *M. sedoides*, *Thlaspi caerulescens*, *Arenaria norvegica*, *Cerastium alpinum*, *C. nigrescens*, *C. fontanum* ssp. *scotica*, *Lychnis alpina*, *Cochlearia micacea* and *Arabis petraea*. Many of the richest examples of *Festuca-Minuartia* vegetation are now protected within candidate SACs.

## Management

Most stands of the *Festuca-Minuartia* community and related vegetation are open to grazing, but the structure and flora of the vegetation owes more to the peculiar and toxic soils than to management (Rodwell 2000). The main threat to most examples of the community is reclamation of mine spoil and conversion to agricultural grassland or other land uses.

# OV38 *Gymnocarpium robertianum-Arrhenatherum elatius* community

OV38 *Gymnocarpium-Arrhenatherum* community (stippled): among well-lit Carboniferous limestone scree and outcrops, in mosaics with woodland, scrub and grassland.

## Description

The *Gymnocarpium-Arrhenatherum* community is a distinctive assemblage of plants which occurs on limestone pavement and scree in the south of Great Britain. The triangular, grey-green fronds of *Gymnocarpium robertianum* are interleaved with the grasses *Arrhenatherum elatius*, *Festuca rubra*, *F. ovina*, *Brachypodium sylvaticum* and, in northern England, *Sesleria caerulea*. Also common here are lush shoots of *Mercurialis perennis*, *Teucrium scorodonia* and *Mycelis muralis*, and fragrant, reddish-green, lacy leaves of *Geranium robertianum*. The mosses *Homalothecium sericeum* and *Ctenidium molluscum* cling to the rock and *Dicranum scoparium* grows in thick tufts among the vascular plants.

## Differentiation from other communities

Stands of the *Gymnocarpium-Arrhenatherum* community containing much *Sesleria* might be confused with *Sesleria-Galium* grassland CG9, which can occur as fragments over rocky ground. However, *Gymnocarpium robertianum*, *Geranium robertianum*, *Arrhenatherum elatius*, *Mercurialis perennis* and *Mycelis muralis* are all rare in *Sesleria-Galium* grassland. *Arrhenatherum-Filipendula* grassland MG2 shares *Arrhenatherum* and *Mercurialis perennis* with the *Gymnocarpium-Arrhenatherum* community, but also has species of damp ground such as *Filipendula ulmaria*, *Geum rivale* and *Valeriana officinalis*, and lacks *Gymnocarpium*. The *Asplenium-Cystopteris* crevice community OV40 can resemble stands of *Gymnocarpium-Arrhenatherum* vegetation containing the ferns *Asplenium viride* and *Cystopteris fragilis*, but can be distinguished by the scarcity of *Gymnocarpium* and tall grasses.

## Ecology

*Gymnocarpium-Arrhenatherum* vegetation occurs in the crevices or grikes of limestone pavement and on limestone scree. The soils are coarse-grained, fragmentary accumulations of humus and limestone gravel in the cracks between the stones. Neither *Gymnocarpium* nor *Arrhenatherum* can tolerate shade, and so this is a community of sunny, open ground. It rarely occurs more than a few hundred metres above sea-level.

Within the upland regions of Great Britain, *Gymnocarpium-Arrhenatherum* vegetation occurs on limestone pavement and limestone scree in the Yorkshire Dales and in Wales. It is also in the lowlands around Morecambe Bay, in Derbyshire and on the Mendips in south-west England. The geographic range has not been thoroughly studied, and no distribution map is available.

There is similar vegetation in mainland Europe.

## Conservation interest

The *Gymnocarpium-Arrhenatherum* community is a scarce vegetation type. It is the main habitat for the scarce calcicole fern *Gymnocarpium robertianum*. A number of other species that are generally uncommon in the British uplands can also grow here, including *Sesleria caerulea*, *Aquilegia vulgaris*, *Rubus saxatilis*, *Origanum vulgare*, *Polygonatum odoratum*, *Arabis hirsuta* and *Galium sterneri*.

## Management

The characteristic species of the *Gymnocarpium-Arrhenatherum* community are eliminated by heavy grazing. Lack of grazing will also lead to the disappearance of the community, as succession to woodland is inevitable. The vegetation would almost certainly be colonised by ash and hazel, eventually becoming a herb-rich woodland of the *Fraxinus-Sorbus-Mercurialis* type W9 in the uplands, or the *Fraxinus-Acer-Mercurialis* type W8 at lower elevations. Light grazing may postpone or prevent succession to woodland. Another threat to the community is quarrying of limestone pavement for the horticultural trade.

# OV39 *Asplenium trichomanes-Asplenium ruta-muraria* community

## Synonyms

*Asplenium trichomanes-Fissidens cristatus* Association *p.p.* (Birks 1973).

Typical habitats of OV39 *Asplenium trichomanes-Asplenium ruta-muraria* community:

1   On mortar on old stone wall
2   On mortar on old stone building
3   On steep artificial exposures of basic rock
4   On steep natural exposures of basic rock

## Description

This community comprises colonies of small ferns on base-rich cliffs and walls. The most common species are *Asplenium trichomanes*, *A. ruta-muraria*, and the bryophytes *Homalothecium sericeum* and *Porella platyphylla*.

There are two sub-communities. In the *Trichostomum crispulum-Syntrichia intermedia* sub-community OV39a, *H. sericeum* and *P. platyphylla* are typically accompanied by a little *Neckera crispa*, *Targionia hypophylla* and a few crustose lichens. The more species-rich *Sedum acre-Arenaria serpyllifolia* sub-community OV39b has an array of small vascular plants including *Festuca ovina*, *Thymus polytrichus*, *Koeleria macrantha*, *Helianthemum nummularium*, *Sedum acre* and *Arenaria serpyllifolia*.

## Differentiation from other communities

The *Asplenium trichomanes-Asplenium ruta-muraria* community can be confused only with the *Asplenium-Cystopteris* community OV40, a related form of rock crevice vegetation in which both *Asplenium trichomanes* and *A. ruta-muraria* also occur. They are easy to separate because *A. viride* and *Cystopteris fragilis* occur only in the *Asplenium-Cystopteris* community.

## Ecology

This is a community of crevices in basic rocks such as limestone and basalt, and of mortared walls at low to moderate altitudes. On walls the plants grow in mortar, but in more natural rocky habitats the soils are patchy, with thin accumulations of humus and sand forming a disintegrating substrate. The *Asplenium trichomanes-Asplenium ruta-muraria* community is widespread especially in the west of Great Britain from south-west England to northern Scotland thinning out eastwards. The geographic range has not been thoroughly studied, and no distribution map is available.

There is similar vegetation elsewhere in Europe.

## Conservation interest

In south-west England and Wales this type of vegetation is the main habitat of the fern *Ceterach officinarum*, a plant that persists in the community as far north as the south and west of Scotland. Upland stands usually lack rare plants, but contribute to the diversity of vegetation structure and flora in the upland margins.

## Management

Most stands of the *Asplenium trichomanes-Asplenium ruta-muraria* community are scarcely affected by grazing. In its natural habitat the community is not threatened except by rock-falls and weathering of the substrate; stands probably disappear and reappear in a cyclical pattern. On mortared walls the community can be a casualty of re-pointing. The ferns are susceptible to herbicides and especially to Asulox which is used to kill bracken. The mortar in walls can be acidified by air pollution; this may result in the community dying out or failing to colonise new sites.

# OV40 *Asplenium viride-Cystopteris fragilis* community

## Synonyms

*Asplenium trichomanes-Fissidens cristatus* Association, Limestone facies *p.p.* and Montane facies (Birks 1973); *Asplenium-Fissidens cristatus* crevice community D4 (Birks and Ratcliffe 1980).

Typical habitats of OV40 *Asplenium viride-Cystopteris fragilis* community:

1   Shaded crevices and ledges among steep outcrops of basic rock on hill slopes
2   Shaded crevices and ledges among steep outcrops of basic rock in ravines
3   Shaded crevices in limestone pavement

## Description

This is a community of small ferns, herbs and mosses in rocky habitats. The main species are *Asplenium viride* with its bright-green fronds and the delicate, pale-grey-green *Cystopteris fragilis*. *Asplenium ruta-muraria* and *A. trichomanes* grow here too, together with a few small herbs such as *Festuca ovina*, *Campanula rotundifolia*, *Thymus polytrichus* and *Hieracium* species. Bryophytes are common, including *Ctenidium molluscum*, *Tortella tortuosa*, *Fissidens dubius*, *Neckera crispa* and *Homalothecium sericeum*. Where the community occurs on limestone pavement there can be much *Phyllitis scolopendrium* and *Polystichum aculeatum*.

## Differentiation from other communities

*Asplenium-Cystopteris* vegetation may be confused with the other crevice community in which ferns dominate: the *Asplenium trichomanes-Asplenium ruta-muraria* community OV39. However, it can generally be distinguished by the presence of *Cystopteris fragilis* and *Asplenium viride*.

## Ecology

The *Asplenium-Cystopteris* community occurs in crevices in cliffs of base-rich rock, in the shallower grikes in limestone pavement, and more rarely in base-rich screes. Its

416

habitat is similar to that of the *Asplenium trichomanes-Asplenium ruta-muraria* community, but it is characteristic of places where the microclimate is damper, more sheltered
and more shaded. Although it occurs just above sea-level on Skye it ascends well into
the montane zone above 600 m on mica-schist ledges on the hills of the central Highlands. The soils are fragmentary accumulations of gravel and humus: the ferns root deep
into the rock crevices and gain access to water far below the surface.

## Conservation interest

Although the stands are usually small, this near-natural community is an important habitat for a range of uncommon upland plants. The most notable include the ferns *Woodsia
alpina* and *Polystichum lonchitis*, and the mosses *Orthothecium rufescens*, *Encalypta
alpina*, *Mnium spinosum*, *Plagiobryum demissum*, *Ptychodium plicatum* and *Blindia
caespiticia*. The rare orchid *Epipactis atrorubens* grows in this community in limestone
pavements on Skye and in the north-west Highlands.

## Management

Most stands of *Asplenium-Cystopteris* vegetation are out of the reach of grazing animals,
although they are accessible to slugs and snails, which can strip plants bare. Limestone

*Asplenium-Cystopteris* vegetation is locally common on base-
rich rocks in Scotland, the Lake
District, the Pennines and Wales.
There are particularly good
examples on the Craven Pennine
hills of Ingleborough, Pen-y-
Ghent and Whernside, and on
many of the Breadalbane cliffs.
It also occurs in the lowlands in
the Peak District of Derbyshire
(not shown on the distribution
map).
　Similar rock crevice vegetation
is known from Scandinavia and
mainland Europe.

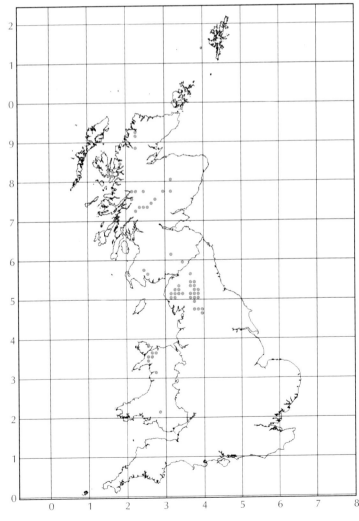

417

pavements have huge populations of snails (Gilbert 2000) and their lichen floras are much poorer than one would expect because of snail grazing. In the south of Skye a few palatable vascular plants seem to have been lost from more accessible pavements as a result of grazing by sheep, deer and invertebrates. Some examples of the community are threatened by the removal of limestone pavement for the horticultural trade; this is a particular problem in the north of England where many pavements are easily accessible. Other stands may be damaged by climbers who remove vegetation from cracks and crevices in order to clear routes up cliffs.

# Brief accounts of other upland habitats and vegetation types

Some types of upland vegetation and habitat are not included in the accounts that comprise the main body of this *Guide*, but are described briefly below. Most of them are not in the NVC. Many have an interesting flora including some notable rarities. Those which may be confused with communities described in the NVC are also mentioned in the key on pp. 57–93. An overview of possible gaps in coverage in the NVC is provided by Rodwell *et al.* (2000).

## Conifer plantation

This is perhaps the most obvious omission from the NVC, considering how much upland ground is now under conifers; Sitka spruce is the most common tree species over large areas of the uplands. Conifer plantations vary in their flora and in their value for nature conservation. Open plantations of native Scots pine can have a ground flora that is indistinguishable from that of native pine woods belonging to the *Pinus-Hylocomium* community W18. Their origin is given away only by the fact that the trees occur in even-aged blocks and are planted in straight lines. Larch plantations vary from dense to quite well-lit with a ground flora similar to that of some deciduous woodlands. At the other extreme are the dense commercial plantations of species such as Sitka spruce *Picea sitchensis*, Norway spruce *Picea abies* and lodgepole pine *Pinus contorta*, with their dark, seemingly lifeless, interiors. However, even these are not entirely without interest. Uncommon lichens such as *Bacidia saxenii* are occasionally recorded on cut stumps (Gilbert 2000), and some uncommon oceanic liverworts have been recorded on conifer bark in a few plantations in the western Highlands.

Open conifer plantations can be very good habitats for birds, especially Black Grouse *Tetrao tetrix*, Capercaillie *T. urogallus*, Scottish Crossbill *Loxia scotica*, Goshawk *Accipiter gentilis*, Redpoll *Carduelis flammea*, Siskin *C. spinus* and Redstart *Phoenicurus phoenicurus*.

## Earthy and gravelly gullies at high altitudes

On the highest hills in the Scottish Highlands there are precipices seamed with almost bare earthy or gravelly gullies. These gullies are often choked with snow in winter. The earth or gravel is unstable, always shifting, and prone to landslips and rock-falls. The vegetation is sparse and may be almost imperceptible. The most noticeable species include bryophytes such as *Oligotrichum hercynicum*, *Pohlia ludwigii*, *Philonotis seriata* and *Conostomum tetragonum*, and there can be a sprinkling of small montane vascular plants including *Gnaphalium supinum* and *Saxifraga stellaris*. A number of rarities grow in places such as this: for example *Gnaphalium norvegicum*, *Veronica alpina*, *Poa flexuosa* and *Carex lachenalii*, and the bryophytes *Bryum dixonii*, *Weissia controversa* var. *wimmeriana*, *Ptychodium plicatum* and *Lophozia opacifolia*. Similar habitats occur on lower Scottish hills and in the Lake District and north Wales. They have a less exacting assemblage of species, although relict montane plants such as *Dryas octopetala*, *Saxifraga nivalis* and *Saussurea alpina* cling on in some unstable, inaccessible places.

## Screes

Scree is produced when rocks are broken into fragments by physical and chemical weathering. It can occur in vast skirts of loose stone on the steep flanks of hills, or as tiny patches half-hidden by vegetation. It is generally a well-drained, coarse substrate, colonised by bryophytes and lichens but with few vascular plants.

In many of the more stable block screes, especially in the west, the most common plant is the moss *Racomitrium lanuginosum*, which smothers the rocks and stones with a thick, silver-grey blanket. In such situations, *R. lanuginosum* can grow in continuous carpets that superficially resemble *Carex-Racomitrium* moss-heath U10, but the scree vegetation usually differs in having deeper moss carpets and lacking *Carex bigelowii*, *Diphasiastrum alpinum*, *Salix herbacea*, and other montane species. The habitat is also very different from that of *Carex-Racomitrium* moss-heath. Eventually, *R. lanuginosum* can help to stabilise some screes so that soil forms and vascular plants are able to colonise. Other typical species in this form of scree include *R. ericoides* and, on rock surfaces, *Andreaea* species. The rocks and stones can be coated with a layer of small crustose lichens, but in general these are common pioneer species; this is not a habitat known for rare lichens.

Of the few vascular plants that can maintain a foothold in these unstable wastes of stone, perhaps the most common is *Vaccinium myrtillus*, the rhizomes of which are able to penetrate between the stones. *Deschampsia flexuosa*, *Festuca ovina*, *F. vivipara* and *Huperzia selago* are common in many stands. *Silene uniflora*, with its blue-grey leaves and stems, can grow here, especially in the west, and a notable rarity of screes at high altitudes in the Highlands is the grass *Poa flexuosa*.

## Cliff faces and small cliff ledges

Much of the vegetation of cliffs cannot be accommodated within existing NVC communities because it is too open and fragmentary, comprising scattered plants growing in rock crevices, and cushions or tufts of smaller plants clinging to tiny ledges or growing on the face of the rock itself. The flora of cliff habitats varies according to altitude and the acidity of the rock. The most common species are usually those that occur in the vegetation of adjacent, less rocky ground. So on acid cliffs typical plants include *Calluna vulgaris*, *Vaccinium myrtillus*, *Erica cinerea*, *Teucrium scorodonia*, *Festuca* species, *Carex* species, various ferns, bryophytes such as *Racomitrium* species, *Andreaea* species and *Diplophyllum albicans*, and a colourful array of crustose lichens. On more basic rocks there may be *Thymus polytrichus*, *Festuca ovina*, *F. rubra*, *Carex* species, the bryophytes *Ctenidium molluscum*, *Tortella tortuosa* and *Hypnum lacunosum*, and many lichens. Some of the vegetation in these habitats belongs to various recognised bryophyte and lichen assemblages (see p. 426–429).

There is little competition in this often unstable habitat, and in some places rare species grow on tiny, narrow ledges and in cracks and crevices. Among these are *Arabis alpina* on Skye; the specialities of the Breadalbanes and the Glen Clova hills *Erigeron borealis*, *Gentiana nivalis*, *Minuartia rubella*, *Myosotis alpestris*, *Veronica fruticans*, *Oxytropis campestris*, *Carex norvegica*, *Sagina nivalis*, *Draba norvegica*, *Potentilla crantzii* and *Saxifraga nivalis*; the Snowdon Lily *Lloydia serotina* in Wales; the rare saxifrages *Saxifraga cernua*, *S. cespitosa* and *S. rivularis*; the ferns *Woodsia alpina*, *W. ilvensis* and *Cystopteris montana*; and the bryophytes *Blindia caespiticia*, *Bryoerythrophyllum caledonicum*, *Grimmia unicolor*, *Timmia austriaca*, *T. norvegica*, *Saelania glaucescens*, *Hygrohypnum styriacum*, *Plagiobryum demissum*, *Mnium ambiguum, M. spinosum, Stegonia latifolia, Hymenostylium insigne, Hypnum bambergeri, Barbilophozia quadriloba* and *Gymnomitrion corallioides*. The most notable plant assemblages in these habitats are on rock faces that are out of the reach of grazing animals and, especially in the past, of plant collectors, who have sadly depleted the montane floras in the more accessible and well-known sites in north Wales, the Lake District and the Breadalbanes.

Although cliffs and rock outcrops are most often thought of as an element of the higher montane landscape, they are also common at low altitudes, both in the open and in woodland. Here they can bear varied assemblages of vascular plants, bryophytes and lichens. In sheltered places in the west they are one of the most important habitats for the filmy ferns *Hymenophyllum wilsonii* and *H. tunbrigense*, and for several uncommon oceanic bryophytes such as the diminutive liverworts *Aphanolejeunea microscopica, Drepanolejeunea hamatifolia, Harpalejeunea molleri* and *Colura calyptrifolia*. The undersides of overhangs – facing the ground – can be home to a rich array of lichens. Few vascular plants are able to grow upside-down in the shade, but the rare Killarney fern *Trichomanes speciosum* is one of them.

## Metalliferous rocks including serpentine

Metalliferous rocks have long been known for their distinctive and peculiar flora. The *Festuca-Minuartia* community OV37 represents vegetation on lead-rich soils in England and Wales and more locally in Scotland. Other Scottish metalliferous habitats are rather different, although they are related to the *Festuca-Minuartia* community, as shown by the presence of *Minuartia verna* in some stands. Most examples are on serpentine rocks and soils. Serpentine rocks occur in the eastern Highlands and on Shetland, many of them forming bare, open, stony fell-fields with a sparse but distinctive flora. Soils derived from serpentine have more magnesium than calcium, and nickel and chromium in amounts that are toxic to most plants.

The species that grow scattered among the stones of serpentine debris in the eastern Highlands include *Cochlearia officinalis, Danthonia decumbens, Thymus polytrichus, Minuartia verna, Huperzia selago, Galium verum, Cerastium fontanum, C. alpinum, Viola riviniana, Potentilla erecta, Deschampsia flexuosa, Carex panicea, C. viridula* ssp. *oedocarpa, Sagina procumbens, Armeria maritima, Festuca ovina, Selaginella selaginoides, Silene uniflora, Arabis petraea, Asplenium viride* and the distinctive serpentine form of *A. adiantum-nigrum* (Averis 1991b). The moss *Grimmia sessitana* is a notable rarity here, and *Antitrichia curtipendula* is locally common in serpentine habitats. On Meikle Kilrannoch in the eastern Highlands, the rare *Lychnis alpina* grows on serpentine debris between 850 m and 870 m with *Cerastium fontanum* ssp. *scoticum, Cochlearia micacea* and *Minuartia sedoides*. *L. alpina* also grows on Hobcarton Crag in the Lake District, where the soils have an unusually large amount of manganese (Ratcliffe 2000).

There are also serpentine soils on Shetland. The most famous locality is the Keen of Hamar, where the endemic *Cerastium nigrescens* and the rare *Arenaria norvegica* ssp. *norvegica* grow on bare, dry serpentine debris at about 90 m with *Arabis petraea, Armeria maritima, Plantago maritima, Anthyllis vulneraria, Silene uniflora, S. acaulis* and *Thymus polytrichus* (Lusby and Wright 1996).

Several bryophytes and lichens are characteristic of metalliferous rocks in the British uplands. These include the mosses *Grimmia atrata, G. donniana, Mielichhoferia mielichhoferiana* and *M. elongata; Grimmia montana* and *G. sessitana* are especially associated with serpentine (Hill *et al.* 1992). Metallophyte lichens that can tolerate lead and zinc include *Bacidia saxenii, B. viridescens, Cladonia cariosa, Gyalidea subscutellaris, Peltigera neckeri, P. venosa, Placynthiella hyporhoda, Sarcosagium campestre, Steinia geophana,*

*Stereocaulon condensatum, S. dactylophyllum, S. delisei, S. glareosum, S. leucophaeopsis, S. nanodes, S. pileatum,* and all six British species of *Vezdaea* (Gilbert 2000). Iron-rich rocks have a distinctive lichen flora, including *Acarospora sinopica, Arthonia lapidicola, Lecanora epanora, L. handelii, L. subaurea, Lecidea silacea, Miriquidica atrofulva, Rhizocarpon furfurosum, R. oederi* and *Stereocaulon leucophaeopsis* (Gilbert 2000).

## Fell-fields

Fell-fields are areas of gravelly ground on exposed summits, ridges and plateaux in the Highlands and the Hebrides, at altitudes from 300 m to over 1100 m. They have been recorded on the Cairngorms and Lochnagar, on the hills of Skye and Mull (Averis and Averis 1995a, 1995b, 1998b, 1999b), on the hills of Morvern and Ardnamurchan, at Invernaver on the north coast of Sutherland (Averis 1997), and on Shetland (Jóhansen 1985; A B G Averis and A M Averis, pers. obs.). Most fell-fields are on basalt or granite, which are particularly prone to weathering to fine gravel. They usually look like nothing more than bare spreads of fine-grained, unstable and often wet gravel, and one has to inspect them quite closely in order to see that they actually contain a scattering of very small plants among the stones. These assemblages can be very distinctive in their species composition. In many fell-fields the flora shows that continual physical and chemical weathering provides a rich supply of nutrients. Most fell-fields are probably the natural consequence of erosion by frost, wind and rain in a severe upland climate, but it seems likely that high concentrations of sheep and deer may cause them to expand. For example, there are ridges on Mull and Skye where fell-fields are pocked by hoof-prints, and the edges of the more continuous vegetation around the fell-fields are obviously being worn away by trampling (Averis and Averis 1999b).

On basalt fell-fields in Skye, Mull and Morvern the most common plants are *Agrostis canina, Festuca vivipara, Thymus polytrichus* and the moss *Oligotrichum hercynicum. Potentilla erecta, Calluna vulgaris, Carex viridula* ssp. *oedocarpa, Galium saxatile* and the mosses *Polytrichum piliferum* and *Racomitrium ellipticum* also occur; basalt fell-field is one of the main habitats of the nationally uncommon *R. ellipticum* in Scotland. There is usually a sprinkling of other species including *Viola riviniana, Plantago maritima, Salix herbacea, Alchemilla alpina, Solidago virgaurea,* the bryophytes *Polytrichum alpinum, Pogonatum urnigerum, Racomitrium lanuginosum, R. fasciculare, Marsupella emarginata, Anthelia julacea, Nardia scalaris* and *Jungermannia gracillima,* and the lichen *Cladonia subcervicornis.* Some low-altitude examples have *Sedum anglicum* or *Armeria maritima,* and at higher altitudes the flora may include *Alchemilla alpina, Juncus triglumis, Luzula spicata, Carex bigelowii, Diphasiastrum alpinum, Oxyria digyna, Salix herbacea, Saxifraga stellaris, Sedum rosea, Silene acaulis* and *Vaccinium vitis-idaea.* The more montane British basalt fell-fields have a flora quite similar to that of fell-fields on the Faroe Islands, where the habitat is very extensive (Hobbs and Averis 1991a). Basalt fell-fields on the Trotternish peninsula of Skye and the Ardmeanach peninsula of Mull are the main habitat in Great Britain of the nationally rare montane annual *Koenigia islandica.* This species also occurs more sparingly in gravelly *Carex-Koenigia* flushes M34 and on exposed soil of erosion scars on hill slopes at these two sites – its only known localities in Great Britain. It is common on basalt fell-fields in the Faroes and Iceland. On the Ardmeanach peninsula and on Beinn Iadainn in Morvern the presence of *Arabis petraea* and *Sedum villosum* in basalt fell-fields provides another floristic link with Faroese fell-fields.

Fell-fields on the granite of the Cairngorm hills and Lochnagar have many species in common with similar habitats on basalt: for example, *Festuca vivipara, Agrostis canina, Galium saxatile, Potentilla erecta, Alchemilla alpina, Thymus polytrichus, Saxifraga stellaris, Luzula spicata,* the mosses *Oligotrichum hercynicum* and *Polytrichum piliferum*, and the lichen *Cladonia subcervicornis*. Other species include *Antennaria dioica, Juncus trifidus* and in some places the rare *Luzula arcuata*.

In the far north of Scotland fell-fields occur at very low altitudes. Some examples on metamorphic rocks on the north coast of Sutherland are enriched by wind-blown calcareous sand, and have an interesting calcicolous flora including *Saxifraga aizoides, S. oppositifolia, Carex capillaris, Draba incana*, the mosses *Ditrichum flexicaule s.l., Tortella tortuosa* and *Encalypta streptocarpa*, and the lichen *Cladonia rangiformis* (Averis 1997). On Shetland there are fell-fields on Devonian Old Red Sandstone, with an acidophilous flora including *Jasione montana*.

## Small-sedge mires

In many places in northern and western Great Britain there are various types of small-sedge mire that are not described in the NVC. They generally have a sward made up of mixtures of *Carex nigra, C. panicea, C. echinata* and *C. pulicaris*; in some stands, *C. flacca, C. hostiana* and *C. viridula* ssp. *oedocarpa* are common. *Anthoxanthum odoratum, Molinia caerulea, Holcus lanatus* and other grasses grow among the sedges. Some of these mires contain a rich array of mesotrophic forbs, including *Angelica sylvestris, Filipendula ulmaria, Caltha palustris, Valeriana officinalis, Geum rivale, Parnassia palustris, Ranunculus acris, Lychnis flos-cuculi, Triglochin palustris, Prunella vulgaris, Hydrocotyle vulgaris, Pinguicula vulgaris, Ranunculus flammula* and *Cardamine pratensis*. They usually have a patchy underlay of bryophytes such as *Calliergonella cuspidata, Rhizomnium punctatum, Campylium stellatum, Brachythecium rutabulum, Chiloscyphus polyanthos* and *Aneura pinguis*. Mires of this type may contain uncommon species such as *Crepis paludosa, Trollius europaeus, Pinguicula lusitanica* and, in the Breadalbane hills, *Carex vaginata*. Other forms of short-sedge mire are less herb-rich but have carpets of base-tolerant Sphagna (*Sphagnum teres, S. warnstorfii, S. contortum, S. squarrosum* and *S. subsecundum*); this is the *Carex nigra-Sphagnum contortum* mire described by Birks and Ratcliffe (1980).

These small-sedge mires have been found widely in upland Great Britain (Cooper and MacKintosh 1996; Rodwell *et al.* 1998) and also in Ireland (O'Criodain and Doyle 1994). They occur from near sea-level up to 650 m or more, where wet soils are flushed with moderately base-rich water. Stands are usually small and occur in mosaics with grasslands, heaths and other types of mire. Many examples appear to have been derived from taller *Molinia*-dominated vegetation as a result of heavy grazing. They differ from *Carex echinata-Sphagnum* mires M6 in having mesotrophic herbs and bryophytes, but little or no *Sphagnum fallax, S. denticulatum* or *S. palustre. Juncus-Galium* rush-pasture M23, which can have a similar assemblage of herbs and bryophytes, is dominated by rushes rather than small sedges. Compared with *Carex-Pinguicula* mire M10 and *Carex-Saxifraga* mire M11, the sward is denser and there are generally more grasses and mesotrophic forbs.

## Bryophyte and lichen assemblages

Bryophytes and lichens are small plants which grow not only on the ground but also on rock and bark surfaces not colonised by vascular plants. It is therefore not surprising that they form distinct assemblages of their own, and that these assemblages vary on a finer scale than those of vascular plants.

Most bryophyte and lichen assemblages on the ground occur in close association with vascular plant vegetation, and are included as integral components of various NVC types. Carpets of Sphagna form a significant part of many mires and some forms of wet woodland. Various bryophytes dominate in upland springs, including *Anthelia julacea, Philonotis fontana, Dicranella palustris, Scapania undulata, Pohlia wahlenbergii* var.

*glacialis* and *Palustriella commutata*. Extensive mats of *Racomitrium lanuginosum* form the bulk of summit moss-heath and some other montane communities. *Rhytidiadelphus loreus* is dominant in certain types of snow-bed vegetation. Low mats of mixed montane liverworts and mosses form late-lying snow-bed communities. *Cladonia* lichens form extensive patches within both montane and sub-montane heaths and bogs.

Among many heavily grazed acid grasslands and mires, especially in England and Wales, the moss *Polytrichum commune* forms large conspicuous dark green patches which are not described in the NVC. Accompanying species are sparse and include *Juncus effusus, J. squarrosus, Nardus stricta, Festuca ovina, Agrostis capillaris, A. canina, Carex nigra, Potentilla erecta, Galium saxatile* and the mosses *Hypnum jutlandicum, Rhytidiadelphus squarrosus* and *Sphagnum fallax*. In these places *P. commune* appears to be encouraged by heavy grazing reducing competition from vascular plants, and perhaps also by nitrogen deposition.

There are many finer-scale bryophyte assemblages, especially on rock and bark, which are independent of the adjacent vascular plant vegetation. Many such assemblages have been described from Great Britain and Ireland, particularly by Richards (1938), Paton (1956), Proctor (1962), Ratcliffe (1968), Birks (1973), Averis (1991a) and Rothero (1991) for bryophytes, by James *et al.* (1977), Purvis and Halls (1996) and Fryday (1997) for lichens, and by Orange and Fryday (1998) for both. Some assemblages described by these authors can be recognised among the European epiphytic communities classified by Barkman (1958), but on the whole that scheme, with its origins in mainland Europe, does not work well in upland Great Britain. Unfortunately, there is no single, unified classification that deals with British bryophyte and lichen assemblages as comprehensively and systematically as the NVC does for vegetation in which vascular plants play a prominent role. Table 4 provides an outline of the types that we have found in upland Great Britain. Although bryophyte assemblages and lichen assemblages have generally been treated separately, and are listed separately in Table 4, it is important to note that bryophytes and lichens commonly grow together and there is some overlap between the assemblages described for each of the two groups. There is a great deal of interest to be said about many lower plant assemblages, but that is beyond the scope of this book, and the following discussion is necessarily rather general.

Bryophyte and lichen assemblages are very well represented in Great Britain and Ireland, especially in rocky habitats and woodlands in the west where they can be very conspicuous. The habitats of bryophytes and lichens overlap considerably, but in general bryophytes are more shade-tolerant or shade-demanding, and lichens more light-demanding. The assemblages vary immensely, from very open 'pioneer' assemblages of small plants on steep rock and bark surfaces, to dense, deep mats of large species where some soil or humus has accumulated on the ground, on banks, on the upper parts of boulders and on the lower parts of tree trunks.

Very often one can trace patterns of succession on steep rock, bark and soil surfaces. Assemblages of small 'pioneer' species initially colonise a bare substrate, after which slightly larger species are able to gain a foothold, and over time increasingly larger species colonise until thick mats have developed and swamped out the original 'pioneer' assemblages. These mats eventually fall off under their own weight to leave a bare surface on which the process begins again. The larger bryophytes and lichens may never be able to grow on the steepest substrates, so the process may be arrested at an early stage. Hence small species tend to predominate on vertical tree trunks, whereas larger species can grow on sloping tree bases. Richards (1938) described these processes in woodland in south-western Ireland, and they have been observed in many places since then by people surveying and monitoring assemblages of bryophytes and lichens.

**Table 4** Main bryophyte and lichen assemblages in upland Great Britain[1]

### Assemblages on acidic soil, peat or humus

*Bryophytes*

*Pogonatum urnigerum/aloides-Jungermannia gracillima-Nardia scalaris-Diplophyllum albicans* (P)
>Scattered patches and shoots of low-grown species on acidic soil, mainly in unwooded habitats

*Pogonatum aloides-Dicranella heteromalla-Fissidens bryoides-Calypogeia* species-*Diplophyllum albicans* (P)
>Scattered patches and shoots of low-grown species on acidic soil, especially in woods

*Mnium hornum-Atrichum undulatum* (P-M)
>Low patches on acidic soil on ground and banks, especially in woodland

*Pseudotaxiphyllum elegans* (P-M)
>Low species-poor patches on acidic soil on ground and banks in woodland

*Dicranum scoparium/majus-Polytrichum formosum-Plagiothecium undulatum-Hypnum jutlandicum* (M-L)
>Typical (M-L)
>>Mats on acidic soil and humus on ground, banks, boulders and logs in woodland

>*Leucobryum glaucum* (M-L)
>>Patches and cushions on acidic soil and humus on ground, banks and logs in woods

>*Hylocomium splendens-Rhytidiadelphus loreus-Thuidium tamariscinum-Pleurozium schreberi* (L)
>>Deep carpets on acidic soil and humus on ground, banks, boulders and logs in woods

>*Hylocomium splendens-Rhytidiadelphus triquetrus* (L)
>>Deep carpets on acidic soil and humus, mainly on ground and banks in woods in the north and north-east

*Pellia epiphylla* (P-M)
>Extensive species-poor sheets on wet acidic soil on shaded banks, especially in woods

*Lichens*

*Cladonion coniocraeae* (P)
>Scattered small lichens on peat

*Lecanorion variae* (P)
>Species-poor mixtures of crustose lichens on acidic soil, especially in polluted areas

### Assemblages on neutral to basic soil

*Bryophytes*

*Fissidens bryoides/taxifolius-Lophocolea bidentata-Calypogeia* species (P)
>Scattered patches and shoots on neutral to basic soil, especially in woods

*Eurhynchium praelongum/striatum-Plagiomnium undulatum-Lophocolea bidentata* (L)
>Typical (L)
>>Fairly deep mats on neutral to basic soil on ground and banks in woods

>*Thuidium tamariscinum-Hylocomium brevirostre/splendens-Rhytidiadelphus triquetrus* (L)
>>Deep carpets on neutral to basic soil on ground, banks and boulders in woods

*Eurhynchium praelongum-Lophocolea bidentata-Hookeria lucens-Plagiochila asplenioides-Trichocolea tomentella* (L)
>Deep mats on moist neutral to basic soil on steep banks in woodland

*Conocephalum conicum* (P-M)
>Extensive species-poor sheets on moist neutral soil on banks, mainly in woods

*Lichens*

*Montane epidiorite* nodum (P)
>Mixtures of mainly crustose species on basic montane rock (epidiorite and mica-schist) and soil

### Assemblages on acidic rock, bark and decorticated wood

*Well-drained habitats: bryophytes*

These assemblages are ordered into two groups. Those in the first group have been found only on rocks. Those in the second group can occur on both rock and bark, or on decorticated wood.

*Andreaea rupestris/rothii* (P)
>Scattered patches on dry acidic rocks in open to lightly shaded places at low to high altitude

**Table 4** (continued)

---

*Diplophyllum albicans* (P)
> Thin mats on shaded acidic rocks at low to high altitude

*Racomitrium sudeticum/heterostichum/fasciculare-Andreaea* species (P-M)
> Scattered patches on dry acidic rocks in open to lightly shaded places at low to high altitude

*Racomitrium lanuginosum* (M) (described under vegetation of screes on p. 420)
> Large, deep patches on acidic scree and block litter at medium to high altitude

*Frullania tamarisci* (P)
> Typical (P)
> > Thin, species-poor mats on shaded acidic rock and bark, especially in woodland
> *Microlejeunea ulicina-Scapania gracilis-Plagiochila punctata* (P)*
> > Thin patches on acidic rock and bark in western woodland
> *Microlejeunea ulicina-Metzgeria temperata* (P)*
> > Thin patches on acidic rock and bark in western woodland

*Lepidozia reptans-Cephalozia bicuspidata/lunulifolia-Lophocolea bidentata* (P)
> Typical (P)
> > Scattered patches or thin mats on shaded acidic rocks, tree bases and rotting logs in woods
> *Nowellia curvifolia-Riccardia palmata-Scapania umbrosa* (P)
> > Scattered patches or thin mats on shaded rotting logs in woods, mainly in the north and west

*Isothecium myosuroides* (P-M)
> Thin to fairly deep, species-poor mats on acidic rock and bark in woods

*Hypnum cupressiforme/andoi* (P-M)
> Thin, species-poor mats on acidic rock and bark in woods

*Hypnum cupressiforme/andoi-Isothecium myosuroides-Frullania tamarisci-Diplophyllum albicans* (M)
> Typical (M)
> > Patches on acidic rock and bark, mainly in woodland
> *Scapania gracilis-Plagiochila spinulosa-Bazzania trilobata-Hymenophyllum wilsonii* (M)*
> > Patches and cushions on acidic rock and bark in sheltered western woodland
> *Dicranum scottianum-Lepidozia cupressina-Bazzania trilobata-Hymenophyllum wilsonii/tunbrigense* (M)*
> > Patches and cushions on acidic rock faces, rocky banks and bark in sheltered western woods
> *Leucobryum glaucum-Bazzania trilobata* (M)
> > Patches and cushions on acidic rocky banks and logs in western woods

*Mnium hornum* (M)
> Low, dense, species-poor patches on acidic rocks and tree bases in woodland

*Well-drained habitats: lichens*

These assemblages are ordered into two groups. Those in the first group occur only on rock. Those in the second group occur on both rock and bark, or on bark only; some occur also on decorticated wood.

*Rhizocarpion alpicolae* (P)
> Crustose lichens on well-lit to rather shaded acidic rocks

*Umbilicarion cylindricae* (P)
> Crustose and foliose lichens on well-lit to lightly shaded acidic rocks

*Lecideion tumidae* (P)
> Crustose lichens on well-lit acidic rocks

*Leprarion chlorinae* (P)
> Crustose lichens on dry underhangs and recesses on acidic rocks

*Parmelion conspersae* (P)
> Crustose and foliose lichens on nutrient-enriched well-lit acidic rocks

*Acarosporion sinopicae* (P)
> Crustose lichens on acidic rocks rich in heavy metals

*Cladonion coniocraeae* (P)
> Scattered small lichens on acidic bark, rock, logs and peat, especially in woods

*Calicion hyperelli* (P)
> Scattered very small lichens on steep surfaces of dry acidic to basic bark or rotting wood

*Lecanorion variae* (P)
> Species-poor mixtures of crustose lichens on acidic bark and rock, especially in polluted areas

**Table 4** *(continued)*

*Parmelion laevigatae* (P-M)
> Mixed crustose, foliose and fruticose lichens on acidic bark and rock in western woods

*Pseudevernion furfuraceae* (P-M)
> Mixed crustose, foliose and fruticose lichens on acidic bark and rock in woods and in open habitats

*Usneion barbatae* (P-M)
> Scattered hanging, fruticose lichens on acidic bark and rock, especially in woods

*Wet habitats: bryophytes*

*Racomitrium aquaticum-Campylopus atrovirens-Marsupella emarginata-Diplophyllum albicans* (P-M)
> Patches on wet acidic rocks in open to shaded situations at low to high altitude

*Brachythecium plumosum-Racomitrium aciculare-Scapania undulata* (P-M)
> Scattered patches on wet acidic rocks, mainly in or by streams and rivers

*Heterocladium heteropterum* (P-M)
> Thin, species-poor sheets on wet, acidic rock faces, especially by streams in western woods

*Hyocomium armoricum* (M)*
> Species-poor mats on wet acidic rocks, especially by streams and rivers in western woods

*Isothecium holtii* (M)*
> Species-poor mats on wet acidic rocks, mainly by streams and rivers in western woods

*Wet habitats: lichens*

*Aspicilietum lacustris* (P)
> Crustose and foliose lichens on acidic to basic rocks by streams, rivers and lochs

*Ephebetum lanatae* (P)
> Crustose and fruticose lichens on acidic rocks by streams, rivers and lochs

*Ionaspidetum suaveolentis* (P)
> Crustose and foliose lichens on acidic rocks by montane streams

*Verrucarietum siliceae* (P)
> Crustose and fruticose lichens on acidic rocks in and by streams, rivers and lochs

## Assemblages on neutral to basic rock, bark and decorticated wood

*Well-drained habitats: bryophytes*

These assemblages are ordered into two groups. Those in the first group have been found only on rocks. Those in the second group occur on both rock and bark, or on bark only.

*Ctenidium molluscum-Tortella tortuosa-Fissidens dubius-Metzgeria furcata* (P-M)
> Typical (P-M)
>> Scattered mats, patches and cushions on dry, well-lit to shaded, basic rock at low to medium altitude
>
> *Anoectangium aestivum-Mnium marginatum-Plagiochila britannica-Scapania aspera* (M)
>> Patches and cushions on moderately shaded basic rocks in northern localities or at higher altitudes

*Frullania tamarisci-Metzgeria furcata* (P)
> Species-poor (P)
>> Thin mats and small cushions on neutral-basic bark and rock in well-lit to shaded, low-altitude sites
>
> *Ulota crispa* (P)*
>> Thin mats and small cushions on neutral-basic bark in well-lit to shaded, low-altitude sites in the west
>
> *Ulota crispa/phyllantha/calvescens-Frullania teneriffae* (P)*
>> Mats, patches or small cushions on neutral-basic bark in well-lit, low-altitude sites in the west

*Lejeunea patens-Aphanolejeunea microscopica-Drepanolejeunea hamatifolia-Harpalejeunea molleri-Colura calyptrifolia-Radula aquilegia-Plagiochila exigua/killarniensis* (P)*
> Thin mats or small patches on neutral-basic rock and bark in shaded, humid, low-altitude sites in the west

*Frullania dilatata-Metzgeria furcata* (P)
> Species-poor (P)
>> Thin mats and patches on dry neutral to basic rock and bark at low altitude

**Table 4** (*continued*)

---

*Ulota crispa* (P)
> Thin mats and small *Ulota* cushions on well-lit to shaded neutral-basic bark at low altitude in the west

*Ulota phyllantha/crispa/calvescens-Cololejeunea minutissima* (P)*
> Thin mats and scattered cushions on moderately shaded, neutral-basic bark at low altitude in the west

*Orthotrichum* species (P)
> Thin mats and small cushions on well-lit, neutral-basic bark in the east

*Homalothecium sericeum-Neckera complanata* (M)
> Scattered patches on steep, well-drained, shaded to well-lit, basic rock and bark at low altitude

*Neckera crispa* (M)
> Large patches on usually shaded basic rocks and bark, especially in woodland

*Well-drained habitats: lichens*

These assemblages are ordered into two groups. Those in the first group occur only on rocks. Those in the second group occur on both rock and bark, on bark only, or on decorticated rotting wood.

*Montane epidiorite* nodum (P)
> Mixtures of mainly crustose species on basic montane rock (epidiorite and mica-schist) and soil

*Rhizocarpon petraeum* nodum (P)
> Mixtures of mainly crustose species on basic rocks

*Aspicilion calcareae* (P)
> Varied mixtures of mainly crustose species on limestone

*Lecanorion subfuscae* (P)
> Scattered crustose lichens on well-lit twigs and branches

*Graphidion scriptae* (P)
> Scattered crustose lichens on smooth, variably shaded, neutral-basic bark, especially in woods

*Calicion hyperelli* (P)
> Scattered very small lichens on steep surfaces of dry acidic to basic bark or rotting wood

*Xanthorion parietinae* (P-M)
> Crustose lichens on nutrient-rich, neutral-basic, usually well-lit rock and bark at low altitude

*Parmelion perlatae* (P-M)
> Mixed foliose and crustose lichens, mainly on neutral bark

*Lobarion pulmonariae* (M)
> Foliose and fruticose lichens on neutral-basic bark and rock, especially in woods in unpolluted western areas

*Wet habitats: bryophytes*

*Gymnostomum aeruginosum-Fissidens adianthoides* (P-M)
> Scattered patches, tufts and cushions on wet basic rocks at low to medium altitude

*Brachythecium plumosum-Schistidium rivulare-Chiloscyphus polyanthos* (P-M)
> Scattered patches on wet neutral-basic rocks, mainly in or by streams and rivers

*Rhynchostegium riparioides* (M)
> Extensive mats on wet, shaded, neutral-basic rock

*Thamnobryum alopecurum* (M)
> Extensive deep mats on wet, shaded, neutral-basic rock, especially by woodland streams

*Wet habitats: lichens*

*Aspicilietum lacustris* (P)
> Crustose and foliose lichens on acidic to basic rocks by streams, rivers and lochs

---

[1]Bryophyte assemblages are defined by Ben Averis and are named after selected characteristic or consistent species; some are divided into sub-types. Lichen assemblages are those described by James *et al.* (1977). Some lichen assemblages are included in more than one place in the table because they span two or more major habitat types. Abbreviations for successional stages: P = pioneer stage; M = medium stage; L = late stage. Western bryophyte assemblages containing oceanic species are indicated by an asterisk (*); this has not been done for lichens because there is insufficient information readily available, especially as there is not yet a systematic phytogeographical classification of all British lichen species.

Several assemblages include many of the oceanic bryophyte and lichen species for which western Scotland and western Ireland are the European headquarters. They are therefore of great importance for nature conservation. Oceanic bryophyte assemblages are indicated in Table 4. An additional distinctive assemblage of northern oceanic leafy liverworts occurs on north-facing to east-facing hill slopes in the western Highlands and western Ireland. This was described by Ratcliffe (1968) as the Northern Atlantic Hepatic Mat, and was studied in detail by A M Averis (1994). It occurs as a component of heathland vegetation (the *Bazzania-Mylia* sub-community of *Vaccinium-Racomitrium* heath H20c and the *Mastigophora-Herbertus* sub-community of *Calluna-Vaccinium-Sphagnum* heath H21b) rather than as a separate bryophyte assemblage. It is of great international importance, being confined in Europe to Great Britain and Ireland.

Many bryophyte and lichen assemblages have been found to be good indicators of the effects of pollution and of past disturbance, such as burning of heathland and removal of trees in woodland. A well-known example is the *Lobarion pulmonariae*. This distinctive and colourful assemblage of large macrolichens, including *Lobaria* species, *Sticta* species, *Nephroma* species, *Pseudocyphellaria* species, *Degelia* species and *Pannaria* species, was once widespread on trees throughout much of Europe but has become greatly reduced by air pollution. The western Highlands is one of its remaining strongholds. Assemblages of small oceanic liverworts, such as *Aphanolejeunea microscopica*, *Drepanolejeunea hamatifolia*, *Harpalejeunea molleri*, *Colura calyptrifolia*, *Radula aquilegia* and *Plagiochila exigua*, also appear to be vulnerable to air pollution. In Great Britain some oceanic bryophytes, especially *Plagiochila atlantica*, *Adelanthus decipiens* and *Sematophyllum micans*, appear to be so dependent on trees to provide the right amount of shade and humidity that they are now strongly associated with 'ancient' woodland that has suffered little disturbance over many years (Ratcliffe 1968). Many lichens are also associated with ancient woodland, although in some cases for different reasons; for example, some species require the specialised bark conditions on very old trees. The Northern Atlantic Hepatic Mat is highly vulnerable to moor-burning which destroys the characteristic liverworts and reduces the shade and shelter of the habitat to such an extent that they are unable to recolonise before their place has been taken by commoner species.

## Streams and rivers

Some streams and rivers contain vegetation that can be classified within the NVC. For example, tiny streamlets running downstream from springs can be filled with bryophyte-dominated springhead vegetation, and various swamp communities can occur in sluggish water in small streams and at the edges of larger streams and rivers.

Rocks within and at the edges of streams and rivers are typically thickly clothed with bryophytes. For example, the *Brachythecium plumosum-Racomitrium aciculare-Scapania undulata* assemblage is characteristic of acidic rocks, and the *Brachythecium plumosum-Schistidium rivulare-Chiloscyphus polyanthos* assemblage occurs on more basic rocks (see Table 4). Bryophyte assemblages on acid rocks can include *Fontinalis antipyretica*, *F. squamosa*, *Hygrohypnum luridum*, *H. ochraceum*, *Jungermannia atrovirens*, *Marsupella emarginata*, *Nardia compressa* and, in the west, *Hyocomium armoricum*, *Isothecium holtii*, *Fissidens curnovii* and the scarce *Rhynchostegium alopecuroides*. *Grimmia hartmanii* and *G. curvata* are locally common on the upper parts of acidic boulders in streams. Neutral to basic rocks can support *Jungermannia atrovirens*, *Fontinalis antipyretica*, *Hygrohypnum luridum*, *H. eugyrium*, *Rhynchostegium riparioides*, *Thamnobryum alopecurum*, *Plagiochila porelloides*, *Cratoneuron filicinum*, *Amblystegium tenax*, *Cinclidotus fontinaloides*, *Rhynchostegiella teneriffae*, *Schistidium strictum* and *Porella cordaeana*. *Dichodontium pellucidum* is common on rocks and gritty soil along stream margins both in woodlands and in the open. The presumably introduced moss *Atrichum crispum* is common in similar habitats in unwooded uplands in parts of Wales, northern England and Devon.

Some aquatic lichen communities have been described by James *et al.* (1977) (see Table 4), and include characteristic riparian species such as *Dermatocarpon* species, *Ephebe lanata, Massalongia carnosa, Polychidium muscicola, Rhizocarpon lavatum* and *Verrucaria* species.

Rarities associated with streamside rocks include the bryophytes *Dicranum subporodictyon, Fissidens rufulus, Grimmia retracta, Hygrohypnum molle, H. smithii, H. duriusculum, Schistidium agassizii, Seligeria carniolica, Dumortiera hirsuta, Porella pinnata* and the lichen *Collema dichotomum*. Steep rock faces just above streams in western ravines are one of the main habitats of the scarce oceanic liverworts *Aphanolejeunea microscopica, Drepanolejeunea hamatifolia, Harpalejeunea molleri, Colura calyptrifolia, Radula aquilegia, R. carringtonii, R. voluta, Acrobolbus wilsonii, Jubula hutchinsiae, Lophocolea fragrans* and *Plagiochila exigua*.

Steep, regularly wetted streamside and riverside banks may have assemblages similar to those described above, or may support large patches of a single species: *Pellia epiphylla* on acidic soil, *Conocephalum conicum* on shaded neutral soil, *Heterocladium heteropterum, Hyocomium armoricum* and *Isothecium holtii* on shaded acidic rocks in the west, or *Thamnobryum alopecurum* on shaded, neutral to basic rock faces. Patches of *T. alopecurum* on well-shaded rock faces in the west are a favoured habitat of the scarce *Jubula hutchinsiae*, in many places accompanied by *Lejeunea patens, L. lamacerina, Riccardia multifida* and *R. chamedryfolia*.

## Aquatic vegetation in standing water

There are many NVC communities of open water, but they are sketchily described both in the NVC and in most survey work, and their distributions are poorly known. Plants characteristic of standing water such as *Lobelia dortmanna, Littorella uniflora, Isoetes lacustris, Menyanthes trifoliata, Nymphaea alba* and *Nuphar lutea* commonly occur as very sparse stands dominated by a single species. There are no accounts of these aquatic vegetation types in this book because at present we cannot add much to the NVC descriptions. The *Nymphaea alba* community A7 is common in the upland fringes of the north and west of Great Britain; in Caithness and Sutherland, the rare *Nuphar pumila* has been recorded in this community. The *Nuphar lutea* community A8 is primarily a lowland type, but also occurs in the uplands, especially in the west Highlands and Inner Hebrides. The *Potamogeton natans* community A9 is common in peaty pools in the uplands, as well as in larger bodies of water and even in moderately fast-flowing streams. The *Littorella uniflora-Lobelia dortmanna* community A22 occurs on stony substrates in shallow water. The presence of the *Isoetes lacustris/setacea* community A23 is often inferred from plants of *Isoetes lacustris* washed up after storms, as it generally occurs in deep water and can rarely be viewed from the shore. It is a community of deep standing water from north Wales northwards. The *Juncus bulbosus* community A24 is common in shallow peaty pools and around the sheltered margins of larger bodies of water. It occurs throughout the uplands from south-west England to Shetland.

## Ruderal vegetation

Even in the harsh climate of the uplands plants can rapidly colonise habitats created by human activities. The most common of these habitats are tracks and abandoned roads. Bryophytes and lichens are usually the first pioneers on these stony substrates. The most common species are disturbance-tolerant bryophytes, such as *Hypnum cupressiforme, Ceratodon purpureus, Polytrichum piliferum, P. juniperinum, Pogonatum urnigerum, P. aloides, Racomitrium ericoides, Oligotrichum hercynicum, Jungermannia gracillima* and *Nardia scalaris*. More exacting species, such as the liverworts *Anastrepta orcadensis* and *Bazzania tricrenata*, have been seen colonising disturbed ground on roadsides in sheltered, humid places in the north-west Highlands (D G Long, pers. comm.). *Juncus effusus* is very common on damp disturbed ground along tracks in the uplands, and mire and springhead species such as *J. bulbosus* and the mosses *Philonotis fontana* and *Dicranella palustris* are reasonably common colonisers of flushed gravel on tracks. *Rubus fruticosus* can grow luxuriantly along disturbed roadsides, even in colder northern areas where it is a scarce plant in other habitats. Some uncommon species have been recorded on tracks and roadsides. For example, the fern *Botrychium lunaria* has been recorded on disturbed ground and in lay-bys beside hill roads, and in at least one place in the north-west Highlands *Salix herbacea* grows on the edges of a track in a carpet of *Racomitrium ericoides* at only about 300 m. Among the more notable lichens that colonise disturbed ground in the uplands is *Lempholemma radiatum*, which is abundant on a well-trodden track on Ben Lawers (Brian and Sandy Coppins, pers. comm.). Several uncommon bryophytes, including *Fossombronia fimbriata*, *F. incurva* and *Aongstroemia longipes*, also occur on tracks in the uplands.

Ground around buildings and abandoned dwellings can become enriched by lime mortar, by dumping of refuse, or simply because sheep and cattle have congregated in the shelter of the walls and left piles of droppings. These conditions can be favourable to nettles and thistles (thickening up in places to form the *Urtica dioica-Galium aparine* and *Urtica dioica-Cirsium arvense* communities OV24 and OV25), together with other weedy species such as *Poa annua*, *Rumex acetosa*, *R. obtusifolius*, *Digitalis purpurea*, *Stellaria media* and *Ranunculus repens*.

Disturbed ground associated with skiing developments seems to take a long time to revegetate, partly because of the severe upland climate and partly because of repeated disturbance by the machines used to shovel snow onto the *pistes* where there is otherwise not enough snow for skiing. Vegetation is also damaged by being squashed under the compacted snow of the ski-runs. In these situations, *Calluna vulgaris* becomes more susceptible to fungal diseases and can be replaced by *Vaccinium myrtillus* and grasses (Bayfield *et al.* 1988), which form conspicuous and visually intrusive green patches when there is no snow on the ground.

Mine workings and quarries are common in the uplands, especially in Wales, northern England and parts of western Scotland. The characteristic vegetation of old lead mines has been comprehensively described and is represented in the NVC by the *Festuca-Minuartia* community OV37. Bryophytes on lead mine spoil include the rarities *Ditrichum plumbicola* and *Weissia controversa* var. *densifolia*. There are also iron and copper mines in the British uplands. Although their vascular flora is generally less distinctive than that of lead mines, they are rich habitats for lichens. For example, the copper mines in Snowdonia and the Lake District are home to the rare *Lecidea inops*, *Psilolechia leprosa*, *Stereocaulon leucophaeopsis* and *S. symphycheilum* (Gilbert 2000).

Worked-out quarries are essentially cliffs, and their bare ledges and crevices can become colonised by small ferns, dwarf shrubs such as *Calluna vulgaris* and *Vaccinium myrtillus*, and trees. In the bare, sheep-bitten uplands of Wales and northern England, disused quarries are among the few places with vegetation out of the reach of grazing animals. They can have an interesting flora, as well as being valuable nesting sites for raven, peregrine and, in north Wales, choughs.

Old stone walls can become clothed with a layer of colourful crustose lichens, generally comprising assemblages of common pioneer species similar to those of dry scree (Brian Coppins, pers. comm.). The bryophyte flora too is generally made up of common species such as *Racomitrium lanuginosum*, *R. heterostichum*, *R. fasciculare*, *Grimmia pulvinata, Hypnum cupressiforme, H. lacunosum* and *Isothecium myosuroides*. The liverwort *Barbilophozia barbata* grows abundantly on old stone walls in some sheltered, shaded situations. More exacting species can gain a foothold locally; for example, the oceanic liverworts *Herbertus aduncus* ssp. *hutchinsiae* and *Plagiochila killarniensis* have been recorded on stone walls in the western Highlands.

## Lowland types of vegetation that can occur in the uplands

There are many types of vegetation that occur predominantly in lowland Great Britain, but which are sufficiently common in the uplands to merit inclusion in this *Guide*. Some are described in the main accounts and also included in the key: for example, *Festuca-Agrostis-Rumex* grassland U1, *Cynosurus-Centaurea* grassland MG5, and *Holcus-Juncus* rush-pasture MG10. Others are not described in the accounts but are included in the key.

# References

Allen, D S (1995) Habitat selection by whinchats: a case for bracken in the uplands. In *Heaths and moorlands: cultural landscapes* (eds D B A Thompson, A J Hester and M B Usher), pp 200–205. HMSO, Edinburgh.

Averis, A B G (1991a) *A survey of the bryophytes of 448 woods in the Scottish Highlands.* Scottish Field Survey Unit, Nature Conservancy Council, Edinburgh.

Averis, A B G (1991b) *A survey of serpentine-influenced vegetation at four sites in Grampian Region, Scotland.* Nature Conservancy Council for Scotland, Aberdeen.

Averis, A B G (1994) *The vegetation of Roineabhal, Bleaval and Chaipaval, South Harris, Scotland.* Scottish Natural Heritage, Battleby.

Averis, A B G (1997) *The vegetation of Druim Chuibhe, Invernaver, Sutherland.* Scottish Natural Heritage, Battleby.

Averis, A B G (1999) *Vegetation survey of Ballyoukan Juniper Wood, Perthshire, 1999.* Scottish Natural Heritage, Battleby.

Averis, A B G and Averis, A M (1995a) *The vegetation of south and east Mull.* Scottish Natural Heritage, Battleby.

Averis, A B G and Averis, A M (1995b) *The vegetation of north Mull.* Scottish Natural Heritage, Battleby.

Averis, A B G and Averis, A M (1996) *The vegetation of Lismore and Kerrera.* Scottish Natural Heritage, Battleby.

Averis, A B G and Averis, A M (1997a) *The vegetation of Cranstackie, Sutherland.* Scottish Natural Heritage, Battleby.

Averis, A B G and Averis, A M (1997b) *The vegetation of the Trotternish Ridge, Skye.* Scottish Natural Heritage, Battleby.

Averis, A B G and Averis, A M (1997c) *The vegetation of the Sgurr na Stri-Druim Hain area, Skye.* John Muir Trust, Edinburgh.

Averis, A B G and Averis, A M (1998a) *Vegetation survey of Beinn Eighe, Wester Ross, 1997.* Scottish Natural Heritage, Battleby.

Averis, A B G and Averis, A M (1998b) *Vegetation survey of the Glamaig-Beinn Dearg-Marsco area, Skye, 1998.* John Muir Trust, Edinburgh.

Averis, A B G and Averis, A M (1999a) *Vegetation survey of Ben Lui, 1998.* Scottish Natural Heritage, Battleby.

Averis, A B G and Averis, A M (1999b) *Vegetation survey of mid Mull, 1998.* Scottish Natural Heritage, Battleby.

Averis, A B G and Averis, A M (1999c). *Vegetation survey of the north-western part of Strathaird estate, Island of Skye, May and September 1999.* John Muir Trust, Edinburgh.

Averis, A B G and Averis, A M (2000a) *A survey of the distribution, flora and condition of limestone and woodland habitats at Strath Suardal, Skye, in September 1999.* Scottish Natural Heritage, Battleby.

Averis, A B G and Averis, A M (2000b) *Vegetation survey of Cadair Idris National Nature Reserve, Gwynedd, Wales, August-September 1999.* Countryside Council for Wales, Bangor.

Averis, A B G and Averis, A M (2000c) *Vegetation survey of Woodland Grant Scheme sites on the Isle of Eigg, 1999–2000.* Scottish Wildlife Trust, Edinburgh.

Averis, A B G and Averis, A M (2003) *Vegetation survey of Glen Coe.* Scottish Natural Heritage, Battleby.

Averis, A B G, Averis, A M, Horsfield, D and Thompson, D B A (2000) The upland vegetation of the Western Isles, Scotland. *Scottish Natural Heritage Research, Survey and Monitoring Report* No 164, Battleby.

Averis, A B G and Coppins, A M (1998) *Bryophytes, lichens and woodland management in the western Highlands.* Scottish Natural Heritage, Battleby.

Averis, A M (1994) *The ecology of an Atlantic liverwort community.* PhD Thesis, University of Edinburgh.

# References

Averis, A M (2001a) *Vegetation survey of selected proposed extensions to the Eryri SAC, Gwynedd, September 2000.* Countryside Council for Wales, Bangor.

Averis, A M (2001b) *Vegetation survey of Lí and Coire Dhorrcail, Knoydart, June 2001.* John Muir Trust, Edinburgh.

Baddeley, J A, Thompson, D BA and Lee, J A (1994) Regional and historical variation in the nitrogen content of *Racomitrium lanuginosum* in Britain in relation to atmospheric nitrogen deposition. *Environmental Pollution, 84: 189–196.*

Barkman, J J (1958) *Phytosociology and ecology of cryptogamic epiphytes.* Van Gorcum, Assen.

Bayfield, N G, Watson, A and Miller, G R (1988) Assessing and managing the effects of recreational use on British hills. In *Ecological change in the uplands.* (eds M B Usher and D B A Thompson), pp 399–414. Blackwell Scientific Publications, Oxford.

Birks, H J B (1973) *Past and present vegetation of the Isle of Skye: a palaeoecological study.* Cambridge University Press, Cambridge.

Birks, H J B (1976) The distribution of European pteridophytes: a numerical analysis. *New Phytologist, 77, 257–287.*

Birks, H J B (1988) Long-term ecological change in the British Uplands. In *Ecological change in the uplands* (eds M B Usher and D B A Thompson), pp 37–56. Blackwell Scientific Publications, Oxford.

Birks, H J B and Ratcliffe, D A (1980) *Classification of upland vegetation types in Britain.* Nature Conservancy Council, Edinburgh.

Birse, E L (1980) Plant communities of Scotland. Revised and additional tables. *Soil Survey of Scotland Bulletin 4.* The Macaulay Institute for Soil Research, Aberdeen.

Birse, E L (1984) The phytocoenonia of Scotland. Additions and revision. *Soil Survey of Scotland Bulletin 5.* The Macaulay Institute for Soil Research, Aberdeen.

Birse, E L and Robertson, J S (1976) *Plant communities and soils of the lowland and southern upland regions of Scotland.* Macaulay Institute for Soil Research, Aberdeen.

Blockeel, T L and Long, D G (1998) *A check-list and census catalogue of British and Irish bryophytes.* British Bryological Society, Cardiff.

Brown, A, Horsfield, D and Thompson, D B A (1993a). A new biogeographical classification of the Scottish uplands. I. Descriptions of vegetation blocks and their spatial variation. *Journal of Ecology,* 81: 207–230.

Brown, A, Birks, H J B and Thompson, D B A (1993b). A new biogeographical classification of the Scottish uplands. II. Vegetation-environment relationships. *Journal of Ecology,* 81: 231–251.

Burt, T P, Thompson, D B A and Warburton, J (eds.) (2002). *The British Uplands: dynamics of change.* Joint Nature Conservation Committee, Peterborough.

Clapham, A R, Tutin, C G and Moore, D (1987) *Flora of the British Isles,* 3rd edn. Cambridge University Press, Cambridge.

Cooper, E A (1997) *Summary descriptions of National Vegetation Classification grassland and montane communities.* UK Nature Conservation No. 14. Joint Nature Conservation Committee, Peterborough.

Cooper, E and MacKintosh, J (1996) *NVC review of Scottish grassland surveys.* Scottish Natural Heritage Review No. 65, Battleby.

Coppins, A M and Coppins, B J (1999) *Lichen survey of the Hill of White Hamars, Orkney.* Scottish Wildlife Trust, Edinburgh.

Coppins, B J (1990) Lichens of the native Scottish pinewoods. *British Lichen Society Bulletin,* **66,** 11–13.

Dargie, T C D (1994) *The sand dune vegetation of Scotland. Coull Links to Golspie (Loch Fleet cSAC).* Scottish Natural Heritage, Edinburgh.

Düll, R (1983) Distribution of the European and Macaronesian liverworts (Hepaticophytina). *Bryologische Beitraege,* 2, 1–114.

Düll, R (1984) Distribution of the European and Macaronesian mosses (Bryophytina). Part 1. *Bryologische Beitraege,* 4, 1–113.

Düll, R (1985) Distribution of the European and Macaronesian mosses (Bryophytina). Part 2. *Bryologische Beitraege,* 5, 110–232.

Edwards, M E and Birks, H J B (1986) Vegetation and ecology of four western oakwoods (*Blechno-Quercetum petraeae* Br.-Bl. et Tx 1952) in North Wales. *Phytocoenologia,* **14,** 237–261.

Elkington, T, Dayton, N, Jackson, D L and Strachan, I M (2001) *National Vegetation Classification: field guide to mires and heaths.* Joint Nature Conservation Committee, Peterborough. www.jncc.gov.uk/communications/pubcat/heathland.htm

Farmer, A M, Bates, J W and Bell, J N B (1992) Ecophysiological effects of acid rain on bryophytes and lichens. In *Bryophytes and lichens in a changing environment* (eds J W Bates and A M Farmer), pp 284–313. Oxford University Press, Oxford.

Forestry Commission (1998) Caledonian Pinewood Inventory. The Forest Authority Scotland, Edinburgh.

Forrest, G I (1971) Structure and production of North Pennine blanket bog vegetation. *Journal of Ecology*, **59**, 453–479.

Fowler, D and Irwin, J G (1989) The pollution climate of Scotland. In *Acidification in Scotland* (eds F T Last and S Warren), pp 10–25. Proceedings of a symposium held in Edinburgh on 8 November 1988. Scottish Development Department, Edinburgh.

Fryday, A M (1997) *Ecology and taxonomy of montane lichen vegetation in the British Isles*. PhD Thesis, University of Sheffield.

Galbraith, H, Duncan, K, Murray, S, Smith, R, Whitfield, D P and Thompson, D B A (1993) Diet and habitat use of the dotterel (*Charadrius morinellus*) in Scotland. *Ibis*, **135**, 148–155.

Gee, A S and Stoner, J H (1988) The effects of afforestation and acid deposition on the water quality and ecology of upland Wales. In *Ecological change in the uplands* (eds M B Usher and D B A Thompson), pp 273–287. Blackwell Scientific Publications, Oxford.

Gilbert, O L (2000) *Lichens*. HarperCollins, London.

Gilbert, O L and Fox, B W (1985) Lichens of high ground in the Cairngorm mountains, Scotland. *Lichenologist*, **17**, 51–66.

Gordon, J E, Thompson, D B A, Haynes, V M, Brazier, V and MacDonald, R (1998) Environmental sensitivity and conservation management in the Cairngorm mountains, Scotland. *Ambio*, **27**, 335–344.

Gordon, J E, Brazier, V, Thompson, D B A and Horsfield, D (2000) Geo-ecology and the conservation management of sensitive upland landscapes in Scotland. *Catena*, **42**, 323–332.

Gregory, S (1954). Accumulated temperature maps of the British Isles. *Transactions of the Institute of British Geographers*, **20**, 59–73.

Greig-Smith, P (1950) Evidence from hepatics on the history of the British flora. *Journal of Ecology*, **38**, 320–344.

Hall, J E, Kirby, K J and Whitbread, A M (2001) *National Vegetation Classification: field guide to woodland*. Joint Nature Conservation Committee, Peterborough

Hill, M O and Preston, C D (1998) The geographical relationships of British and Irish bryophytes. *Journal of Bryology*, **20**, 127–226.

Hill, M O, Evans, D F, Bell, S A (1992) Long-term effects of excluding sheep from hill pastures in North Wales. *Journal of Ecology*, **80**, 1–13.

Hill, M O, Preston, C D and Smith, A J E (eds) (1991) *Atlas of the bryophytes of Britain and Ireland. Volume 1. Liverworts*. Harley Books, Colchester.

Hill, M O, Preston, C D and Smith, A J E (eds) (1992) *Atlas of the bryophytes of Britain and Ireland. Volume 2. Mosses (except Diplolepidae)*. Harley Books, Colchester.

Hill, M O, Preston, C D and Smith, A J E (eds) (1994) *Atlas of the bryophytes of Britain and Ireland. Volume 3. Mosses (Diplolepideae)*. Harley Books, Colchester.

Hillaby, J (1970) *Journey through Britain*. Paladin, London.

Hobbs, A M and Averis, A B G (1991a) *The vegetation of the Faroe Islands in relation to British plant communities*. CSD Note No. 53. Nature Conservancy Council, Peterborough.

Hobbs, A M and Averis, A B G (1991b) *The vegetation of the Mourne Mountains in relation to plant communities of Great Britain*. CSD Note No. 54. Nature Conservancy Council, Peterborough.

Hobbs, R J (1984) Length of burning rotation and community composition in high-level *Calluna-Eriophorum* bog in N. England. *Vegetatio*, **57**, 129–136.

Hobbs, R J and Gimingham, C H (1987) Vegetation, fire and herbivore interactions in heathlands. *Advances in Ecological Research*, **16**, 87–173.

Horsfield, D, Hobbs, A M, Averis, A B G and Kinnes, L H (1991) *The vegetation of Connemara in relation to plant communities of Great Britain*. Nature Conservancy Council, Edinburgh.

James, P W, Hawksworth, D L and Rose, F (1977) Lichen communities in the British Isles: a preliminary conspectus. In *Lichen Ecology* (ed M R D Seaward), pp 295–413. Academic Press, London.

Jóhansen, J (1985) *Studies in the vegetational history of the Faroe and Shetland Islands*. Føroya Fróðskaparfelag, Tórshaven.

Lavin, J C and Wilmore, G T D (1994) *The West Yorkshire Plant Atlas*. City of Bradford Metropolitan Council, Bradford.

# References

Lawton, J H (1976) The structure of the arthropod community on bracken. *Botanical Journal of the Linnean Society*, **73**, 187–216.

Lee, J A, Tallis, J H and Woodin, S J (1988) Acidic deposition and British upland vegetation. In *Ecological change in the uplands* (eds M B Usher and D B A Thompson), pp 151–162. Blackwell Scientific Publications, Oxford.

Legg, C J (1995) Heathland dynamics: a matter of scale. In *Heaths and moorland: cultural landscapes* (eds D B A Thompson, A J Hester and M B Usher), pp 117–134. HMSO, Edinburgh.

Legg, C J (2000) *Review of published work in relation to monitoring of trampling impacts and change in montane vegetation.* Scottish Natural Heritage Review No. 131, Battleby.

Lindsay, R A (1995) *Bogs: the ecology, classification and conservation of ombrotrophic mires.* Scottish Natural Heritage, Battleby.

Lindsay, R A, Charman, D T, Everingham, F, O'Reilly, R M, Palmer, M A, Rowell, T A and Stroud, D A (1988) *The Flow Country: the peatlands of Caithness and Sutherland.* Nature Conservancy Council, Peterborough.

Lavin, J C and Wilmore, G T D (1994) *The West Yorkshire Plant Atlas.* City of Bradford Metropolitan Council, Bradford.

Lusby, P and Wright, J (1996) *Scottish wild plants. Their history, ecology and conservation.* The Stationery Office, Edinburgh.

MacDonald, A, Stevens, P, Armstrong, H, Immirzi, P and Reynolds, P (1998) *Upland habitats: surveying land management impacts.* 2 vols. Scottish Natural Heritage, Battleby.

MacKenzie, N (1999) *The native woodland resource of Scotland.* Forestry Commission Technical Paper No. 30. Forestry Commission, Edinburgh.

Mackey, E C, Shewry, M C and Tudor, G J (1998) *Land cover change: Scotland from the 1940s to the 1980s.* The Stationery Office, Edinburgh.

MacKintosh, E J (1988) *The woods of Argyll and Bute.* Research and survey in nature conservation series: report no. 10. Nature Conservancy Council, Edinburgh. (First produced in 1985 as NCC Scottish Field Survey Unit report S28).

MacKintosh, E J (1990) *A botanical survey of the semi-natural deciduous woods of Lochaber District.* Scottish Field Survey Unit Report S39. Nature Conservancy Council, Edinburgh.

MAFF (1992) *Heather and grass burning code.* Ministry of Agriculture, Fisheries and Food, London.

Mardon, D K (1990) Conservation of montane willow scrub in Scotland. *Transactions of the Botanical Society of Edinburgh*, **45**, 427–436.

Matthews, J R (1937) Geographical relationships of the British flora. *Journal of Ecology*, **25**, 1–90.

Matthews, J R (1955) *Origin and distribution of the British flora.* Hutchinson, London.

McLeod, C R, Yeo, M, Brown, A E, Burn, A J, Hopkins, J J and Way, S F (eds) (2002) The Habitats Directive: selection of Special Areas of Conservation in the UK. 2nd edn. Joint Nature Conservation Committee, Peterborough. www.jncc.gov.uk/SACselection.

McPhail, P and Taylor, N (1997) *Management plan for Ballyoukan juniper wood SSSI for the period 1997/8 to 2001/02.* Scottish Natural Heritage, Battleby.

McVean, D N (1961) Post-glacial history of juniper in Scotland. *Proceedings of the Linnean Society of London*, **172**, 53–55.

McVean, D N and Ratcliffe, D A (1962) *Plant communities of the Scottish Highlands.* Monograph no. 1 of the Nature Conservancy. HMSO, London.

Meteorological Office (1952) *Climatological atlas of the British Isles.* Meteorological Office, London.

Mitchell, F J G and Averis, A B G (1988) *Atlantic bryophytes in three Killarney woods.* Report for Royal Irish Academy and Office of Public Works, Dublin.

Moorland Working Group (1998) *Good practice for grouse moor management.* Scottish Natural Heritage, Battleby.

Murray, B (1988) The genus *Andreaea* in the British Isles. *Journal of Bryology*, **15**, 1–17.

Nagy, L, Grabherr, G, Körner, C and Thompson, D B A (eds.) (2003a). *Alpine Biodiversity in Europe.* Springer, Berlin.

Nagy, L, Thompson, D B A, Grabherr, G and Körner, C (2003b). *Alpine Biodiversity in Europe: an introduction.* Joint Nature Conservation Committee, Peterborough.

National Expert Group on Transboundary Air Pollution (2001) *Transboundary air pollution: acidification, eutrophication and ground-level ozone in the UK.* NEGTAP, DEFRA, London. www.nbu.ac.uk/negtap/finalreport.htm

Nethersole-Thompson, D and Watson, A (1981) *The Cairngorms: their natural history and scenery.* Melven, Inverness.

O'Criodain, C and Doyle, G J (1994) An overview of Irish small-sedge vegetation: syntaxonomy and a key to communities belonging to the Scheuchzerio-Caricetea nigrae (Nordh. 1936) Tx. 1937. *Biology and Environment: Proceedings of the Royal Irish Academy*, **94**, 127–144.

Orange, A and Fryday, A (1998) *A survey of the saxicolous lichen and bryophyte communities of Eryri cSAC and Moel Hebog SSSI*. Countryside Council for Wales, Bangor.

Page, C N (1982) *The ferns of Britain and Ireland*. Cambridge University Press, Cambridge.

Pakeman, R J and Marrs, R H (1992) The conservation value of bracken *Pteridium aquilinum* (L.) Kuhn-dominated communities in the UK; an assessment of the ecological impact of bracken expansion or its removal. *Biological Conservation*, **62**, 101–114.

Paterson, S, Marrs, R H and Pakeman, R J (1997) Efficacy of bracken *Pteridium aquilinum* (L.) Kuhn control treatments across a range of climatic zones in Great Britain. A national overview and regional examination of treatment effects. *Annals of Applied Biology*, **130**, 283–303.

Paton, J A (1956) Bryophyte succession on the Wealden sandstone rocks. *Transactions of the British Bryological Society*, **3**, 103–114.

Pearsall, W H (1968) *Mountains and moorlands*. Collins, London.

Pearsall, W H and Pennington, W (1973) *The Lake District*. Collins, London.

Peterken, G F (1981) *Woodland conservation and management*. Chapman & Hall, London.

Phillips, J, Watson, A and MacDonald, A (1993) *A muirburn code*. Scottish Natural Heritage, Battleby.

Pigott, C D (1956) The vegetation of Upper Teesdale in the North Pennines. *Journal of Ecology*, **44**, 545–586.

Poore, M E D and McVean, D N (1957) A new approach to Scottish mountain vegetation. *Journal of Ecology*, **45**, 401–439.

Preston, C D and Hill, M O (1997) The geographical relationships of British and Irish vascular plants. *Botanical Journal of the Linnean Society*, **124**, 1–120.

Proctor, M C F (1962) A sketch of the epiphytic bryophyte communities of the Dartmoor oakwoods. *Report and Transactions of the Devon Association for the Advancement of Science*, **94**, 531–554.

Prosser, M V (1990a) *A botanical survey of hay meadows in Teesdale, Lunedale and Baldersdale, Durham, 1986–1988*. England Field Unit Project No. 94, Nature Conservancy Council.

Prosser, M V (1990b) *A botanical survey of hay meadows in West Allendale and South Tynedale, Northumberland, 1986–1988*. England Field Unit Project Nos 93 and 96, Nature Conservancy Council.

Purvis, O W and Halls, C (1996) Review of lichens in metal-enriched environments. *Lichenologist*, **28**, 571–601.

Purvis, O W, Coppins, B J, Hawksworth, D L, James, P W and Moore, D M (1992) *The lichen flora of Great Britain and Ireland*. The British Museum (Natural History), London.

Ratcliffe, D A (1959) The vegetation of the Carneddau, North Wales. I. Grasslands, heaths and bogs. *Journal of Ecology*, **47**, 371–413.

Ratcliffe, D A (1968) An ecological account of Atlantic bryophytes in the British Isles. *New Phytologist*, **67**, 365–439.

Ratcliffe, D A (ed) (1977) *A nature conservation review*. 2 vols. Cambridge University Press, Cambridge.

Ratcliffe, D A (2000). *In search of nature*. Peregrine Books, Leeds.

Ratcliffe, D A and Thompson, D B A (1988) The British uplands: their ecological character and international significance. In *Ecological change in the uplands* (eds M B Usher and D B A Thompson), pp 9–36. Blackwell Scientific Publications, Oxford.

Rawes, M and Hobbs, R (1979). Management of semi-natural blanket bog in the northern Pennines. *Journal of Ecology*, **69**, 651–669.

Richards, P W (1938) The bryophyte communities of a Killarney oakwood. *Annals of Bryology*, **11**, 108–130.

Robertson, H J and Jefferson, R G (2000) *Monitoring the condition of lowland grassland SSSIs. I. English Nature's rapid assessment method*. English Nature, Peterborough.

Robertson, J S (1984) *A key to common plant communities of Scotland*. Soil Survey of Scotland Monograph. The Macaulay Institute for Soil Research, Craigiebuckler, Aberdeen.

Rodwell, J S (ed) (1991a) *British plant communities. Volume 1. Woodlands and scrub*. Cambridge University Press, Cambridge.

Rodwell, J S (ed) (1991b) *British plant communities. Volume 2. Mires and heaths*. Cambridge University Press, Cambridge.

# References

Rodwell, J S (ed) (1992) *British plant communities. Volume 3. Grasslands and montane communities.* Cambridge University Press, Cambridge.

Rodwell, J S (ed) (1995) *British plant communities. Volume 4. Aquatic communities, swamps and tall-herb fens.* Cambridge University Press, Cambridge.

Rodwell, J S (1997) Scottish vegetation in a European context. *Botanical Journal of Scotland,* **49**, 177–190.

Rodwell, J S (ed) (2000) *British plant communities. Volume 5. Maritime communities and vegetation of open habitats.* Cambridge University Press, Cambridge.

Rodwell, J S, Dring, J C, Averis, A B G, Proctor, M C F, Malloch, A J C, Schaminée, J H J and Dargie, T C D (2000) *Review of coverage of the National Vegetation Classification.* JNCC Report, No. 302. www.jncc.gov.uk/publications/jncc302.

Roper-Lindsay, J and Say, A M (1986) Plant communities of the Shetland Islands. *Journal of Ecology,* **74**, 1013–1030.

Rothero, G P (1991) *Bryophyte-dominated snow-beds in the Scottish Highlands.* MSc Thesis, University of Glasgow.

Salisbury, E J (1932) The East Anglian flora: a study in comparative plant geography. *Transactions of the Norfolk and Norwich Naturalists' Society,* **13**, 191–263.

Schuster, R M (1983) Phytogeography of the Bryophyta. In *New manual of bryology, Volume 1* (ed R M Schuster), pp 463–626. Hattori Botanical Laboratory, Nichinan, Miyazaki, Japan.

Scotland's Moorland Forum (2003). *Scotland's Moorland Forum: Principles of moorland management.* Scottish Natural Heritage, Battleby.

Shaw, S C and Wheeler, B D (1990) Comparative survey of habitat conditions and management characteristics of herbaceous poor-fen vegetation types. Contract surveys series No. 129. Nature Conservancy Council, Peterborough.

Smith, M A (2000) *Borders grassland survey.* Scottish Natural Heritage, Battleby.

Stace, C A (1997) *New flora of the British Isles*, 2nd edn. Cambridge University Press, Cambridge.

Steven, H M and Carlisle, A (1959) *The native pinewoods of Scotland.* Oliver & Boyd, Edinburgh.

Stroud, D A, Chambers, D, Cook, S, Buxton, N, Fraser, B, Clement, P, Lewis, P, McLean, I, Baker, H, and Whitehead, S (eds) (2001) *The UK SPA network: its scope and content.* Joint Nature Conservation Committee, Peterborough.

Tansley, A G (1939) *The British Islands and their vegetation.* Cambridge University Press, Cambridge.

Thompson, D B A (2002). The importance of nature conservation in the British uplands: nature conservation and land use changes. In *The British uplands: dynamics of change.* (eds. T P Burt, D B A Thompson and J Warburton), pp. 36–40, Joint Nature Conservation Committee, Peterborough.

Thompson, D B A and Baddeley, J (1991) Some effects of acidic deposition on montane *Racomitrium lanuginosum* heaths. In *The effects of acid deposition on nature conservation in Great Britain* (eds S J Woodin and A M Farmer), pp 17–28. Nature Conservancy Council, Peterborough.

Thompson, D B A and Brown, A (1992) Biodiversity in montane Britain: habitat variation, vegetation diversity and some objectives for conservation. *Biodiversity and Conservation,* **1**, 179–208.

Thompson, D B A, Galbraith, H and Horsfield, D (1987) Ecology and resources of Britain's mountain plateaux: conflicts and land-use issues. In *Agriculture and conservation in the hills and uplands* (eds M Bell and R G H Bunce), pp 22–31. Institute of Terrestrial Ecology, Merlewood.

Thompson, D B A, Gordon, J E and Horsfield, D (2001). Montane landscapes in Scotland: are these natural artefacts or complex relics. In *'Earth Science and the Natural Heritage: interactions and integrated management'.* (eds. J E Gordon and K F Leys), pp. 105–119. The Stationery Office, Edinburgh.

Thompson, D B A, MacDonald, A J, Marsden, J H and Galbraith, C A (1995). Upland heather moorland in Great Britain: a review of international importance, vegetation change and some objectives for nature conservation. *Biological Conservation,* 71: 163–178.

Tidswell, R J (1988) *A botanical survey of the semi-natural deciduous woods of Badenoch and Strathspey District.* Scottish Field Survey Unit Report S34. Nature Conservancy Council, Edinburgh.

Tidswell, R J (1990) *A botanical survey of the semi-natural deciduous woods of Angus District.* Scottish Field Survey Unit Report S44. Nature Conservancy Council, Edinburgh.

Tipping, R (2003). Woods and people in prehistory to 1000 BC. In *People and Woods in Scotland* (ed T C Smout), pp 14–39. Edinburgh University Press, Edinburgh.

Tyler, S J and Ormerod, S J (1994) *The Dipper*. T & A D Poyser, London.

UK Biodiversity Group (2001) *Sustaining the variety of life: 5 years of the UK Biodiversity Action Plan*. DETR, London.

UK Government (1994) *Biodiversity: the action plan*. HMSO, London.

Watson, A and Birse, E L (1990). Lichen-rich pinewood, *Cladonia ciliata-Pinus sylvestris* community in north-eastern Scotland. *Botanical Journal of Scotland*, **46**, 73–88.

Wheeler, D and Mayes, J (eds) (1997) *Regional climates of the British Isles*. Routledge, London.

White, J and Doyle, G (1982) The vegetation of Ireland: a catalogue raisonné. *Royal Dublin Society Journal of Life Sciences*, **3**, 289–368.

Woods, R G (1993) *Flora of Radnorshire*. National Museum of Wales, Cardiff.

Woolgrove, C E (1994) *Impacts of climate change and pollutants in snowmelt on snowbed ecology*. PhD Thesis, University of Aberdeen.

Woolgrove, C E and Woodin, S J (1995) Current and historical relationships between the tissue nitrogen content of a snowbed bryophyte and nitrogenous air pollution. *Environmental Pollution*, **93**, 283–288.

Wormell, P (2000) Pinewoods of the Black Mount – a progress report. *Native Woodlands Discussion Group Spring Newsletter*, **25**, 17–19.

# Glossary

Although we have tried to avoid unnecessarily complicated terminology and phytosociological jargon, the world of vegetation science includes a few words which may be unfamiliar to some people. These words are defined here.

**acid, acidic** Rocks and soils that are deficient in the alkaline constituents calcium, magnesium, sodium and potassium, and that are generally poor in plant nutrients too. Examples of acid rocks are granite and quartzite. Water percolating through or running over such soils or rocks may also be acid.

**acidophilous** Growing exclusively or usually on acid substrates.

**aftermath** (In relation to hay-meadows) plant growth after a hay cut has been taken.

**anthropogenic** Caused by human activities (e.g. woodland clearance) or by the activities of domestic animals (e.g. grazing).

**basic, base-rich** Rocks and soils with large amounts of alkaline elements, such as calcium, magnesium, sodium and potassium. Examples of base-rich rocks are limestone and calcareous mica-schist. Water percolating through or running over such soils or rocks may also be base-rich, carrying these elements in solution.

**basiphilous** Growing exclusively or usually on base-rich substrates.

**blanket mire** Bog vegetation on deep peat that covers the terrain in a continuous layer, like a blanket.

**bog** This term is used here for the vegetation of deep peat (blanket bogs, raised bogs and valley bogs) where the source of water and nutrients is either rain or a high water-table, and where there is little or no lateral movement of water through the peat.

**bryophytes** A division of the plant kingdom, comprising mosses, liverworts and hornworts.

**calcareous** (Of a substrate) containing calcium carbonate.

**calcicole** A plant species that grows mainly or exclusively on calcareous substrates.

**calcifuge** A plant species that grows mainly or exclusively on non-calcareous substrates.

**canopy** The highest tier or layer of plants in the vegetation; most often used of the tree tops in woodland but also used to refer to the layer of dwarf shrubs in heathland.

**carr** A wooded fen.

**community** An assemblage of different plant species growing together; the second of the three levels of classification used in the National Vegetation Classification (*see* NVC).

**coppice** A traditional method of woodland management in which trees are cut down to the stump so that they re-grow producing multiple stems.

**crustose** (Of a lichen) growing as a very thin crust closely attached to the substrate.

**epiphyllous** (Of a bryophyte or lichen) growing on the leaf of a vascular plant.

**epiphytic** Growing on the bark of trees and shrubs.

**ericoid** (Of a dwarf shrub) belonging to the genera *Calluna, Vaccinium, Arctostaphylos, Phyllodoce, Loiseleuria, Andromeda* or *Erica* in the family Ericaceae.

**eutrophic** Soil that is rich in plant nutrients. Also used of water and soil enriched with abnormally high levels of nitrogen, potassium and phosphorus by run-off from agricultural land.

**fen** A mire with neutral or basic water and a tall, dense sward of plants.

**field** 'In the field' means outside, in the real world (rather than in the office or laboratory); not necessarily in enclosed fields!

**field layer** The layer or tier of grasses, shrubs and herbs in a wood, above the ground layer (see below), but below the canopy.

**flush** A mire irrigated by water moving laterally through the soil; the same as a soligenous mire (see below).

**fluvioglacial** Referring to landscapes, or elements in a landscape, produced by the action of glacial melt-water.

**forb** A herbaceous (non-woody) vascular plant that is not a grass, rush or sedge, for example primrose and bluebell.

**foliose** (Of a lichen) with a flattened, leaf-like thallus that is not closely attached to the substrate.

**fruticose** (Of a lichen) with a shrubby thallus loosely attached to the substrate; the thallus is commonly erect and branched.

**graminoid** Collective term for grasses, rushes, sedges and similar grass-like species.

**ground layer** That part of terrestrial vegetation comprising mosses, liverworts and lichens. The associated vascular vegetation is referred to as the field layer (see above), but the term 'ground flora' refers to the vascular and non-vascular vegetation combined.

**hag** A block of peat with bare eroded sides; the result of erosion in blanket bogs.

**hepatic** Synonymous with liverwort (see below).

**herb** A non-woody vascular plant.

**lazy-bed** A form of cultivation, once common in the west Highlands for growing potatoes and, less, commonly, other crops. They were made by turning over the soil in a long narrow strip about a metre wide and covering it with manure and seaweed. This led to a sequence of raised ridges or beds separated by channels, which served to drain off excess water. Now disused, the outlines of lazy-beds can often be seen under grassland or bracken, especially when the sun is low in the sky.

**leaching** The gradual washing out of nutrients by water flowing over rocks or through soil.

**lichens** Small plants consisting of two components – an alga and a fungus – growing closely together.

**liverwort** One of the three divisions of bryophytes. Leaf, stem and capsule cells are less differentiated than in mosses; leaves are in two or three distinct ranks along stem; many species have lobed leaves, or a thallus not separated into a stem and leaves.

**mesotrophic** Soils with a moderate amount of plant nutrients; also refers to plant species that need moderate amounts of nutrients.

**metallophyte** A plant that is able to grow on soils enriched with metal ores.

**mire** Vegetation of wet ground, generally with a short to medium sward of shrubs, sedges, rushes and forbs. Interchangeable with 'bog' (see above) but more often used for vegetation in places where there is some lateral movement of water through the ground.

**montane** The land above the potential limit of woodland (the tree-line) or the plants and types of vegetation that occur here.

**moss** One of the three divisions of bryophytes. Leaves are distinct from stems, and not in two or three obvious ranks; leaf, stem and capsule cells are more differentiated than in liverworts; leaves are not lobed, in contrast to many liverworts.

**nodum (plural = noda)** A vegetational point of reference. The NVC communities are noda: abstract units based on the common features of a number of samples. They represent assemblages of plants that tend to occur together repeatedly and consistently across the country.

**NVC** National Vegetation Classification – the current standard system of classification used in Great Britain, and adopted by this book. Vegetation is classified at three levels: a) broad vegetation types such as heaths, mires etc; b) communities; and c) sub-communities.

**occult deposition** The deposition of pollution directly from mist and clouds, rather than from rain.

**oceanic** (Of vegetation types or species) restricted mainly to areas where the climate is wet with equable temperatures throughout the year, and in Europe showing a markedly western distribution.

**oligotrophic** Soils with small amounts of plant nutrients.

**ombrogenous** (Of a mire) receiving most of its water from rain.

**phytosociology** The study of plant communities.

**pleurocarpous** Mosses with stems that are usually well branched, and with capsules arising from lateral branches or branchlets.

**quadrat** Sample of vegetation recorded in a (usually) square plot. In continental texts the term relevé is generally used.

**raised mire** A type of bog, originating as a body of open water that becomes filled with peat. Over millennia the peat accumulates until it builds a raised dome or mound above the level of the surrounding land.

**ruderal** Growing on waste ground or among rubbish.

**SAC** Special Area of Conservation – a site of European importance for habitats, plants or animals, designated under the Habitats Directive.

**saxicolous** Growing on rock surfaces.

**snow-bed** The characteristic vegetation of places where snow accumulates over winter and lies late far into spring.

**soakway** A narrow, vegetated track of slowly moving water within a heath or bog.

**solifluction** Creep of soil down a slope.

**soligenous** (Of a mire) irrigated by ground water moving down a slope.

**SPA** Special Protection Area – a site of European importance for birds, designated under the Birds Directive.

**SSSI** Site of Special Scientific Interest – a site of national importance for wildlife conservation.

**stand** An area or patch of a particular vegetation type.

**sub-community** A sub-division of a National Vegetation Classification community (*see* NVC).

**sub-montane** The land below the potential limit of woodland (the tree-line) or the plants and vegetation types that occur here.

**substrate** A surface on which plants grow, e.g. soil, peat, sand, humus, rock, living bark or rotting wood.

**swamp** Vegetation growing on substrates that are covered by water for all or most of the year, typically dominated by tall grasses or sedges.

**taxon (plural = taxa)** A named group of organisms: includes species, sub-species and so on.

**thallus** The body of a lichen or liverwort consisting of a plate or branched/lobed mass of tissue, not differentiated into a stem and leaves.

**thermophilous** Adapted to warm temperatures.

**topogenous** (Of a mire) with a high water-table maintained by poor drainage; usually in basins or on valley floors.

**trap landscape** The typical stepped terrain of basalt landscapes.

**tree-line** The potential upper limit of woodland.

**valley mire** A type of bog where a deep layer of peat has accumulated on the floor of a valley; in many places this is on either side of a stream. Unlike raised mires and most blanket mires, which receive water only from rain, valley mires receive water that drains down the surrounding slope and seeps outwards through the peat from the central stream or streams.

**vascular plant** A plant with specialised conducting vessels through which water and nutrients are internally transported. Vascular plants include ferns, trees, shrubs and all herbaceous species. Bryophytes and lichens lack these conductive tissues and are non-vascular plants.

**wet heath** Heath vegetation consisting of mixtures of dwarf shrubs, including *Erica tetralix*, sedges, grasses and other plants, commonly including *Sphagnum* mosses, on shallow peat.

**wether** A castrated male sheep.

# Index

# Index